Frontiers in Dairy Cows Nutrition and Milk Quality

◎ Edited by Jiaqi WANG Shengguo ZHAO Nan ZHENG

China Agricultural Science and Technology Press

图书在版编目(CIP)数据

奶牛营养与牛奶质量科学研究前沿 = Frontiers in dairy cows nutrition and milk quality / 王加启，赵圣国，郑楠主编 . --北京：中国农业科学技术出版社，2022. 11

ISBN 978-7-5116-5984-2

Ⅰ. ①奶…　Ⅱ. ①王…②赵…③郑…　Ⅲ. ①乳牛-家畜营养学-研究②牛奶-质量管理-研究　Ⅳ. ①S823. 9②TS252. 2

中国版本图书馆 CIP 数据核字(2022)第 204229 号

责任编辑　金　迪
责任校对　李向荣
责任印制　姜义伟　王思文

出 版 者　中国农业科学技术出版社
北京市中关村南大街 12 号　　邮编：100081
电　　话　(010) 82106625 (编辑室)　　(010) 82109702 (发行部)
(010) 82109709 (读者服务部)
网　　址　https://castp.caas.cn
经 销 者　各地新华书店
印 刷 者　北京建宏印刷有限公司
开　　本　185 mm×260 mm　1/16
印　　张　19. 25
字　　数　578 千字
版　　次　2022 年 11 月第 1 版　2022 年 11 月第 1 次印刷
定　　价　80. 00 元

Frontiers in Dairy Cows Nutrition and Milk Quality

Editorial Board

Forword

Dairy science is an important driving force to support the development of dairy industry. In order to strengthen international cooperation and exchange in dairy science, we have set up an academic exchange platform "International Symposium on Dairy Cow Nutrition and Milk Quality". Since 2009, it has been held for 7 times and 13 years. At each time, internationally renowned scientists from the United States, Canada, New Zealand, United Kingdom, Netherlands and other developed dairy countries gather together to exchange and discuss the international frontier progress in the basic research of dairy cow nutrition, milk quality and safety research, new dairy cow feeding technology and so on. The symposium reaches the theoretical consensus and the willingness to cooperate in research. We invite some internationally renowned scientists to write articles and exchange academic views and research results on dairy cow nutrition and milk quality. This is a collection of 23 key articles, mainly divided into two aspects: dairy cow nutrition progress, milk quality and safety progress. We hope that the book will provide you with cutting-edge research advances and encourage you to undertake more extensive study of these topics.

Contents

Session 2 Advances in Milk Quality and Safety

Introduction

Emerging Trends in Agriculture

Robert J. Collier[1], Todd R. Bilby[2] and Benjamin R. Renquist[1]

[1] University of Arizona, Tucson, AZ 85745

[2] Merck Animal Health, One Merck Dr., PO Box 100 Ws3Ab-05, Whitehouse Station, NJ 08889

Abstract

Population growth rate peaked in 1961 and has been falling dramatically worldwide since that time which will result in stabilization of world population around 2100. Stabilization of the world population will have very big impacts on economic growth, agricultural research programs and political systems. However, worldwide food production must double by 2050 in order to meet demands for the current growing population. In addition, the developing countries where most of the future population growth will occur and also will require a major increase in high quality protein in their diets to match the diets of developed countries. The primary forces in agriculture that have provided the increased food required for population growth has been improved agricultural efficiency and maximizing use of arable land. Currently there is no indication that available arable land will increase in the future and projections for future rates of improved agricultural efficiency indicate that the projected rate of efficiency improvements will fall below that required for meeting future food demands. This drop in the rate of improved agricultural efficiency will be due primarily to impacts of climate change which will adversely affect food output in current primary food production areas, and move the primary grain growing regions northward in the northern hemisphere and southward in the southern hemisphere. The current global climate change models predict a reduction in food production in both the U.S. and China associated with increased environmental temperatures. The primary opportunities to address this shortfall include major investments in reducing impacts of climate change on food production, reducing waste in food production, processing and consumption, increased use of precision agriculture and reducing impact of agriculture on the environment. Special reference is made to improved reproductive strategies to increase reproductive performance during global warming.

Introduction

The world population has more than doubled since 1960, from approximately 2.9 billion to more than 7.5 billion today. This increase in population has required a tripling of agricultural production because of a simultaneous demand for more and better food in less developed countries (Wik et al., 2008). Global agriculture has been successful in meeting this increase in demand. Steady growth in agricultural output and a long-term decline in real commodity prices attest to this success (Wik et al., 2008).

Population Growth and Food Demand

By 2050, world population is projected to grow to between 9 and 10 billion people. Most of the growth is expected to occur in poor developing countries, where income elasticity of demand for food continues to be high. The population increase, combined with moderately high income growth, could result in a more than 70% increase in demand for food and other agricultural products by 2050 (UN, 2011). However, population growth is slowing dramatically. Before 1800 the world population growth rate was always well below 1% (Figure 1). While in the course of the first fifty years of the 20th century annual growth increased to up to 2.1%—the highest annual growth rate in history, recorded in 1962. After this point, it has been systematically going down with projections estimating an annual rate of growth of 0.1% for 2100 (Figure 1).

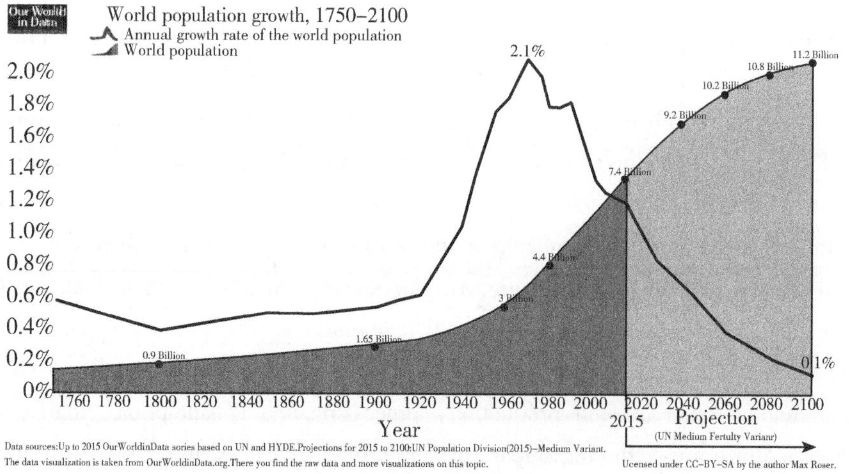

Figure 1 The world population growth rate

Agricultural Productivity

Meeting the increasing demand for more food will require improvements in agricultural productivity. Indeed, improving agricultural productivity has been the primary driver for growth in American and world agriculture since 1950 (Figure 2).

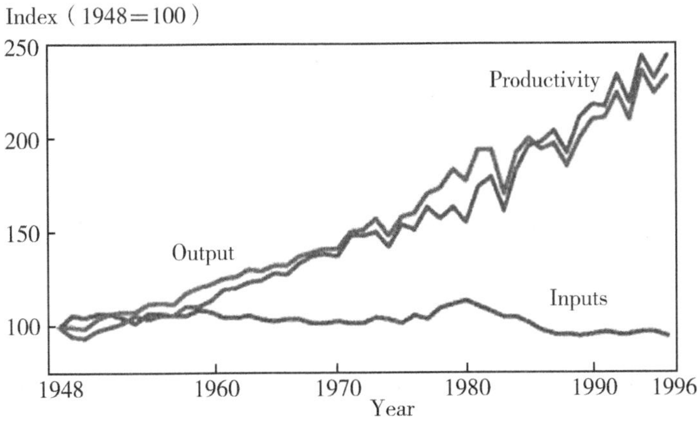

Figure 2 Agricultural productivity in the united states (1948–1996)

(Source: Economic Research Service, USDA)

This has also been true for the rest of the world. The value of total agricultural output (all food and non - food crop and livestock commodities) has almost trebled since 1961 (Figure 3), an average increase of 2.3% per year, always keeping ahead of global population growth rates (1.7% p.a.). Much of this growth originated in developing countries (3.4%–3.8% p.a.).The high growth rates of the latter reflected, among other things, developments in some large countries, most importantly China. Without China, the rest of the developing countries grew at 2.8%–3.0% p. a. These figures also reflect the rising share of high value commodities such as livestock products in the total value of production; interms of quantities (whether measured in tons or calorie content), the growth rates have been lower (FAO, 2006 a).

However, there appears to be a growing gap in the current global rate of agricultural productivity.

Climate Change

The actual agricultural output needs to double by 2050 in order to meet population growth and food demands associated with increased consumption of higher quality protein. However, the actual rate of total food production (TFP) growth in agricultural productivity is currently

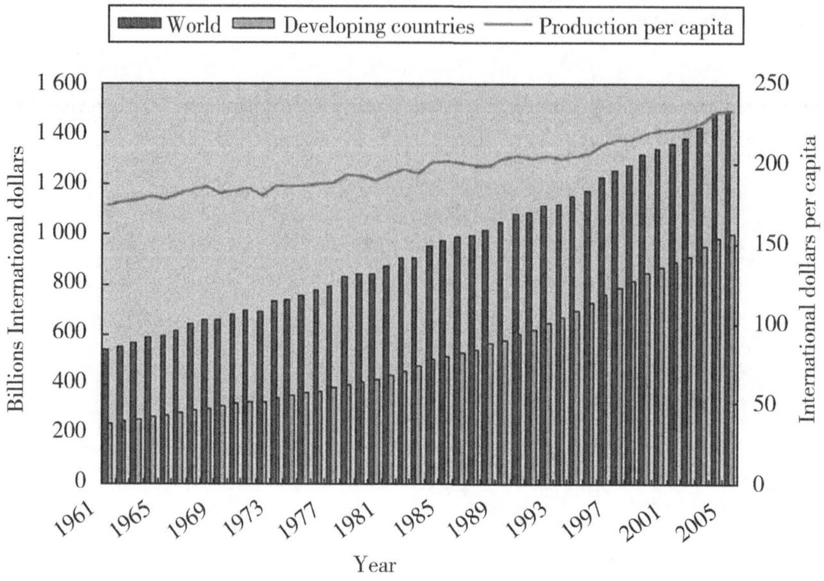

Figure 3 Total and per capita agricultural production (Wik et al., 2008)

falling below the required rate of growth. The causes of this lag in agricultural are multifactorial and include increased survival at birth, increased longevity, rising world affluence and increased consumer demand for high quality protein. However, one additional major potential culprit is climate change. Indeed, agriculture is the most sensitive component of an economy to climate change. Currently, all global climate models predict a rise in surface temperature of the earth (IPPC, 2014).

There is now a strong scientific consensus that climate change is occurring and it is projected that the global average temperature will likely rise an additional 1.1–5.4℃ over the next century (CCSP, 2008). These changes will have large and measurable impacts on animal productivity worldwide (Klinedinst et al., 1993) through a variety of routes including changes in food availability and quality, changes in pest and pathogen populations, alteration in immunity and both direct and indirect impacts on animal performance such as growth, reproduction and lactation. As productivity increases with continued genetic, nutritional and management improvement the sensitivity of high producing animals to heat stress increases. Currently, heat stress causes an estimated $ 1.7 billion (US) in losses each year in animal production (St-Pierre et al., 2003).

Genetic selection for milk production as well as nutritional and management improvements have all contributed to increased milk yield and therefore increased metabolic heat output per cow. This has considerably increased the susceptibility of the lactating dairy cows to heat stress. The objectives of intensive livestock systems are to take advantage of scale effects to maximize profitability, provide a uniform thermoneutral environment and consistent nutrition in order to maximize production output and to reduce impacts of adverse environmental conditions. The use of intensive livestock systems is increasing and will continue to do so for the immediate future because they are essential to achieving increases in animal productivity. However, proper construction and management of these systems present several-challenges to producers, who must consider several factors including management of the micro environments inside the facility, maximizing efficiency of labor, capital and nutrients required as well as waste disposal in the form of waste water and manure.

In modern dairy facilities the objective is to achieve consistent high milk production, feed efficiency and reproductive efficiency while maintaining the health of the dairy cow. Heat stress reduces intake, milk production, health and reproduction of dairy cows. Therefore, a large component of capital expenditure for dairy facilities involves addressing the cooling needs of lactating dairy cows. In many parts of the U.S., milk production and reproductive performance drastically decline during periods of heat stress despite large capital investments in cooling equipment. Heat stress also has a negative impact on a dairy farm's future by reducing the peak milk production of cows that go through the transition period during periods of heat stress. The impact of reduced peak milk production often lingers into late fall or early winter. Cows can be managed and cooled to minimize the impact of heat stress. The method used will vary depending on the severity of the climate and the ambient relative humidity. In modern dairy facilities it is essential to minimize variation in the cow's core body temperature during periods of heat stress to maximize milk production and reproductive performance.

Metabolic Impacts

Heat stress imposes dramatic challenges to the metabolism of high producing animals and forces changes in many metabolic pathways as the animal attempts to maintain homeostasis while adjusting its productive output. The process of acclimation to thermal stress is homeorhetic and involves changes in tissue sensitivity to homeostatic regulators of the endocrine system as well as changes in secretion rate of some of these regulators as well. Also, in some cases acclimation involves use of alternate metabolic pathways not yet fully described (Baumgard and Rhoads, 2013). The maintenance of a homeothermic state within an organism has high metabolic costs. Environmental factors such as high temperature with or without increased levels of humidity, wind, and solar exposure can impair the ability to maintain homeostasis and increase the need for metabolic fuel.

Rising body temperature leads to a range of metabolic effects on lactating dairy cows. Cows pant when heat stressed and panting leads to respiratory alkalosis in addition to rapidly reducing feed intake. Spain et al. (1998) showed that lactating cows under heat stress decreased feed intake 6% – 16% as compared to thermal neutral conditions. Holter et al. (1996) reported heat stress depressed intake of cows more than heifers. Other studies have reported similar results. In addition to a reduction in feed intake, there is also a 30%–50% reduction in the efficiency of energy utilization for milk production (McDowell et al., 1969). Lactating dairy cows subjected to heat stress go into negative energy balance as feed intake drops more quickly than milk yield and several studies indicate that lactating dairy cows losing greater than 0.5 units body condition score within 70 d postpartum had longer calving to first detected estrus and (or) ovulation interval (Butler, 2000; Beam and Butler, 1999). Garnsworthy and Webb (1999) reported that cows displaying the lowest conception rates had lost more than 1.5 BCS units between calving and insemination. Butler (2000) also reported that conception rates range between 17% and 38% when BCS decreases 1 unit or more, between 25% and 53% if the loss is between 0.5 and 1 unit, and is > 60% if cows do not lose more than 0.5 units or gain weight.

Metabolic acclimation to chronic thermal stress was recently reviewed (Baumgard and Rhoads, 2013). They point out that domestic animals alter their metabolic and fuel selection priorities independent of nutrient intake or energy balance during acclimation to thermal stress. These alterations include a shift in carbohydrate metabolism including changes in basal and stimulated circulating insulin levels. The production of glucose from liver and metabolism of glucose by muscle also demonstrate differences in glucose production and use during heat stress. Their studies also demonstrated an apparent lack of fat mobilization from adipose tissue coupled with a reduced responsiveness to lipolytic stimuli despite a pronounced negative energy balance demonstrating marked changes in metabolic fuel availability during acclimation to thermal stress. Collectively, their work demonstrates that acclimation to heat stress is a homeorhetic process involving alteration of carbohydrate, lipid, and protein metabolism independently of reduced feed intake. This process is mediated by coordinated changes in metabolic fuel supply and utilization by multiple tissues. This is a rapidly evolving area of thermal biology and we do not yet sufficiently understand the genomic regulatory process involved to identify potential pathways to pursue to improve metabolic performance during environmental stress.

Another potential heat stress-related deterrent to dairy cow fertility is increased circulating plasma urea nitrogen concentrations. In terms of effects on fertility, most research has focused on the urea produced as a result of protein metabolism within the rumen. However, elevated urea concentrations are also a consequence of increased skeletal muscle breakdown. The end result of these physiological changes that occur during heat stress are elevated plasma urea nitrogen concentrations in heat stress cows compared to pair fed thermoneutral cows (Baumgard and Rhoads, 2013). Therefore, elevated plasma urea nitrogen concentrations may be exacerbating the decrease in fertility that is frequently observed during periods of heat stress.

Estrous Activity, Hormone Function, and Follicular Development

Heat stress reduces the duration and intensity of estrus, increasing the difficulty of detecting cows in estrus for artificial insemination. Motor activity and behavioural estrus are reduced during summer months (Hansen and Arechiga, 1999) while incidence of anestrous and silent ovulations are increased (Gwazdauskas et al., 1981). Holsteins in estrus during the summer had 4.5 vs. 8.6 mounts/estrus for those in winter (Nebel et al., 1997). Undetected estrous events were estimated at 76% to 82% during June through September compared to 44% to 65% during October through May on a commercial dairy in Florida (Thatcher and Collier, 1986).

Heat stress impairs follicle selection and increases the length of follicular dominance while reducing its degree thereby increasing the length of follicular waves; thus reducing the quality of oocytes, modulating follicular steroidogenesis, and reducing fertility (Mihm et al., 1994; Roth et al., 2001). Summer heat stress has been shown to increase the number of subordinate follicles; while reducing the degree of dominance of the dominant follicle and decreasing inhibin and estrogen levels (Wolfenson et al., 1995; Wilson et al., 1998). These cumulative changes in ovarian follicular dynamics result in increased twinning rates in dairy cows during summer vs. winter (Ryan and Boland, 1991).

Oocytes and Early Developing Embryos

Wolfenson et al. (2000) demonstrated that summer heat stress reduces pregnancy and conception rates, which can carry-over into the fall months. This carryover effect occurs because the early developing follicle which emerges 40-50 d prior to ovulation is adversely affected by heat stress. Thus, oocytes that undergo maturation near the end of the summer are damaged by heat stress months before they ovulate. Numerous studies both *in vitro* and *in vivo* have clearly demonstrated the devastating effects of heat stress on the development and competence of oocytes (Edwards and Hansen, 1997; Putney et al., 1988; Rocha et al., 1998; Rutledge et al., 1999).

Heat stress can also affect the early developing embryo post-fertilization. When heat stress was applied from d 1 to d 7 after estrus, embryo quality, stage and competance on d 7 was reduced (Putney et al., 1989; Monty and Racowsky, 1987). Early developmental embryonic stages are more susceptible to the deleterious effects of HS. When embryos are exposed to heat stress at the 2-to 4-cell stage there was a larger reduction in embryo cell number and development than heat stress at the morula stage (Edwards and Hansen, 1997; Ju et al., 1999; Paula-Lopes and Hansen, 2004). This adverse effect of heat stress on the early developing embryo can be ameliorated by transferring embryos produced under thermoneutral conditions to heat stress cows (as opposed to inseminating HS cows; Drost et al., 1999) and thereby increasing pregnancy rates.

Latter Stages of Embryo Development and Embryo Loss

Embryonic growth is also reduced by heat stress up to d 17, which is the period critical for maternal recognition of pregnancy. Sufficient interferon-tau must be secreted from the embryo in order to prevent luteolysis and to maintain pregnancy. However, heat stress has been shown to reduce the size of embryos recovered on d 17 after insemination (Biggers et al., 1987) and also to reduce interferon-tau secretion (Putney et al., 1988). Furthermore, *in vitro* endometrial secretion of the luteolytic hormone, PGF2α, increases in response to heat stress (Putney et al. 1988). Circulating concentrations of metabolic hormones and nutrients necessary for the early embryonic growth leading up to the maternal recognition of pregnancy (such as insulin, IGF-1, and glucose) are reduced by heat stress. Bilby et al. (2006) proposed that exogenous supplementation of these factors during heat stress may improve fertility (Bilby et al., 2006). Thus, heat stress induces effects on the embryo as well as the uterine environment that can disrupt the events necessary for maternal recognition of pregnancy, thereby reducing pregnancy rates.

Other studies have shown that embryonic loss following maternal recognition of pregnancy is also elevated during periods of heat stress. Dairy cows carrying singletons or twins are 3.7 and 5.4 times more likely suffer embryonic loss, during the hot versus cool season (Lopez-Gatius et al., 2004). In addition, the likelihood of pregnancy loss has been shown to increase by a factor of 1.05 for each unit increase in mean maximum temperature-humidity index (THI) from d 21-30 of gestation (Figure 4).

Improving Reproduction and Milk Production by Cooling Dry Cows

Traditionally, dry pregnant cows are provided little protection from heat stress because they are not lactating; and it is incorrectly assumed they are less prone to heat stress. Additional stressors are imposed during this period due to abrupt physiological, nutritional, and environmental changes. These changes can increase the cows' susceptibility to heat stress and have a critical influence on postpartum cow health, milk production, and reproduction. The dry period is particularly crucial since it involves mammary gland involution and subsequent development, rapid fetal growth, and induction of lactation. Heat stress during this time period can affect endocrine responses that may increase fetal abortions, shorten the gestationlength, lower calf birth weight, and reduce follicle and oocyte maturation associated with the postpartum reproductive cycle. Collier et al. (1982 a, b) demonstrated that heat stress during late gestation reduced both calf birth weight and subsequent milk yield and altered the postpartum uterine environment. Other investigators have confirmed these effects and have extended them to include development and growth of the neonate.

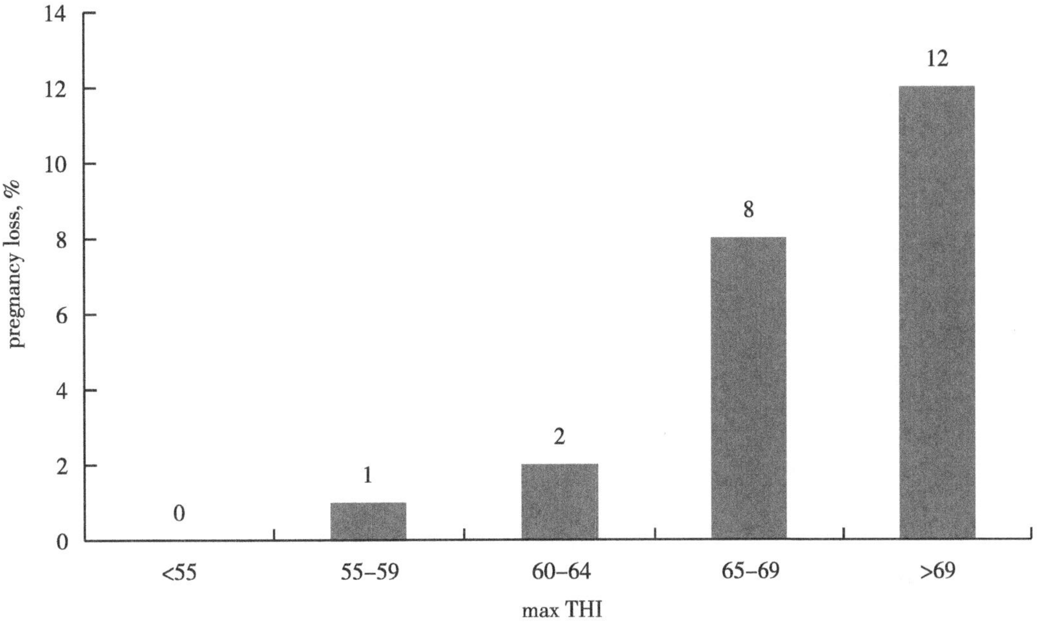

Figure 4 Pregnancy loss rates for different maximum temperature-humidity indices (THI) during d 21-30 of gestation (adapted from Garcia-Ispierto et al., 2006)

Many studies reporting subtle effects of HS on subsequent fertility were published when the average milk yield was much lower than it is today and it is well established that both heat stress and high production impact reproductive performance. In addition, our knowledge of cooling systems and their proper use (when, where, and to what extent) to reduce HS has increased substantially. For example, Wiersma and Armstrong (1988) conducted a study in Saudi Arabia on 3 different farms and observed an improvement in peak milk production (90.9 vs. 87.2 lb.), decreased services per conception (3.1 vs. 3.7 services), and reduced culling for reproductive failure (7.7% vs. 19%) for dry cows evaporative cooled vs. shade only. More recently, Avendano-Reyes et al. (2006) concluded that cooling dry cows with shades, fans, and water spray vs. cows with only shade decreased services per conception and days open, while milk yield increased during the postpartum period. In 2006, Urdaz et al. (2006) observed that dry cows with feed line sprinklers, fans, and shade compared to cows with only feed line sprinklers had an increased 60 d milk yield with no difference in body condition score body condition score changes, incidence of post parturient disorders, or serum nonesterified fatty acid (NEFA) concentrations. Although reproductive parameters were not measured, cooling dry cows with shades, fans, and sprinklers compared with only sprinklers improved total 60 d milk production by 185.5 lb./cow, and increased estimated annual profits by $ 8.92/cow based on milk only (Table 1). The problem of carry over effects from summer HS to fall fertility may be accentuated due to HS during the dry period. Cooling dry cows may reduce HS effects on the antral follicle destined to ovulate 40-50 d later, which coincides with the start of most breeding periods, and possibly increases first service conception rates.

Table 1 Projected economic returns from dry cow pen fans, sprinklers, and shades vs. sprinklers only based on marginal milk production for the first 60 days of lactation for dry multiparous holsteins enrolled une through october 2002[1]

Period	5 years
Fans used, no.	7
Cows cooled, no.	139
Interest rate (cost capital)	7.00%
Cows culled during first 60 days in milk	10%
Median days in milk at culling	25
Net number of cow-days to benefit	13,504
Capital costs	
Fansshade clothframeetc.	$ 7,040.00
Residual value of capital equipment after 5 years	$ 1,500.00
Annual capital costs	$ 1,456.15
Operating costs per year	
Maintenance and electricity	$ 450.54
Marginal feed for dry cows	$ 326.24
Annual operating costs	$ 776.78
Total annual costs	$ 2,232.93
Revenues	
Additional milk during first 60 days postpartum	1.4kg/d
Marginal milk price for additional milk	$ 0.23/xx
Total annual benefit (milk returns)	$ 4,363.66
Profit per year (based on milk alone)	$ 2,130.72
Profitper cow	$ 8.92
Percentage profit per dollar spent per year	95%

[1]Adanted from Urdaz et al. (2006)

The greatest opportunity to reduce the negative effects of HS during both the pre-and postpartum periods is through aggressive use of cooling. Cooling dry cows with shade, feed line sprinklers and fans or evaporative cooling has been demonstrated to be beneficial for reducing services per conception, reproductive culls, and days open; as well as increasing milk yield with a significant return on investment compared to cows with either shades alone or feed line sprinklers alone (Wiersma and Armstrong, 1988; Avendano-Reyes et al., 2006; Urdaz et al., 2006). In addition to proper cooling, changing management decisions may help reduce the severity of HS in areas of intermittent heat waves. For instance, Bilby et al. (2009) pointed out that at dry-off, many cows receive vaccines that can cause a fever spike which, when coupled with HS, can cause body temperature to rise above normal (101.3–102.8℉). Therefore, during severe heat waves, it might prove beneficial to delay vaccinations at dry-off, if the dry pen does not contain adequate cooling.

Reducing Negative Effects of Postpartum Heat Stress

Current and past research has resulted in dramatic improvements in dairy cow management in hot environments. Two primary strategies are to minimize heat gain by reducing solar heat load and maximize heat loss by reducing air temperature around the animal or increasing evaporative heat loss directly from animals. Following are several strategies to potentially help reduce the negative impacts of HS on reproduction in lactating dairy cows.

Cow Comfort and Cooling

Locating where heat stress is occurring on the dairy facility is central to implementing the proper cooling or management strategy to eliminate these hot spots. Temperature monitoring equipment can be employed to monitor core body temperatures in cows by attaching a thermal data logger to a blank continuous intravaginal drug release (CIDR®, Pfizer Animal Health, New York, NY) device for practical on-farm use. The data logger mounted on the CIDR is inserted into the cow's vagina, measuring core body temperature every minute for up to 6 d. This allows monitoring of the cow's body temperature and identification of where the cow is experiencing heat stress.

Providing enough shade and cow cooling is vital for proper cow comfort. There should be at least 38 to 45 sq. ft. of shade/mature dairy cow to reduce solar radiation. Spray and fan systems should be used in the holding pen, over feeding areas, over the feeding areas in some free stall barns, and under shades on dry lot dairies in arid climates. Exit lane cooling is an inexpensive way to cool cows as they leave the parlor. Providing enough access to water during heat stress is critical. Water needs increase 1.2 to 2 times during heat stress conditions. Lactating cattle require 35 to 45 gal of water/d. Access to clean water troughs when cows leave the parlor, at 2 locations in dry lot housing, and at every crossover between feeding and resting areas in free stall housing is recommended. Keep in mind milk is approximately 90% water; therefore water intake is vital for milk production and to maintain thermal homeostasis.

The bedding material used can also be source of stress to animals. Data recently produced at the University of Arizona demonstrated that cows exposed to heat stress and bedded on sand had lower rectal temperatures compared to cows bedded on dried manure (Figure 5, Ortiz et al., 2014).

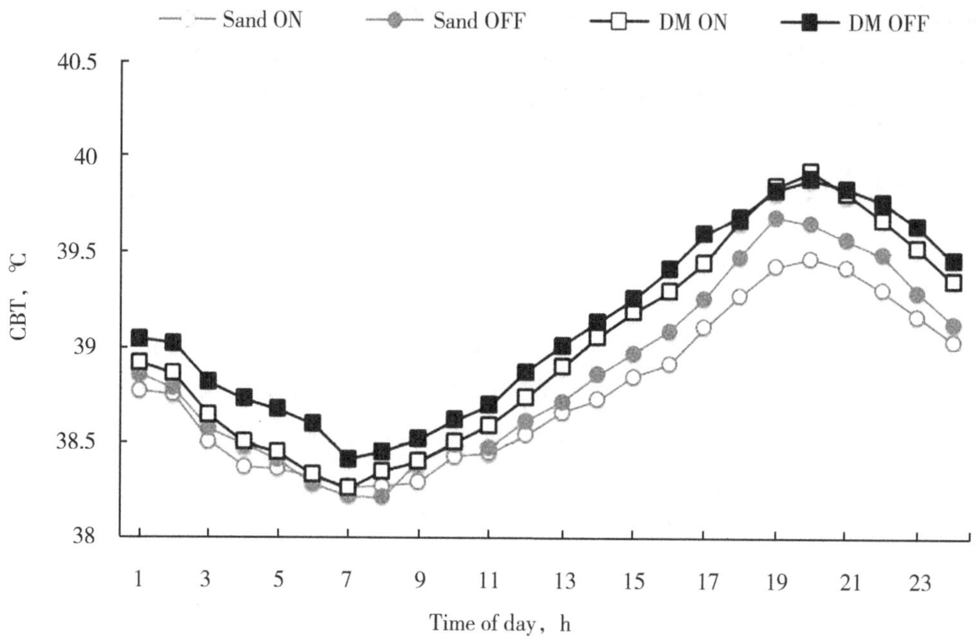

Figure 5 Effect of bedding type on continuous core body temperature (CBT, ℃) in a hot dry environment. DM = dried manure. Sand ON and Sand OFF refer to whether a buried heat exchanger below the bedding material has water circulating through it or not. the same is true of dried manure (DM) ON or OFF (from ortiz et al., 2014)

The holding pen is often an area of elevated HS conditions. Cows are crowded into a confined area for several minutes to hours. Cows should not spend more than 60 to 90 min in the holding area and shade, fans, and sprinklers should be provided in the holding pen. An Arizona study showed a 3.5°F drop in body temperature and a 1.76 lb. increase in milk/(cow · d) when cows were cooled in the holding pen with fans and sprinklers (Wiersma and Armstrong, 1983). Cattle handling such as sorting, adding cattle to the herd, vet checks, and lock-up times should be completed in the early morning since the cow's warmest body temperature occurs between 6 p. m. and midnight. Reducing lock-up times can also reduce heat stress, especially in facilities with little or no cooling above head locks.

Reproductive Management Strategies during Periods of Thermal Stress

Reproductive performance of lactating cows is greatly reduced during thermal stress but non-lactating heifers generally show no seasonal trend in reproductive performance even in the humid Southeastern United States. Both expression of estrus and fertility are reduced in heat stressed lactating cows. Tools and methods for timed inseminations have been developed that reduce the need for detection of estrus. The use of fixed timed artificial insemination (TAI) to avoid the deleterious effects of reduced estrous detection has been well documented. Utilizing some type of TAI (i.e. Ovsynch, Cosynch 72, or Ovsynch 56), either coupled with or without estrous detection, can improve fertility during the summer (De la Sota et al., 1998). When timed artificial insemination programs were used, pregnancy rates were improved under thermal stress conditions compared with AI without timing. Recent reports indicate that calving in summer months might reduce the success of a timed first insemination between 60 d and 66 d postpartum, although other researchers have reported increased conception rates under similar conditions. Follicular dynamics are altered by thermal stress and oocyte quality is reduced for an extended interval after thermal stress is removed (Wolfenson et al., 1995, 2000). This is believed to be the reason for decreased fertility of dairy cows during cooler autumn months. Removal of the low viability cohort of follicles led to earlier emergence of healthy follicles and higher quality oocytes. Embryo transfer has also been reported to improve pregnancy rates during warm summer months by removing the need to detect estrus. Embryo transfer can significantly improve pregnancy rates during the summer months (Drost et al., 1999). Embryo transfers can by-pass the period (i. e. before d 7) in which the embryo is more susceptible to heat stress. Nevertheless, embryo transfer is not a widely adopted technique. Improvements need to be made in the *in vitro* embryo production techniques, embryo freezing, timed embryo transfer, and lowering cost of commercially available embryos before this becomes a feasible solution. Recent developments in improving embryo resistance to HS through the use of both genotype manipulation and addition of survival factors, such as insulin-like growth factor-1 which protects cells from a variety of stresses may further improve pregnancy rates with embryo transfer (Block and Hansen, 2007). Use of AI during warm summer months avoids the negative effects of HS on bull performance on dairies. Additional work is warranted to improve reproductive performance of cattle in hot environments as well as effects of heat stress during the transition period on subsequent lactation.

Surface temperature is projected to rise over the 21st century under all assessed emission scenarios. It is very likely that heat waves will occur more often and last longer, and that extreme precipitation events will become more intense and frequent in many regions. The ocean will continue to warm and acidify, and global mean sea level will rise. These changes are predicted to have major effects on food production. Rising carbon dioxide levels are projected to have some beneficial effect on crop yields, however, these benefits are overcome by adverse weather events in current scenarios. Under these scenarios the main growing regions for cereal crops will move further north where you can see gains in food production. However, current grain growing regions will suffer large losses in production. Additionally, the regions associated with the highest population growth in the next 75 years will suffer the greatest losses in food production capability.

The purpose of this paper is not to argue the causes of the documented increases in global atmosphereic carbon dioxide (CO_2) or the documented increases in average global temperature. Our focus instead is to look at the impacts of this increase in global temperature if it continues and what steps might be taken to reduce the negative impacts of global warming. The ability to adapt to population growth and global warming is not uniform across the world (Godbar and Wall, 2014).Livestock production contributes significantly to sustainable food security for many nations, particularly in low income areas and marginal habitats that are unsuitable for crop production. Animal products account for approximatelyone-third of global human protein consumption. Godbar and Wall (2014) utilized a range of indicators, derived from FAOSTAT and World Bank statistics, are used to model the relative vulnerability of nations at the global scale to predicted climateand population changes, which are likely to impact on their use of grazing livestock for food.

These investigators identified nations within sub-Saharan Africa, particularly in the Sahel region, and some Asian nations as the most likely to be vulnerable. They state in their paper that "Livestock-based food security is already compromised in many areas on these continents and suffers constraints from current climate in addition to the lack of economic and technical support allowing mitigation of predicted climate change impacts. Governance is shown to be a highly influential factor and, paradoxically, it issuggested that current self-sufficiency may increase future potential vulnerability because trade networks are poorly developed. This may be relieved through freer trade of food products, which is also associated with improved governance. Policy decisions, support and interventions will need to be targeted at the most vulnerable nations, but given the strong influence of governance, to be effective, any implementation will require considerable care in the management of underlying structural reform."

In addition to differences in world vulnerability to climate change there is also relatively large differences in emissions of greenhouse gases. Improved agricultural productivity will decrease the output of greenhouse gases and reduce impact of animals on the environment. An example of this can be seen in Table 2. If you compare the number of dairy animals in the United States and Pakistan for the year 2013, the U.S. had 8.7 million dairy cattle providing 196 million tons of milk to feed a population of 318 million people and still had sufficient milk and hay to export both. Pakistan had a population of 38 million dairy cows and 45 million water buffalo producing 29 million tons of milk for 176 million people and were chronically short of milk and forage. Improving productivity of dairy animals in Pakistan will increase their total milk output and reduce the number of animals to feed leading to sufficient forage supplies and sufficient milk to feed their people.

Table 2 Comparison of dairy and human populations and milk output for the united states and pakistan in 2013

Impact of Technology	
United States	Pakistan
○ Land: 3,539,224 sq miles ○ 318 million people ○ 8.7 million dairy cattle in 23 states ○ 196.3 million tons of milk ○ Exporting milk and hay	○ Land: 307,374 sq miles ○ 176 million people ○ 38 million dairy cattle ○ 45 million water buffalo ○ 29 million tons of milk ○ Milk shortage ○ Forage shortage

Not all impacts of global warming are negative. To a large degree, the impact of climate change on agriculture will be greatly influenced by the "carbon fertilization" effect (Stern, 2007). Since CO_2 is a basic requirement for plant growth, increasing the concentration in the atmosphere may enhance plant growth and produce yield offsetting reductions in yield due to heat and water stress.

Several models have been developed examining the links between climate change and crop-growth to project future changes in crop yields and food supply across the globe (Parry et al., 2005; Fischer et al., 2005). These predictions from these models, and those developed by the IPCC, predict that reduced crop yields associated with increased temperatures and drought will be offset by the direct fertilization effect of rising CO_2 concentration.

Currently, at its full effect the carbon fertilization effect may even result in a net rise in crop yields when averaged across the globe. However, the majority of studies indicate a slight to moderate (0-5%) negative impact on simulated worldcrop yields (Parry et al., 2004; Parry et al., 2005). These studies were based on data from responses of crops to elevated carbon dioxide obtained from studies in greenhouses and laboratory controlled-environment chambers (Stern, 2007). However, recent analyses from field grown crops grown under commercial conditions indicates that the carbon fertilization effect is potentially less than 50% of that typically included in the crop models (Long et al., 2005, 2006). This is likely because crops grown in commercial green houses have a much more stable environment than crops grown in the field. When a weak carbon fertilization effect isused in the models, global cereal production declines by 5% for a 2℃ increase intemperature and 10% for a 4℃ increase (Stern, 2007).

Another major factor in the long term agricultural productivity curve will be the discovery and implementation of new science – based agricultural tools. The application of these new tools hold great promise for tackling the world's growing population and food demands. These would include improved seeds and crop fertilization, protection, and cropping techniques, use of modern animal nutrition and reproductive programs, use of precision agriculture to reduce crop and animal production, mobile technology for farmers in the fields, to making foods fresher, safer, and healthier along the food chain, improved infrastructure to increase water availability and access to markets.

As pointed out by Wik et al., "Research on Total Factor Productivity (TFP) is pointing in the same direction. Output growth during the 1960s and 1970s was mainly a result of increased input use, and TFP growth rates were negative for most developing regions. It is encouraging to note that TFP growth rates have become positive in the 1981–2000 period. Even Asia and Sub – Saharan Africa were showing solid TFP growth rates for this period. Increased growth in TFP might offer the only feasible solution to the problem of ensuring continued productivity growth. This will require investments in research and development as well as institutional reforms and infrastructure development."

References

Avendano-Reyes, L., Alvarez-Valenzuela, F. D., Correa-Calderon, A., et al., 2006. Effect of cooling Holstein cows during the dry period on postpartum performance under heat stress conditions [J]. Livest. Sci., 105: 198-206.

Baumgard, L. H., Rhoads, R. P., 2013. Effects of heat stress on postabsorptive metabolism and energetics [J]. Ann. Rev. Anim. Biosci., 1: 311-337.

Beam, S. W., Butler W. R., 1999. Effects of energy balance on follicular development and first ovulation in postpartum dairy cows [J]. J. Reprod. Fertil., 54: 411-424.

Biggers, B. G., Geisert, R. D., Wettemann R. P., et al., 1987. Effect of heat stress on early embryonic development in the beef cow [J]. J. Anim. Sci., 64: 1512-1518.

Bilby, T. R., Sozzi, A., Lopez, M. M., et al., 2006. Pregnancy, bovine somatotropin, and dietary n-3 fatty acids in lactating dairy cows: I. Ovarian, conceptus and growth hormone-Insulin-like growth factor system responses [J]. J. Dairy Sci., 89: 3375-3385.

Bilby, T. R., Baumgard, L. H., Rhoads, M. L., et al., 2009. Pharmacological, nutritional and managerial strategies to improve fertility during heat stress in lactating dairy cows [R]. Proceeding of XIII Curso Novos Enfoques na Producao e Reproducao de Bovinos. pg. 59-71. Uberlandia, Brazil. March 12-13th.

Block, J., Hansen, P. J., 2007. Interaction between season and culture with insulin-like growth factor-1 on survival of *in vitro* produced embryos following transfer to lactating dairy cows [J]. Theriogenology, 67: 1518-1529.

Butler, W. R., 2000. Nutritional interactions with reproductive performance in dairy cattle [J]. Anim. Reprod. Sci., 60-61: 449-457.

CCSP, 2008. The effects of climate change on agriculture, land resources, water resources, and biodiversity in the United States [R]. A report by the U.S. Climate Change Science Program and the Subcommitee on Global Change Research. Backlund, P., Janetos, A., Schimael, D., et al., R.

Shaw. U.S. Department of Agriculture, Washington, D. C., USA, 362 pp.

Collier, R. J., Beede, D. K., Thatcher, W. W., et al., 1982a. Influences of environment and its modification on dairy animal health and production [J]. J. Dairy Sci., 65: 2213-2227.

Collier, R. J., Doelger, S. G., Head, H. H., et al., 1982b. Effects of heat stress during pregnancy on maternal hormone concentrations, calf birth weight and postpartum milk yield of Holstein cows [J]. J. Anim. Sci., 54: 309-319.

De la Sota, R. L., Burke, J. M., Risco, C. A., et al., 1998. Evaluation of timed insemination during summer heat stress in lactating dairy cattle [J]. Theriogenology, 49: 761-770.

Drost, M., Ambrose, J. D., Thatcher, M. J., et al., 1999. Conception rates after artificial insemination or embryo transfer in lactating dairy cows during summer in Florida [J]. Theriogenology, 52: 1161-1167.

Edwards, J. L., Hansen, P. J., 1997. Differential responses of bovine oocytes and preimplantation embryos to heat shock [J]. Mol. Reprod. Dev., 46: 138-145.

FAO, WFP, IFAD, 2012. The State of Food Insecurity in the World 2012. Economic Growthis Necessary But Not Sufficient to Accelerate Reduction of Hunger and Malnutrition [EB/OL]. FAO, Rome.

FAOSTAT, 2013. FAO statistical databases [EB/OL]. Available at: http: //faostat. fao. org/.

IPCC, 2014. Climate Change 2014: Synthesis Report. Contribution of Working Groups I, II and III to the Fifth Assessment Report of the Intergovernmental Panel on Climate Change [Core Writing Team, R. K. Pachauri and L. A. Meyer (eds.)] [R]. IPCC, Geneva, Switzerland, 151 pp.

Fischer, G., Shah, M., Tubiello, F. N., et al., 2005. Socio-economic and climate change impacts on agriculture: an integrated assessment, 1990-2080 [J]. Philos. T. Roy. Soc. B., 360: 2067-2083.

Garnsworthy, P. C., Webb, R., 1999. The influence of nutrition on fertility in dairy cows [D]. In: P. C. Garnsworthy and J. Wiseman (eds), Recent Advances in Animal Nutrition, 1999, Nottingham University Press, UK), pp. 39-57.

Gwazdauskas, F. C., Thatcher, W. W., Kiddy, C. A., et al., 1981. Hormonal patterns during heat stress following PGF2α-tham salt induced luteal regression in heifers [J]. Theriogenology, 16: 271-285.

Hansen, P. J., 2004. Physiological and cellular adaptations of zebu cattle to thermal stress. Animal Reprod [J]. Science, 82-83: 349-360.

Hansen, P. J., Arechiga, C. F., 1999. Strategies for managing reproduction in the heat-stressed dairy cow [J]. J. Anim. Sci., 77 (Suppl. 2): 36-50.

Holter, J B, West, J. W., McGillard, M. L., et al., 1996. Predicating ad libitum dry matter intake and yields of Jersey cows [J]. J. Dairy Sci., 79: 912-921.

Ju, J-C., Parks, J. E., Yang, X., 1999. Thermotolerance of IVM-derived bovine oocytes and embryos after short-term heat shock [J]. Mol. Reprod. Dev., 53: 336-340.

Klinedinst, P., Wilhite, D. A., Leroy Hahn, G., et al., 1993. The potential effects of climate change on summer season dairy cattle milk production and reproduction [J]. Climatic Change, 23: 21-36.

Lohmar, B., 2015. "Will China Import More Corn?" [EB/OL]. Choices. Quarter 2. Available online: http: //choicesmagazine. org/choices-magazine/theme-articles/2nd-quarter-2015/will-china-import-more-corn.

Long, S. P., Ainsworth, E. A., Leakey, A. D. B., et al., 2005. Global food insecurity. Treatment of major food crops with elevated carbon dioxide or ozone under large-scale fully open-air conditions suggests recent models may have over estimated future yields [J]. Philos. T. Roy. Soc. B., 360: 2011-2020.

Long, S. P., Ainsworth, E. A., Leakey, A. D. B., et al., 2006. Food for Thought: Lower-Than-Expected Crop Yield Stimulation with Rising CO_2 Concentrations [J]. Science, 312: 1918-1921.

López-Gatius, F., Santolaria, P., Yániz, J. L., et al., 2004. Timing of early foetal loss for single and twin pregnancies in dairy cattle [J]. Reprod. Domest. Anim., 39: 429-433.

McDowell, R. E., Moody, E. G., Van Soest, P. J., et al., 1969. Effect of heat stress on energy and water utilization of lactating dairy cows [J]. J. Dairy Sci., 52: 188.

Mihm, M., Bagnisi, A., Boland, M. P., et al., 1994. Association between the duration of dominance of the ovulatory follicle and pregnancy rate in beef heifers [J]. J. Reprod. Fertil., 102: 123-130.

Monty, D. E., Racowsky, C., 1987. *In vitro* evaluation of early embryo viability and development in summer heat-stressed, superovulated dairy cows [J]. Theriogenology, 28: 451-465.

Nebel, R. L., Jobst, S. M., Dransfield, M. B. G., et al., 1997. Use of radio frequency data communication system, HeatWatch ®, to describe behavioral estrus in dairy cattle [J]. J. Dairy Sci., 80 (Suppl. 1): 179. (Abstr.)

Ortiz, X. A, Smith, J. F., Rojano, F., et al., 2014. Conductive cooling as an alternative to cool down dairy cows. J. Anim. Sci., 92, E-Suppl. 2/J. Dairy Sci., 97, E-Suppl. 1. (Abstr.).

Parry, M., Rosenzweig, C., Iglesias, A., et al., 2004. Effects of climate change on global food production under SRES emissions and socio-economic scenarios [J]. Global Environ. Chang., 14: 53-67.

Parry, M., Rosenzweig, C., Livermore, M., 2005. Climate change, global food supply and risk of hunger [J]. Philos. T. Roy. Soc. B., 360: 2125-2138.

Paula-Lopes, F. F., Chase, C. C., Al-Katanani, Y. M. C. E., et al., 2003. Genetic divergence in cellular resisitance to heat shock in cattle: differences between breeds developed in temperate versus hot climates in responses of preimplantation embryos, reproductive tract tissues and lymphocytes to increased culture temperatures [J]. Reprod., 125: 285-294.

Putney, D. J., Drost, M., Thatcher, W. W., 1988. Embryonic development in superovulated dairy cattle exposed to elevated ambient temperature between days 1 to 7 post insemination [J]. Theriogenology, 30: 195-209.

Putney, D. J., Drost, M., Thatcher, W. W., 1989. Influence of summer heat stress on pregnancy rates of lactating dairy cattle following embryo transfer or artificial insemination [J]. Theriogenology, 31: 765-778.

Rocha, A., Randel, R. D., Broussard, J. R., et al., 1998. High environmental temperature and humidity decrease oocyte quality in Bos taurus but not in Bos indicus cows [J]. Theriogenology, 49: 657-665.

Roth, Z., Arav, A., Bor, A., et al., 2001. Improvement of quality of oocytes collected in the autumn by enhanced removal of impaired follicles from previously heat-stressed cows [J]. Reproduction, 122: 737-744.

Rutledge, J. J., Monson, R. L., Northey, D. L., et al., 1999. Seasonality of cattle embryo production in a temperate region [J]. Theriogenology, 51 (Suppl. 1): 330. (Abstr).

Ryan, D. P., Boland, M. P., 1991. Frequency of twin births among Holstein X Friesian cows in a warm dry climate [J]. Theriogenology, 36: 1-10.

Spain, J. N., Spiers, D. E., Snyder, B. L., 1998. The effects of strategically cooling dairy cows on milk production [J]. J. Anim. Sci., 76 (Suppl. 1): 103.

Spencer, J. D., Gaines, A. M., Berg, E. P., et al., 2005. Diet modifications to improve finishing pig growth performance and pork quality attributes during periods of heat stress [J]. J. Anim. Sci., 83: 243-254.

Stern, N., 2007. The economics of climate change—The stern review [M]. Cambridge, UK: Cambridge Univcrsity Prcss.

St-Pierre, N. R., Cobanov, B., Schnitkey, G., 2003. Economic losses from heat stress by US livestock industries [J]. J. Dairy Sci., 86: E52-E77.

Thatcher, W. W., Collier, R. J., 1986. Effects of climate on bovine reproduction [M]. In: D. A. Morrow (Ed.) Current Therapy in Theriogenology 2. W. B. Saunders, Philadelphia.301-309.

United Nations (UN), Department of Economic and Social Affairs, Population Division, 2015. World Population Prospects: The 2015 Revision, Key Findings and Advance Tables [EB/OL], Working Paper No. ESA/WP241.

Urdaz, J. H., Overton, M. W., Moore, D. A., et al., 2006. Technical Note: Effects of adding shade and fans to a feed bunk sprinkler system for preparturient cows on health and performance [J]. J. Dairy Sci., 89: 2000-2006.

Wiersma, F., Armstrong, D. V., 1988. Evaporative cooling dry cows for improved performance [R]. ASAE paper no. 88-4053, St. Joseph, MI.

Wilson, S. J., Marion, R. S., Spain, J. N., et al., 1998. Effects of controlled heat stress on ovarian function of dairy cattle [J]. J. Dairy Sci., 81: 2139-2144.

Wolfenson, D., Thatcher, W. W., Badinga, L., et al., 1995. Effect of heat stress on follicular development during the estrous cycle in lactating dairy cattle [J]. Biol. Reprod., 52: 1106-1113.

Wolfenson, D., Roth, Z., Meidan, R., 2000. Impaired reproduction in heat-stressed cattle: basic and applied aspects [J]. Anim. Reprod. Sci., 60/61: 535-547.

Zeigler, M. Steensland, A., 2016. Global Agricultural Productivity Report [R]. Global Harvest Initiative, Washington, D. C.

Session 1

Advances in Dairy Cow Nutrition

1

Modifying Gut Microbiota to Enhance Gut Health in Dairy Cows

J. C. (Kees) Plaizier[1], Hooman Derakhshani[1], Ehsan Khafipour[1,2]

[1]Department of Animal Science, University of Manitoba, Winnipeg, MB, Canada;

[2]Department of Medical Microbiology, University of Manitoba, Winnipeg, MB, Canada

Abstract

The symbiosis between cows and their commensal gut bacteria is critical for their health and production. High yielding dairy cows are commonly fed high-grain diets in order to meet their high energy requirements, and it has been assumed that this negatively affects the gut microbiota of these cows. Until recently the technology to determine these effects comprehensively was not available. However, high-throughput genetic sequencing techniques are now available and they have been used to determine the impact of dietary changes on the composition and functionality of entire gut microbiota. This research has shown that core microbiota are robust to large increases in high grain feeding at the phylum and genus level, but that this results in changes in this composition at the species level. However, these changes vary greatly among studies. Most of these changes at the species level reflect changes in the availability of substrates and niches in digesta. Most studies agree that high grain feeding reduces the richness and diversity of gut microbiota, which is considered adverse to their functionality. Several supplements, including as yeasts, yeast culture products, probiotics, and polyphenols, have shown promise to overcome the negative effects of high-grain feeding on gut microbiota. The effects of these supplements also vary among experiments, but differences in the design and technology among these experiments contributed to this variation of these effects. In order to develop efficient supplementation strategies, the mechanisms behind the effects of these supplements on the composition and functionality of gut microbiota need to be better understood. Obtaining this understanding is held back by challenges in the assessment of the functionality of gut microbiota.

Introduction

The survival of ruminants, including dairy cows, relies on symbiotic relationships with their-commensal gut microbiota (Russell and Rychlik, 2001). Without this symbiosis, ruminants would not be able to utilizeplant fiber, convert non-protein nitrogen into microbial protein, detoxify various toxins in the digesta, produce several vitamins, and prevent the establishment of several pathogenic microorganisms in the digestive tract (Russell and Rychlik, 2001; Plaizier et al., 2008; Penner, 2016). Hence, optimizing of the composition and functionality of the gut microbiota is critical for the health and production of dairy cows.

Current feeding practices can jeopardize the symbiosis between the cows and its gut microbiota. The high energy requirements of high yielding dairy cows necessitate the feeding of high energy diets. These diets are commonly high in grain and low in forage, which increases the starch content of digesta and fermentation in the foregut, whilereducing chewing, saliva production and rumen buffering (Plaizier et al., 2008; Plaizier et al., 2012; Penner, 2016). Excessive grain feeding can also affect the microbiome and increase fermentation in the hindgut, most likely via the increasein by-pass starch that escapes rumen fermentation and small intestine digestion (Khafipour et al., 2009; Mao et al., 2012; Petri et al., 2013).

The dietary changes caused by raising the dietary grain content include increases of the acidity, concentrations of organic acids, and osmolality, and alterationsof the availability of readily fermentable carbohydrates for the various microorganisms in the digestive tract. In addition, Ametaj et al. (2010) and Saleem et al. (2012) demonstrated that excessive grain feeding increased the concentrations of many metabolites in rumen digesta. Many of these metabolites were substrates for beneficial microorganisms, but some of which were toxic and inflammatory (Khafipour et al., 2016). Hence, high grain feeding changes the environment and the availability of substrates and niches of microorganisms in digesta of cows greatly. As the microorganisms in this digesta vary in their sensitivity to these changes (Russell and Rychlik, 2001), high grain feeding could alter the composition and functionality of these microbiota. Despite of this, Weimer (2015) concluded that rumen microbiota have a high resilience to dietary changes, which may be explained by the existence of overlapping genes and functional redundancy among microbial taxa (Russell and Rychlik, 2001; Firkins and Yu, 2015).

Until recently, the technology was not available to monitor the effects of dietary changes on the composition and functionality of gut microbiota comprehensively, as most of these microorganisms cannot be cultured, and quantitative PCR techniques do not target all microorganisms (Krause et al., 2013; Khafipour et al., 2016). Recent advances in molecular - based sequencing technologies offer methodologies that can investigate microbial communities as a whole (Krause et al., 2013). These new techniques are either based on high-throughput sequencing of the hypervariable regions of theconserved and universal 16S rRNA genes for bacterial and archaeal communities (Woese and Fox, 1977; Pace et al., 1985), and 18S or the internal transcribed spacer (ITS) regions of rRNA genes for fungal and protozoal communities (Firkins and Yu, 2015). In addition, massive shotgun sequencing of total DNA (metagenomics) and RNA (metatranscriptomics) from microbial communities have become available (Desai et al., 2012). These technologies commonly utilize "operational taxonomic units" (OTUs) instead of "species" (Krause et al., 2013). A unique OTU is defined as a cluster of sequence reads with a given similarity that can be assigned to a taxonomical level. Sequences with 97% and 99% similarity approximately correspond to the genus and species, respectively (Krause et al., 2013; Khafipour et al., 2016). As changes in the composition of microbiota do not necessarily change the functionality of these microbiota, assessment of changes in functionality requires functional metagenomics, metatranscriptomic, and possibly proteomic and metabolomics analyses (Firkins and Yu, 2015), and ultimate animal performance studies (Khafipour et al., 2016).

Different ecological measures, such as richness, relative abundance, evenness, and diversity, are used to describe and compare microbiotas (Gotelli and Colwell, 2010). Richness refers to the number of different OTUs that are present in a community. In addition to the number of OTUs, diversity also takes the distributionand evenness of unique OTUs into account. Gut microorganisms differ in their functionality and their ability to use different substrates in the digestive tract (Levine and D'antonio, 1999; Henderson et al., 2015). Hence, high microbiota richness and diversity, via providing a more comprehensive genetic pool, enable efficient use of these resources, and enhance the stability of these microbiota (Russell and Rychlik, 2001; Ley et al., 2006a). As a result, high richness and diversity of gut microbiota are most often considered as a beneficial trait for their host animal.

Nutritional Challenges to Gut Health

High grain feeding practices can jeopardize the functionality of gut microbiota. It is expected that increases in grain feeding and then accompanying reduction in fiber feeding reduce the relative abundance of fibrolytic bacteria, and increase that of amylolytic and lactic acid utilizing bacteria (Russell and Rychlik, 2001; Henderson et al., 2015; Khafipour et al., 2016). However, large increases in grain feeding could also increase populations of opportunistic and pathogenic bacteria (Russell and Rychlik, 2001; Khafipour et al., 2009; Plaizier et al., 2012). Khafipour et al. (2009), Mao et al. (2013), and Petri et al. (2013) observed that the induction of subacute ruminal acidosis (SARA) by high grain feeding reduced bacterial richness and diversity in the rumen and led to a decline in Bacteroidetes and an increase in Firmicutes in the rumen. High ratios between Firmicutes to Bacteroidetes in the digestive tract of human and mice have been associated with increased energy harvesting and obesity (Ley et al., 2006b). Jami et al., (2014) showed that high ratios between these phyla in the rumen of dairy cows were correlated with milk fat yields. This is surprising, as grain-induced SARA is commonly associated with milk fat depression. This suggests that the Firmicutes-to-Bacteroidetes ratio has traditionally been considered as a biomarker for metabolic potential of the rumen microbiota in cows, but that the optimal value of this ratio has yet to be determined. These phyla include many of the important fibrolytic, amylolytic, and proteolytic rumen bacteria, and the proper balance between these bacteria to allow efficient rumen function may depend on the contents of the rumen digesta.

Plaizier et al. (2016) induced SARA in dairy cows by excessive grain feeding (GBSC), and by feeding pellets of ground forage (APSC). Both SARA inductions reduced the bacterial richness and diversity in rumen fluid, but the GBSC had the larger effects. The microbiota of GBSC also clustered differently from control feeding in cecal digesta and feces. The microbiota of APSC also clustered differently from control feeding in feces. Only GBSC reduced bacterial richness and diversity in feces. The abundances of Bacteroidetes and Tenericutes in rumen fluid were decreased by GBSC, but not by APSC. Effects of the SARA challenges on most major bacterial phyla and genera in the rumen, cecum and feces were absent or limited. Both SARA challenges increased the abundances of several bacterial species that utilize non-structural carbohydrates and their metabolites in the rumen, cecum, and feces. These increases were larger than the increases in the abundances of the genera and phyla to which thesespecies belonged. Both challenges decreased the abundance of *Streptococcus bovis*, and only GBSC increased that of *Megasphaera elsdenii* in the rumen. Differences in the starch content of rumen and hindgut digesta between the GBSC and the APSC may have contributed to the dissimilarities in the gut microbiomes between these challenges.

Plaizier et al. (2017) induced SARA in lactating dairy cows by excessive grain feeding, and also observed that this induction reduced the richness and diversity of the microbiota in the rumen fluid and feces, without affecting the relative abundances of the most abundant phyla. The SARA challenge affected the abundances of many lower taxa in the rumen fluid and feces, but many of these effects were difficult to explain by the change in the chemical composition of the diet and digesta.

Tun et al. (2017) predicted the effects of grain-induced SARA on the functional capacity of rumen bacterial community based on 16S rRNA-based predictions using Phylogenetic Investigation of Communities by Reconstruction of Unobserved States (PICRUSt). This showed that metabolism of terpenoids and polyketides, biosynthesis of other secondary metabolites, amino acid metabolism, signalling molecules and interaction as well as transport and catabolism were under-represented during SARA, whereas transcription, signal transduction, membrane transport, and cell motility were over-represented. The SARA challenge also affected 39 of 254 level-3 KEGG Orthology groups. It is yet unclear to what extend the shifts in composition and functionality of the rumen ecosystem during SARA were responsible for the symptoms of this disorder that were observed in the cows. However, it is unlikely that these shifts were solely responsible for these systems.

Large increases in grain in the diet of cattle have also been shown to increase the content of bacterial endotoxins, such as lipopolysaccharides (LPS) in the digesta of the rumen and hindgut (Plaizier et al., 2008; Plaizier et al., 2012; Saleem et al., 2012; Plaizier et al., 2014).

Most studies agree that high grain feeding reduces the richness, diversity, stability, and affectsfunctionality of microbiota in the rumen and the large intestine, but the results vary greatly among studies. High grain feeding can benefit some microorganisms by providing more substrate and specific niches, but it can be detrimental to microorganisms that rely on fiber and are sensitive to a low pH of digesta or any of the gut metabolites and compounds whose concentrations are increased by grain feeding. Differences among studies may be the result of complexity of the microbiota in the digestive tract of cow, and differences between the designs, sizes, inductions of SARA, and sequencing techniques used in these studies (Mao et al., 2013; Petri et al., 2013; Weimer, 2015). As a result, excessive grain feeding jeopardizes the health and production of the animals as well as the environmental sustainability of ruminant production systems. A challenge in determining how much grain feeding is excessive is that the grains and grain-processing techniques vary in their impact on gut health and the ruminants vary in their susceptibility to the adverse effects of high-grain feeding (Weimer, 2015; Khafipour et al., 2016)

Solutions to the Effects of High Grainfeeding on Gut Microbiota

Several feed supplements have been considered to attenuate the adverse effects of high grain feeding on gut microbiota. Alzahal et al. (2014) determined the effects of feeding active dry *Saccharomyces cerevisiae* yeast on the rumen microbiome during high grain using quantitative PCR (qPCR). Supple-mentation with the yeast increased the populations of *Fibrobacter succinogenes*, *Anaerovibrio lipolytica*, *Ruminococcus albus*, *S. bovis* and anaerobic fungi 9, 2, 6, 1.3, 2.3, and 8 fold, respectively, and decreased the populations of *Prevotella albensis* and *M. elsdenii* 2.2 and 12 times, respectively. The authors concluded from these results that the yeast supplementation increased cellulolytic bacteria and reduced lactate production, which can attenuate the decreases in cellulolytic bacteria and the increases in lactate production that are commonly observed during increase in grain feeding. During the high grain feeding, the yeast supplementation increased dry matter intake, fat corrected milk yield, and the rumen fluid concentrations of total VFA and propionate.

Chiquette et al. (2015) investigated effects of supplementation of *Enterococcus faecium* alone or in combination with *S. cerevisiae* yeast or *Lactococcus lactis* during control feeding and a subacute ruminal acidosis (SARA) challenge on rumen microbiota. During SARA, the supplementation of *E. faecium* and *S. cerevisiae* resulted in the highest protozoa populations. Bacterial profiles clustered by feeding regimen (control or SARA). These profiles were more scattered during SARA than during control, which shows that ruminal instability and variability among animals was greater during SARA than during control.

De Nardi et al. (2016) induced SARA in heifers by high grain feeding and determined the effects offumarate - malate and polyphenol - essential oil on the rumen microbiota of these animals. Both supplements altered rumen microbiota including a reduction in *Prevotella brevis* and an increase in Christensenellaceae. Both supplements also increased the richness and diversity of rumen microbiota, with polyphenol–essential oil having the larger effect. This effect was attributed to specific antimicrobial activities of polyphenol–essential oil.

Tun et al., (2017) determined the effects of *S. cerevisiae* fermentation product (SCFP, Original XPC, Diamond V, Cedar Rapids, IA) in lactating dairy cows during control feeding and SARA induced by excessive grain feeding. This product does not contain active yeast, but does contain dead yeast, yeast growth media, and fermentation products that are rich in vitamins, minerals, oligosaccharides, organics acids, amino acids, peptides and β - glucans (Callaway and Martin, 1997). The study from Tun et al. (2017) showed that, based on pyrosequencing, several of the adverse impacts of the SARA challenge, including the reductions of the richness of diversity of the rumen microbiome, the reductions in the abundance of Bacteroidetes and the reductions in the populations of ciliate protozoa were attenuated by SCFP. However, the effects of SCFP on the richness and diversity and the relative abundances of the major phyla determined by shotgun metagenomics were less than those determined by pyrosequencing. Across diets and based on qPCR results, SCFP increased the populations of several of the major fibrolytic and amylolytic microorganisms, including *P. brevis*, *Ruminococcus flavefaciens*, ciliate protozoa, and *Bifidobacterium* spp. The SCFP did not affect feed intake, yields of milk, milk fat, and milk protein, but it tended to reduce ruminal LPS endotoxin, and increased milk fat percentage.

Tun et al. (2017), also used PICRUSt as a predictive tool for the functional properties of rumen microbiota and this did not show effects of SCFP supplementation on these properties However, in this study the annotation of the metagenome by shotgun sequencing showed that SCFP increased secondary metabolism and stress response. During SARA, SCFP reduced cell wall and capsule, nitrogen metabolism, and sulfur metabolism, and increased phosphorus metabolism and nucleosides and nucleotides. Hence, this study highlighted that the technology used to assess the effects of SCFP on the composition and functionality of microbiota has very substantial impacts on the observed effects.

Conclusions

These adverse effects of high grain diets on gut microbiota in dairy cows can be attenuated by the use of supplements, such as yeasts, yeast culture products, probiotics, and polyphenols. However, in order to develop efficient strategies, the effects of these supplements on the composition and functionality of gut microbiota in the digestive tract need to be better understood. This requires high-throughput sequencing, as entire microbiomes, and not only several taxa need to be assessed. Even then, the predictions of changes in this functionality are still difficult to predict from changes in the composition of these microbiota due to challenges with the currently available technology.

References

AlZahal, O., Dionissopoulos, L., Laarman, A., et al., 2014. Active dry *Saccharomyces cerevisiae* can alleviate the effect of subacute ruminal acidosis in lactating dairy cows [J]. J. Dairy Sci., 97 (12): 7751-7763.

Ametaj, B. N., Zebeli, Q., Saleem, F., et al., 2010. Metabolomics reveals unhealthy alterations in rumen metabolism with increased proportion of cereal grain in the diet of dairy cows [J]. Metabolomics, 6 (4): 583-594.

Callaway, E., Martin, S., 1997. Effects of a *Saccharomyces cerevisiae* culture on ruminal bacteria that utilize lactate and digest cellulose [J]. J. Dairy Sci., 80 (9): 2035-2044.

Callaway, T. R., Dowd, S. E., Edrington, T. S., et al., 2010. Evaluation of bacterial diversity in the rumen and feces of cattle fed different levels of dried distillers grains plus solubles using bacterial tag-encoded FLX amplicon pyrosequencingl [J]. J. Anim. Sci., 88 (12): 3977-3983.

Chiquette, J., Lagrost, J., Girard, C., et al., 2015. Efficacy of the direct-fed microbial *Enterococcus faecium* alone or in combination with *Saccharomyces cerevisiae* or *Lactococcus lactis* during induced subacute ruminal acidosis [J]. J. Dairy Sci., 98 (1): 190-203.

De Nardi, R., Marchesini, G., Li, S., et al., 2016. Metagenomic analysis of rumen microbial population in dairy heifers fed a high grain diet supplemented with dicarboxylic acids or polyphenols [J]. BMC, 12 (1): 1-9.

Desai, S., Eu, Y. J., Whyard, S., et al., 2012. Reduction in deformed wing virus infection in larval and adult honey bees (*Apis mellifera* L.) by double strained RNA ingestion [J]. Insect Mol. Biol., 21 (4): 446-455.

Firkins, J. L., Yu, Z., 2015. Ruminant Nutrition Symposium: How to use data on the rumen microbiome to improve our understanding of ruminant nutrition [J]. J. Anim. Sci., 93 (4): 1450-1470.

Gotelli, N. J., Colwell, R. K., 2011. Estimating species richness [M]. Biological Diversity: Frontiers in Measurement and Assessment, 12: 39-54.

Henderson, G., Cox, F., Ganesh, S., et al., 2015. Rumen microbial community composition varies with diet and host, but a core microbiome is found across a wide geographical range [J]. Sci. Rep., 5: 14567.

Jami, E., White, B. A., Mizrahi, I., 2014. Potential role of the bovine rumen microbiome in modulating milk composition and feed efficiency [J]. PLoS One, 9 (1): e85423.

Khafipour, E., Li, S., Plaizier, J. C., et al., 2009. Rumen microbiome composition determined using two nutritional models of subacute ruminal acidosis [J]. Appl. Environ. Microb., 75 (22): 7115-7124.

Khafipour, E., Li, S., Tun, H., et al., 2016. Effects of grain feeding on microbiota in the digestive tract of cattle [J]. Anim. Front, 6 (2): 13-19.

Krause, D., Nagaraja, T., Wright, A., et al., 2013. Board-invited review: Rumen microbiology: Leading the way in microbial ecology [J]. J. Anim. Sci., 91 (1): 331-341.

Levine, J. M., D'Antonio, C. M., 1999. Elton revisited: a review of evidence linking diversity and invasibility [J]. Oikos, 87: 15-26.

Ley, R. E., Peterson, D. A., Gordon, J. I., et al., 2006a. Ecological and evolutionary forces shaping microbial diversity in the human intestine [J]. Cell, 124 (4): 837-848.

Ley, R. E., Turnbaugh, P. J., Klein, S., et al., 2006b. Microbial ecology: human gut microbes associated with obesity [J]. Nature, 444: 1022-1023.

Li, S., Yoon, I., Scott, M., et al., 2016. Impact of *Saccharomyces cerevisiae* fermentation product and subacute ruminal acidosis on production, inflammation, and fermentation in the rumen and hindgut of dairy cows [J]. Anim. Feed Sci. Tech., 211: 50-60.

Mao, S., Zhang, R., Wang, D., et al., 2012. The diversity of the fecal bacterial community and its relationship with the concentration of volatile fatty acids in the feces during subacute rumen acidosis in dairy cows [J]. BMC, 8 (1): 1.

Mao, S. Y., Zhang, R. Y., Wang, D. S., et al., 2013. Impact of subacute ruminal acidosis (SARA) adaptation on rumen microbiota in dairy cattle using pyrosequencing [J]. Anaerobe, 24: 12-19.

Pace, N. R., Marsh, T. L., 1985. RNA catalysis and the origin of life [J]. Origins Life Evol. B., 16 (2): 97-116.

Penner, G. B., 2016. Influence of microbial ecology in the rumen and lower gut on production efficiency of dairy cows [R]. Pages 75-81 in Proc. Tri-State Dairy Nutrition Conference, 18-20 April 2016, Fort Wayne, Indiana, USA. 25th Anniversary. The Ohio State University.

Petri, R. M., Schwaiger, T., Penner, G. B., et al., 2013. Characterization of the Core Rumen Microbiome in Cattle during Transition from Forage to Concentrate as Well as during and after an Acidotic Challenge [J]. PLoS One, 8 (12): e83424.

Plaizier, J., Khafipour, E., Li, S., et al., 2012. Subacute ruminal acidosis (SARA), endotoxins and health consequences [J]. Anim. Feed Sci. Tech., 172 (1): 9-21.

Plaizier, J. C., Krause, D. O., Gozho, G. N., et al., 2008. Subacute ruminal acidosis in dairy cows: the physiological causes, incidence and consequences [J]. Vet. J., 176 (1): 21-31.

Plaizier, J. C., Li, S., Danscher, A. M., et al., 2017. Changes in Microbiota in Rumen Digesta and Feces Due to a Grain-Based Subacute Ruminal Acidosis (SARA) Challenge [J]. Microb. Ecol., 74 (2): 485-495.

Plaizier, J. C., Li, S., Tun, H. M., et al., 2016. Nutritional models of experimentally-induced subacute ruminal acidosis (SARA) differ in their impact on rumen and hindgut bacterial communities in dairy cows [J]. Front. Microbiol., 7: 2128.

Russell, J. B., Rychlik, J. L., 2001. Factors that alter rumen microbial ecology [J]. Science, 292 (5519): 1119-1122.

Saleem, F., Ametaj, B. N., Bouatra, S., et al., 2012. A metabolomics approach to uncover the effects of grain diets on rumen health in dairy cows [J]. J. Dairy Sci., 95 (11): 6606-6623.

Weimer, P. J., 2015. Redundancy, resilience, and host specificity of the ruminal microbiota: implications for engineering improved ruminal fermentations [J]. Front. Microbiol., 6: 296.

Woese, C. R., Fox, G. E., 1977. Phylogenetic structure of the prokaryotic domain: the primary kingdoms [J]. P. Natl. Acad. Sci., 74 (11): 5088-5090.

Maximizing Fiber Digestion by Understanding and Manipulating the Ruminal Microbiome

T. R. Callaway[1] and B. D. Rooks[2]

[1]Food and Feed Safety Research Unit, USDA/ARS, 2881 F&B Rd. College Station, TX, USA, 77845. [2]BAN Consultants, Simpsonville, SC, USA

Introduction

The ruminant animal and its resident gastrointestinal microbial ecosystem is a marvel of evolution and symbiosis (Hungate, 1966). The pre-gastric fermentation process that is carried out by the microbial consortium allows cattle to be "factories … converting sunlight to meat, milk, and fiber" (Hungate, 1966). The ruminal fermentation is best characterized by its ability to catabolize fiber (including cellulose and hemicellulose) to produce Volatile Fatty Acids (VFA) and cell protein that provide energy and N to the mammalian host (Yokoyama and Johnson, 1988). One of the most critical factors underpinning the mutualistic relationship between microbes and the host ruminant animal is clearly the ability of the ruminal microbiome to degrade fiber (especially cellulose) and produce a source of energy for the animal.

Fibrolysis by the ruminal microbiome is an anaerobic fermentation process that is carried out by a mixed population of bacteria, protozoa, and fungi and was originally studied by Aristotle and others throughout the course of scientific history (Van Tappeiner, 1884). The anaerobic and highly reduced nature of the ruminal environment means that catabolic processes produce far less ATP per glucose than do aerobic catabolism (2-5 ATP versus 38 ATP) (Russell, 2002). Thus the ruminal fermentation is almost always energy limiting, and because the rumen is a highly reduced environment interspecies hydrogen transfer is an important step in maximizing ATP yield (Thiele and Zeikus, 1988; Russell, 2002; McAllister and Newbold, 2008). Recent years have seen the development of novel techniques (e.g., pyrosequencing, Next-Generation Sequencing, Whole Genome sequencing) that have vastly increased our ability to understand the dynamics of the ruminal microbial ecosystem (Firkins and Yu, 2015). However, despite the exponential increase in our knowledge of the bacterial populations, we still have not fully elucidated how the ruminal microbial population interacts with diet, host physiology, and other members of the microbial consortium.

In the present manuscript, we will describe some of the current theories of how we can increase fiber digestion in the rumen by manipulating the ruminal microbiome. While we will focus most closely on fiber degradation, the degradation and fermentation of protein and non-structural carbohydrates cannot be ignored when considering the holistic nutrition of the dairy or beef cow because of the interactions within the microbial ecosystem. Therefore when attempting to manipulate the ruminal fermentation to enhance efficiency, care must be taken to optimize the use of nutrient by the ruminal microbiome.

Fibrolysis: an Overview

Ruminal fiber catabolism is a complex process that is still not fully understood, nor are all the players involved in degradation of fiber known or characterized (Russell, 2002; Krause et al., 2003). Generally, when discussing fiber degradation, ruminant nutritionists tend to focus primarily upon the polymers cellulose, hemicellulose, and pectins which are fermentable. However, lignin plays an important role in fiber degradation because of its recalcitrant nature and the inhibitory impacts of lignin crosslinking (Akin and Benner, 1988; Jung, 1989; Deblois and Wiegel, 1990). Fiber is neither simple nor consistent in its composition, which varies between plant species (or strains), growth environment, and with plant maturity (Bach-Tuyet Lam et al., 1990).

Ruminal fungi, bacteria, and protozoa collectively and systematically catabolize fiber into increasingly smaller oligomers. Degradation of fiber is initially dependent upon colonization and physical contact between degrading organisms and the fiber (Flint et al., 2008). Mastication by the ruminant is a critical first step in puncturing the cutin layer, increasing surface area, and exposing the interior of the grass or hay to microbial attack and degradation (Yokoyama and Johnson, 1988). Forages are initially colonized by fungi which extend rhizoid into the forage, thereby opening the forage to degradation by fungi and cellulolytic bacteria (Bauchop, 1979b, a; Akin and Rigsby, 1987; Borneman et al., 1989). Fungal populations and the time needed for fungal rhizome growth can impact the rate of total fibrolysis (Bauchop, 1979b; Akin and Rigsby, 1987; Borneman and Akin, 1990). Colonization of fibrous feedstuffs by bacteria can be detected within 5 minutes of entry to the rumen, and is largely completed within 15 min (Edwards et al., 2007). It has been proposed that the degradation of forage follows a first - order rate constant, and thus the initial rate of colonization and breakdown plays a large impact on the total extent and efficiency of fiber degradation (Mourino et al., 2001) (Figure 1).

Fibrolytic bacteria that are associated with the particulate phase of the ruminal fluid (Craig et al., 1984; Wallace, 2008) utilize glycosyl hydrolases, such as endoglucanases, exoglucanases, and glucosidases; whereas hemicellulose are degraded by xylanases and other carbo-

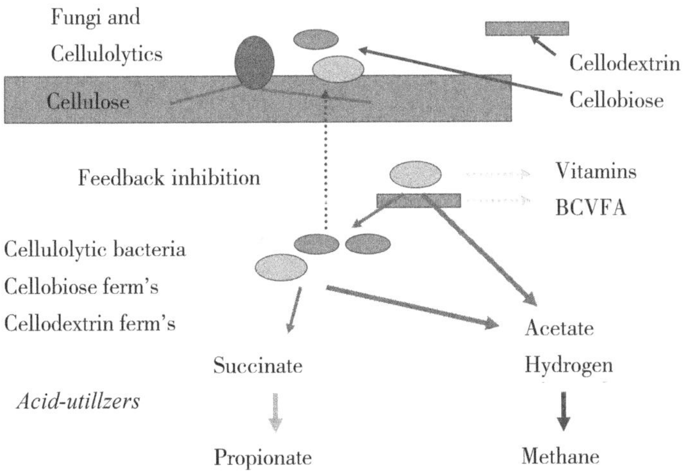

Figure 1 Colonization and distribution of endproducts of ruminal fiber fermentation

hydrate esterases to degrade the structural carbohydrates of forage (Krause et al., 2003; Ransom-Jones et al., 2012). Dextrins and oligosaccharides of a variety of sizes are released by the action of endo-/exo-glucanases (e.g., cellodextrins and cellobiose, along with oligomers of hemicellulose and pectin) which are further catabolized and fermented by protozoa and particle - and fluid - associated ruminal bacteria (Wells et al., 1995; Brulc et al., 2009). As dextrins, oligomers, and monomers are utilized by other members of the "food web", endproduct feedback inhibition is removed allowing fibrolytic bacteria to proceed breaking down cellulose and hemicellulose (Wells et al., 1995).

Lignin poses a separate case when considering ruminal degradation because a complex polymeric structural compound found in forages that makes them more difficult to degrade because of their chemical structure (Jung, 1989). Lignin is recalcitrant to microbial degradation because it forms ester and ether bonds to cellulose and hemicellulose. This occurs because lignin is comprised of p-coumaric, vanillic and ferulic acid which are toxic to some ruminal bacteria (Akin and Rigsby, 1987; Akin and Benner, 1988). As forages mature their proportion of lignin increase and therefore their digestibility decreases, resulting in a reduced rate and extent of degradation (Jung, 1989).

As the ruminal fermentation process continues the degradation of intermediates from fiber degradationand structural CHO monomers, interspecies hydrogen transfer occurs between cellulolytic /hemicellulolytic bacteria and methanogenic archaea (Thiele and Zeikus, 1988). Unfortunately, the highly reduced nature of the rumen means that CH_4 production is inextricably linked with fiber degradation as methane serves as a reducing equivalent sink (Johnson and Johnson, 1995; McAllister and Newbold, 2008; Martin et al., 2010). Furthermore, other interactions and crossfeeding of nutrients (such as pantothenic acid and branched chain fatty acids) amongst the members of the celluloytic consortium and the non-cellulolytic proportion of the population (e. g., providing NH_3 and possibly detoxifying tannic acids) are critical to the most efficient degradation of dietary fiber in the rumen (Allison et al., 1961; Scott and Dehority, 1965; Allison, 1978; Romero - Pérez et al., 2011). Thus the breakdown of fiber is not a simple enzymatic process, but is rather a dynamic, multi-factorial process that is reflective of several populations of microorganisms performing discrete functions that are interconnected in several ways to form a true "food web" throughout the ruminal ecosystem. While the fungi and protozoa play a large role throughout the process of fiber degradation and fermentation (Bauchop, 1979b; Joblin, 1990; Mosoni et al., 2011), in the present manuscript we will focus largely upon the bacterial contribution.

Who is Involved in Rumen Fiber Degradation?

The ruminal microbiome is comprised of a wide variety of bacteria, protozoa, fungi and viruses. To date, there is no estimate of the total diversity of the rumen, but a cursory meta-analysis of published pyrosequencing and NGS datasets indicates that well over 10, 000 bacterial species have been isolated from cattle rumens (Callaway, unpublished analysis). While not all of these bacteria are involved with fibrolysis directly, many are capable of fermenting the intermediate products (e.g., dextrins, cellobiose, glucose, or pentoses) of fiber catabolism (Russell, 2002; Krause et al., 2003). However it appears that a core population of particle-associated bacteria exists in cattle rumens, and this population is less plastic (less changeable) than the fluid-associated bacteria (Wallace, 2008; Henderson et al., 2015).

The best-studied and most well-known cellulolytic bacteria are *Fibrobactersuccinogenes*, *Ruminococcusalbus*, *Ruminococcusflavefaciens*, *Buyrivibriofibrosolvens*, *Pseudobutyrivibrio*, and some species of *Prevotella* are equipped with glycosyl hydrolases of various forms (Hungate, 1950; Hungate, 1966; Stewart and Bryant, 1988; Russell, 2002). Some strains of *F. succinogenes*, *R. flavefaciens*, and *R. albus* adhere simultaneously to the surface of plant materials (Miron et al., 2001; Shinkai and Kobayashi, 2007). Results demonstrated that there was a great deal of variation in the ability of *Fibrobacter* strains in their ability to adhere to and degrade fiber, but this was not correlated with total extent of fiber degradation (Denman and McSweeney, 2006; Shinkai et al., 2009). It was later demonstrated that *F. succinogenes* population increased after feeding but *R. flavefaciens* populations did not (Denman and McSweeney, 2006). Researchers further indicated that when using hay stems as a fiber source for ruminal fluid *in vitro* fermentations *F. succinogenes* could come to dominate a consortium, but that *Butyrivibriofibrisolvens*, *Pseudobutyrivibrioruminis*, *Clostridium* sp., *Prevotellaruminicola*, unclassified *Bacteroides*, *Treponemabryantii*, *Acinetobacter* sp, and *Wolinellasuccinogenes* were all suggested as prominent members of the fiber-degrading consortium (Shinkai et al.,

2010). It was suggested that the ability of the spirochaetes (such as *Treponema*) to move through the ruminal fluid may benefit the fibrolytic consortium (Shinkai et al., 2010), possibly by being chemotactically attracted to areas of high nutrient concentration. As more information is mined out of the genome of these cellulolytics, it is expected that we will understand more of how these organisms survive in the rumen and compete for nutrients (Suen et al., 2011a; Suen et al., 2011b).

Thus it appears that while the traditional "cellulolytic/hemicellulolytic" bacteria are still important, there are other organisms that also take part in the physical colonization of forages. Furthermore, the fluid-associated portion of the ruminal microbiome plays an indirect role in degradation of fiber through endproduct removal and crossfeeding of nutrients and co-factors to the cellulolytic/hemicellulolytic bacteria. The roles of ruminal protozoa and fungi are also critical in the process of nutrient cycling and initial forage colonization/catabolism, respectively (Figure 2).

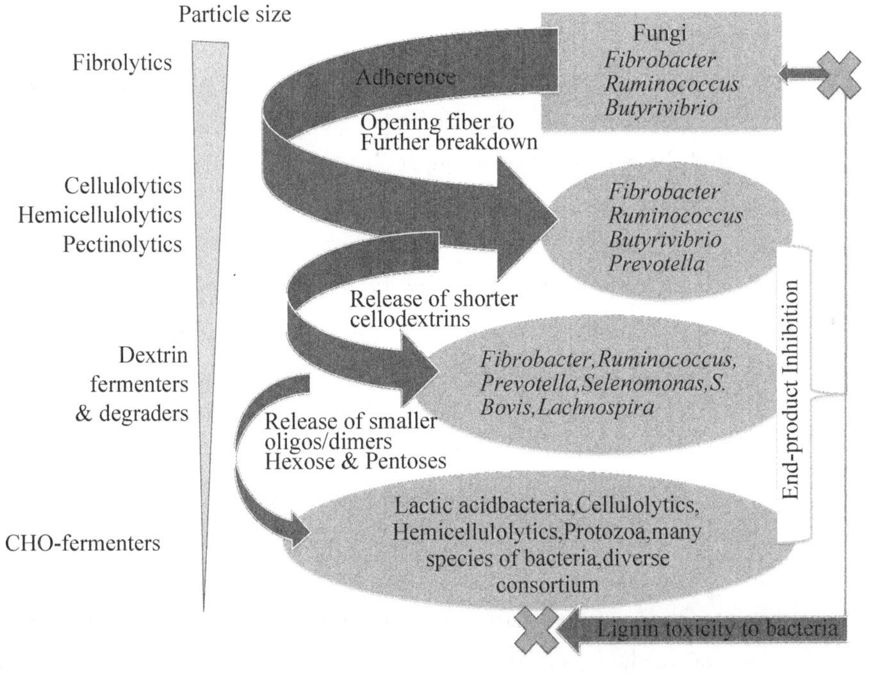

Figure 2 Cascade of degradation of fiber amongst ruminal microbiome

How Can We Alter the Microbiome?

There has been an explosion of information available about the microbiome following the development of NGS, and researchers are struggling to come to grips with how to utilize the data "overload" to best understand the ruminal microbial community (Wright and Klieve, 2011;

Henderson et al., 2015; McAllister et al., 2015; Shabat et al., 2016). This has allowed us to observe impacts of changes, and to determine some of the unintended consequences of changes made to the diet and microbiome. However, collectively it is clear that diet and feed additives can make large changes in the ruminal microbiome.

Microbial seeding or inoculation of the rumen

Calves are inoculated by maternal contact beginning immediately after contamination, and ruminal cross-inoculation between adult cattle has been long used by veterinarians and animal scientists to restart a static rumen (often following a displaced abomasum) (Hungate, 1966). However it was noted that while many facets of rumen function could be improved by transfer of rumen contents in mature animals, fiber digestion was not one of them (Hungate, 1966). While it has been shown that ruminal inoculation does improve performance as mature/growing cattle are stepped up from a high forage diet to a high grain diet to reduce ruminal acidosis, the converse is not true (Allison et al., 1964). Rumen contents were swapped between two mature cattle with different rumen microbiome profiles, and both microbiomes reverted back to the original host population within 14 d for one cow but 61 d for the other cow (Weimer et al., 2010). However when two other cowswere exchanged, they found the microbiomes remained in the new host even after 62 d, suggesting that not all components of the ruminal populations are host specific (Weimer et al., 2010).

The addition of pure cultures of fiber degrading bacteria to the gnotobiotic rumen have been unsuccessful (Mann and Stewart, 1974). It was shown that increasing the complexity of the ruminal population increased the fibrolytic bacteria's ability to colonize the rumen (Fonty et al., 1983). This is likely due to the need of the cellulolytic bacteria for branched chain VFA, and/or NH_3 from protein fermentation by the non-cellulolytic population. However, dosing of the rumen with pure cultures or mixed cultures of fiber-degrading bacteria has not been successful to date (Krause et al., 2003).

In some specific cases, however, forage utilization was improved by ruminal inoculation with bacteria that provides a detoxification function. *Leucaena* is a leguminous forage found in tropical climates that produces mimosine (3-hydroxy-4 pyridone) which is toxic to ruminants. *Synergistesjonesii* is a bacteria that is capable of degrade this toxin which was originally isolated from goats in Hawaii (Allison et al., 1990; Allison et al., 1992). When inoculated into the rumen of animals grazing *Leucaena* this organism made the animals resistant to the toxic effects of mimosine.

Addition of lignolytic fungi is a simple approach, but the most active lignolytic fungi are from the group known as white rot fungi (Akin et al., 1995). Unfortunately, these fungi grow at a rate that means they would only be transient members of the ruminal ecosystem, and would have to be supplemented daily. Utilization of these fungi is feasible for use in a pre-feeding treatment of forages, but this is only feasible at large scale rather than at an individual farm level.

Ruminal pH

Ruminal pH can change dramatically based upon diet and host factors (Fernando et al., 2010; Palmonari et al., 2010; Hook et al., 2011) as well as time of day and time after feeding (Nocek et al., 2002; Palmonari et al., 2010). Many species of ruminal bacteria are sensitive to the physiological impacts of low pH on growth and enzymatic activity, and cellulolytic bacteria are susceptible to low ruminal pH (Hungate, 1950; Russell et al., 1979; Russell and Wilson, 1996; Mourino et al., 2001), as is encountered when cattle eat a high grain ration due to the production of lactic acid from starch fermentation. However, grain-induced acidosis did not reduce the populations of cellulolytic bacteria until ruminal pH became severe (Khafipour et al., 2009). Ruminal pH is determined by flow rate, fermentation rate, dietary components (starch concentration and digestibility), buffering capacity, and microbiome composition (Hook et al., 2011; Romero-Pérez et al., 2011). However ruminal pH does not dictate the microbiome composition entirely, cattle with very different pH profiles (6.5 vs. 6.1) can have very similar microbiome profiles (Palmonari et al., 2010) and cattle with identical pH and diets can have very different microbial profiles (Callaway, 2015, unpublished data). Thus pH is but one factor that can have a widespread impact on fiber degradation by ruminal bacteria.

Dietary impactors

Diet quality varies world wide based on local conditions, and some compounds found in feedstuffs are anti-nutritional for cattle. Tannins are polyphenolic compoundsthat are somewhat similar to lignin found in plants that can be toxic to ruminal bacteria (Nelson et al., 1997; Nelson et al., 1998; Cowan, 1999). When tannins are included in a ration they can alter the ruminal fermentation rate and extent, the composition of the ruminal microbiome, as well as end-products such as methane (Patra and Saxena, 2009; Martin et al., 2010; Romero-Pérez et al., 2011). Typically as tannin level increases the quality of the ration to the animal decreases. In recent years there has been an increase in interest in bacteria capable of degrading (or at least tolerating) tannins (Nelson et al., 1998; Patra and Saxena, 2009), and this finding may lead to the development of specific treatments for cattle fed forages with high levels of tannin. Essential oils are another antimicrobial compound that is found in some forages and may play a role in altering the ruminal microbiome and impact forage digestibility (Kim et al., 1995; Schelz et al., 2006). Furthermore, some of the essential oil-containing by-product feeds can be used to modify the rumen fermentation and potentially enhance food safety in the live animal (Jacob et al., 2009; Patra and Saxena, 2009; Doyle and Erickson, 2012). Cellulolytic bacteria and fungal counts increased when inorganic sulfur was supplemented to animals fed low S diets (McSweeney and Denman, 2007).

Dietary additives

Some probiotic (Direct Fed Microbial) feeding approaches have shown promise in increasing forage digestibility (Nocek and Kautz, 2006), probably due to provision of limiting co-factors for cellulolytic bacteria. However the benefits of DFM on fiber digestion have been inconsistent (Martin and Nisbet, 1992; McAllister et al., 2011). Collectively, DFM feeding improves fermentation efficiency, potentially by enhancing growth factors needed by bacteria that cross-fed nutrients with fibrolytic bacteria.

Dicarboxylic organic acids (e.g., malate, fumarate, and oxaloacetate) stimulate the growth and lactate uptake of the ruminal bacteria *Megasphaeraelsdenii* and *Selenomonasruminantium* (Nisbet and Martin, 1990; Nisbet and Martin, 1993; Martin and Streeter, 1995). The lactilytic bacteria consume lactate and produce propionate, shifting the electron flow in the rumen increasing pH while also reducing the acetate: propionate ratio and methane production in a manner similar to monensin inclusion (Nisbet and Martin, 1990; Callaway and Martin, 1996). Changing the ruminal parameters (especially ruminal pH) can increase fiber digestion (Martin and Streeter, 1995; Martin et al., 1999; Zhou et al., 2012). Additionally increased levels of *Megasphaeraelsdenii* have been correlated to a reduction in milk fat depression in high producing dairy cattle (Palmonari et al., 2010). Disodium fumarate fed to sheep on high forage diets stimulated the populations of *F. succinogenes* and *R. albus* and *B.fibrosolvens*, but decreased methanogens and *R. flavefaciens* (Zhou et al., 2012). Interestingly, fumarate addition increased populations of *Prevotellaruminicola* and *Clostridium* sp., and these species are non-structural carbohydrate fermenting species and are responsible for protein fermentation (Zhou et al., 2012).

Enzyme treatment of forages

The exogenous application of enzymes to forages before feeding has been examined for more than a half century (Burroughs et al., 1960). In this first study the cattle gained significantly more weight and were more efficient than cattle fed a non-enzymatically treated feed (Burroughs et al., 1960). It should be noted that the entire ration was treated with amylolytic, proteolytic, and cellulolytic enzymes, and that the ration was significantly different than current rations that contain higher levels of starch. As research in the 1960s expanded, there was a growing inconsistency in effect of enzymes, based upon diet composition, enzyme choice, and animal factors (Krause et al., 2003). As research in this area has progressed (McAllister et al., 2000; Sujani and Seresinhe, 2015), many of these areas have been elucidated and solutions found, but there is still much room to improve the digestibility of fiber using pre-feeding enzyme treatment.

Microbiome diversity and focus

Our knowledge of the microbiome has grown extensively in recent years, and some researchers have begun attempting to link animal performance measures with microbial populations of individuals or with environmental niches. This ecologically-based jobs analysis is an important step in understanding what is important to rumen function and how can we improve animal growth efficiency and maximize fiber utilization. One study that was recently published begins to dig away at this and found that contrary to microbial ecology theory, that species richness and diversity was negatively correlated with efficiency of dairy cattle (Shabat et al., 2016). The research indicated that a lower diversity, more "job focused" microbiome that produced fewer types of intermediate products that were not used directly by the animal (Shabat et al., 2016). It must be pointed out that this study used cows fed a 70% concentrate ration, so the results cannot be wholesale extrapolated to forage diets, but it indicates that ecological niches and a well-(niche) adapted microbiome can enhance animal efficiency (Shabat et al., 2016).

Conclusions

The ruminal microbiome plays a critical role in the degradation and utilization of forages by ruminant animals. We have only recently begun to understand some of the layers of complexity and subtle interactions that underlie the degradation of forage to produce energy and carbon for cattle. The microbial consortium involves bacteria, fungi, and protozoa to most efficiently degrade forage by breaking forage into smaller "pieces" that can be degraded by a broad cross-section of the ruminal microbial population, thus ensuring that the reaction is pulled along by mass action. Furthermore, the bacteria that benefit from the "crumbs" produce co-factors and NH_3 that are needed by the cellulolytic and hemicellulolytic species. While several compounds do occur in forages that can inhibit bacterial growth (e.g., lignin), there are specialist microorganisms that can degrade these compounds, making most forages utilizable by ruminants. While there is currently no magic bullet to instantly improve forage digestibility, exogenous enzymes specific for forage type and quality offer a possibility of removing the inhibitory factors before feeding. As we delve further into understanding the ruminal microbiome and how it impacts animal health and productivity, we begin to understand how specific members of the cellulolytic community depend on other species. This information can be used to further create conditions in the rumen that are more ideal for forage degradation.

Note: Proprietary or brand names are necessary to report factually on available data; however, the USDA neither guarantees nor warrants the standard of the product, and the use of the name by the USDA implies neither approval of the product, nor exclusion of others that may be suitable.

The U.S. Department of Agriculture (USDA) prohibits discrimination in all its programs and activities on the basis of race, color, national origin, age, disability, and where applicable, sex, marital status, familial status, parental status, religion, sexual orientation, genetic information, political beliefs, reprisal, or because all or part of an individual's income is derived from any public assistance program (Not all prohibited bases apply to all programs).

References

Akin, D. E., Benner, R., 1988. Degradation of polysaccharides and lignin by ruminal bacteria and fungi [J]. Appl. Environ. Microbiol., 54: 1117-1125.

Akin, D. E., Rigsby, L. L., 1987. Mixed fungal populations and lignocellulosic tissue degradation in the bovine rumen [J]. Appl. Environ. Microbiol., 53: 1987-1995.

Akin, D. E., Rigsby, L. L., Sethuraman, A., et al., 1995. Alterations in structure, chemistry, and biodegradability of grass lignocellulose treated with the white rot fungi *Ceriporiopsis subvermispora* and *Cyathus stercoreus* [J]. Appl. Environ. Microbiol., 61 (4): 1591-1598.

Allison, M. J., 1978. Production of branched-chain volatile fatty acids by certain anaerobic bacteria [J]. Appl. Environ. Microbiol., 35: 872-877.

Allison, M. J., Bryant, M. P., Doetsch, R. N., 1961. Studies on the metabolic function of branched-chain volatile fatty acids, growth factors for Ruminocci [J]. J. Gen. Microbiol., 5: 869-879.

Allison, M. J., Bucklin, J. A., Dougherty, R. W., 1964. Ruminal changes after overfeeding with wheat and the effect of intraruminal inoculation on adaptation to a ration containing wheat [J]. J. Anim. Sci., 23: 1164-1171.

Allison, M. J., Hammond, A. C., Jones, R. J., 1990. Detection of ruminal bacteria that degrade toxic dihydroxypyridine compounds produced from mimosine [J]. Appl. Environ. Microbiol., 56 (3): 590-594.

Allison, M. J., Mayberry, W. R., McSweeney, C. S., et al., 1992. *Synergistes jonesii*, gen. nov., sp. nov.: a rumen bacterium that degrades toxic pyridinediols [J]. Syst. Appl. Microbiol., 15 (4): 522-529.

Bach-Tuyet Lam, Iiyama, T., K., Stone, B. A., 1990. Primary and secondary walls of grasses and other forage plants: taxonomic and structural considerations [M]. D. E. Akin, ed. Taxonomy and Structure.

Bauchop, T., 1979a. Rumen anaerobic fungi of cattle and sheep [J]. Appl. Environ. Microbiol., 38: 148-158.

Bauchop, T., 1979b. The rumen anaerobic fungi: colonizers of plant fibre [J]. Ann. Resch. Vet., 10: 246-248.

Borneman, W. S., Akin, D. E., 1990. Lignocellulose degradation by rumen fungi and bacteria: ultrastructure and cell wall degrading enzymes [M]. D. E. Akin, ed. Elsevier Science Publishing Co., Inc, Athens.

Borneman, W. S., Akin, D. E., Ljungdahl, L. G., 1989. Fermentation products and plant cell wall-degrading enzyme produced by monocentric and polycentric anaerobic ruminal fungi [J]. Appl. Environ. Microbiol., 55: 1066-1073.

Brulc, J. M., Antonopoulos, D. A., Berg Miller, M. E., et al., 2009. Gene-centric metagenomics of the fiber-adherent bovine rumen microbiome reveals forage specific glycoside hydrolases [J]. Proc. Nat. Acad. Sci. (USA), 106 (6): 1948-1953.

Burroughs, W., Woods, W., Ewing, S. A., et al., 1960. Enzyme additions to fattening cattle rations [J]. J. Anim. Sci., 19: 458-464.

Callaway, T. R., Martin, S. A., 1996. Effects of organic acid and monensin treatment on *in vitro* mixed ruminal microorganism fermentation of cracked corn [J]. J. Anim. Sci., 74 (8): 1982-1989.

Cowan, M. M., 1999. Plant products as antimicrobial agents [J]. Clin. Microbiol. Rev., 12 (4): 564-582.

Craig, W. M., Hong, B. J., Broderick, G. A., et al., 1984. *In vitro* inoculum enriched with particle-associated microorganisms for determining rates of fiber digestion and protein degradation [J]. J. Dairy Sci., 67: 2902-2909.

Deblois, S., Wiegel, J., 1990. Diversity of hemicellulosic material [M]. Pages 275 - 287 in Hemicellulases in lignocellulose degradation. D. E. Akin, ed. Elsevier Publishing Co.

Denman, S. E., McSweeney, C. S., 2006. Development of a real-time PCR assay for monitoring anaerobic fungal and cellulolytic bacterial populations within the rumen [J]. FEMS Microbiol. Ecol., 58 (3): 572-582.

Doyle, M. P., Erickson, M. C., 2012. Opportunities for mitigating pathogen contamination during on-farm food production [J]. Int. J. Food Microbiol., 152 (3): 54-74.

Edwards, J. E., Huws, S. A., Kim, E. J., et al., 2007. Characterization of the dynamics of initial bacterial colonization of nonconserved forage in the bovine rumen [J]. FEMS Microbiol Ecol., 62 (3): 323-335.

Fernando, S. C., Purvis, H. T., Najar, F. Z., et al., 2010. Rumen microbial population dynamics during adaptation to a high-grain diet [J]. Appl. Environ. Microbiol., 76 (22): 7482-7490.

Firkins, J. L., Yu, Z., 2015. How to use data on the rumen microbiome to improve our understanding of ruminant nutrition [J]. J. Anim. Sci., 93 (4): 1450-1470.

Flint, H. J., Bayer, E. A., Rincon, M. T., et al., 2008. Polysaccharide utilization by gut bacteria: potential for new insights from genomic analysis [J]. Nature Rev. Microbiol., 6 (2): 121-131.

Fonty, G., Jouany, J. P., Thivend, P., et al., 1983. A descriptive study of rumen digestion in meroxenic lambs according to the nature and complexity of the microflora [J]. Reprod. Nutr. Dev., 23: 857-873.

Hook, S. E., Steele, M. A., Northwood, K. S., et al., 2011. Impact of subacute ruminal acidosis (SARA) adaptation and recovery on the density and diversity of bacteria in the rumen of dairy cows [J]. FEMS Microbiol. Ecol., 78 (2): 275-284.

Hungate, R. E., 1950. The anaerobic mesophilic cellulolytic bacteria [J]. Bacterial Rev., 14: 1-49.

Hungate, R. E., 1966. The Rumen and its Microbes [M]. New York, NY: Academic Press.

Jacob, M. E., Callaway, T. R., Nagaraja, T. G., 2009. Dietary interactions and interventions affecting *Escherichia coli* O157 colonization and shedding in cattle [J]. Foodborne Path. Dis., 6 (7): 785-792.

Joblin, K. N., 1990. Bacterial and protozoal interactions with ruminal fungi [M]. D. E. Akin, ed. Elsevier Science Publishing Co., Inc.

Johnson, K. A., Johnson, D. E., 1995. Methane emissions from cattle [J]. J. Anim. Sci., 73: 2483-2494.

Jung, H. G., 1989. Forage lignins and their effects on fiber digestibility [J]. Agron. J., 81: 33-38.

Khafipour, E., Li, S., Plaizier, J. C., et al., 2009. Rumen microbiome composition determined using two nutritional models of subacute ruminal acidosis [J]. Appl. Environ. Microbiol., 75 (22): 7115-7124.

Kim, J., Marshall, M. R., Wei, C. I., 1995. Antibacterial activity of some essential oil components against five foodborne pathogens [J]. J. Agric. Food Chem., 43 (11): 2839-2845.

Krause, D. O., Denman, S. E., Mackie, R. I., et al., 2003. Opportunities to improve fiber degradation in the rumen: microbiology, ecology, and genomics [J]. FEMS Microbiol. Rev., 27: 663-693.

Mann, S. O., Stewart, C. S., 1974. Establishment of a limited rumen flora in gnotobiotic lambs fed on a roughage diet [J]. J. Gen. Microbiol., 84 (2): 379-382.

Martin, C., Morgavi, D. P., Doreau, M., 2010. Methane mitigation in ruminants: From microbe to the farm scale [J]. Animal, 4 (3): 351-365.

Martin, S. A., Nisbet, D. J., 1992. Effect of direct-fed microbials on rumen microbial fermentation [J]. J. Dairy Sci., 75: 1736-1744.

Martin, S. A., Streeter, M. N., 1995. Effect of malate on *in vitro* mixed ruminal microorganism fermentation [J]. J. Anim. Sci., 73: 2141-2145.

Martin, S. A., Streeter, M. N., Nisbet, D. J., et al., 1999. Effects of DL-Malate on ruminal metabolism and performance of cattle fed a high-concentrate diet [J]. J. Anim. Sci., 77: 1008-1015.

McAllister, T. A., Beauchemin, K. A., Alazzeh, A. Y., et al., 2011. Review: The use of direct fed microbials to mitigate pathogens and enhance production in cattle [J]. Can. J. Anim. Sci., 91 (2): 193-211.

McAllister, T. A., Hristov, A. N., Beauchamin, K. A., et al., 2000. Enzymes in ruminant diets [M]. Pages 273-298 in Enzymes in farm animal nutrition. M. Bedford and G. Partridge, ed. CAB International, Wallingford.

McAllister, T. A., Meale, S. J., Valle, E., et al., 2015. Ruminant nutrition symposium: Use of genomics and transcriptomics to identify strategies to lower ruminal methanogenesis [J]. J. Anim Sci., 93 (4): 1431-1449.

McAllister, T. A., Newbold, C. J., 2008. Redirecting rumen fermentation to reduce methanogenesis [J]. Aust. J. Exp. Agric., 48: 7-13.

McSweeney, C., Denman, S., 2007. Effect of sulfur supplements on cellulolytic rumen micro-organisms and microbial protein synthesis in cattle fed a high fibre diet [J]. J. Appl. Microbiol., 103 (5): 1757-1765.

Miron, J., Ben-Ghedalia, D., Morrison, M., 2001. Adhesion mechanisms of rumen cellulolytic bacteria [J]. J. Dairy Sci., 84: 1294-1309.

Mosoni, P., Martin, C., Forano, E., et al., 2011. Long-term defaunation increases the abundance of cellulolytic ruminococci and methanogens but does not affect the bacterial and methanogen diversity in the rumen of sheep [J]. J. Anim Sci., 89 (3): 783-791.

Mourino, F., Akkarawongsa, R., Weimer, P., 2001. Initial pH as a determinant of cellulose digestion rate by mixed ruminal microorganisms *in vitro* [J]. J. Dairy Sci., 84 (4): 848-859.

Nelson, K. E., Pell, A. N., Doane, P. H., et al., 1997. Chemical and biological assays to evaluate bacterial inhibition by Tannins [J]. J. Chem. Ecol., 23 (4): 1175-1194.

Nelson, K. E., Thonney, M. L., Woolston, T. K., et al., 1998. Phenotypic and phylogenetic characterization of ruminal tannin-tolerant bacteria [J]. Appl. Environ. Microbiol., 64 (10): 3824-3830.

Nisbet, D. J., Martin, S. A., 1990. Effect of dicarboxylic acids and *Aspergillus oryzae* fermentation extract on lactate uptake by the ruminal bacterium *Selenomonas ruminantium* [J]. Appl. Environ. Microbiol., 56 (11) 3515-3518.

Nisbet, D. J., Martin, S. A., 1993. Effects of fumarate, L-malate, and an *Aspergillus oryzae* fermentation extract on D-lactate utilization by the ruminal bacterium *Selenomonas ruminantium* [J]. Curr. Microbiol., 26: 136-136.

Nocek, J. E., Kautz, W. P., 2006. Direct-fed microbial supplementation on ruminal digestion, health, and performance of pre-and postpartum dairy cattle [J]. J. Dairy Sci., 89 (1): 260-266.

Nocek, J. E., Kautz, W. P., Leedle, J. A. Z., et al., 2002. Ruminal supplementation of direct-fed microbials on diurnal pH variation and *in situ* digestion in dairy cattle [J]. J. Dairy Sci., 85 (2): 429-433.

Palmonari, A., Stevenson, D., Mertens, D., et al., 2010. pH dynamics and bacterial community composition in the rumen of lactating dairy cows [J]. J. Dairy Sci., 93 (1): 279-287.

Patra, A. K., Saxena, J., 2009. Dietary phytochemicals as rumen modifiers: a review of the effects on microbial populations [J]. Antonie van Leeuwenhoek, 39: 1-13.

Ransom-Jones, E., Jones, D. L., McCarthy, A. J., et al., 2012. The Fibrobacteres: An Important Phylum of Cellulose-Degrading Bacteria [J]. Microb. Ecol., 63 (2): 267-281.

Romero-Pérez, G. A., Ominski, K. H., McAllister, T. A., et al., 2011. Effect of environmental factors and influence of rumen and hindgut biogeography on bacterial communities in steers [J]. Appl. Environ. Microbiol., 77 (1): 258-268.

Russell, J. B., Sharp, W. M., Baldwin, R. L., 1979. The effect of pH on maximum bacterial growth rate and its possible role as a determinant of bacterial competition in the rumen [J]. J. Anim. Sci., 48: 251-255.

Russell, J. B., Wilson, D. B., 1996. Why are ruminal cellulolytic bacteria unable to digest cellulose at low Ph [J]. J. Dairy Sci., 79: 1503-1509.

Schelz, Z., Molnar, J., Hohmann, J., 2006. Antimicrobial and antiplasmid activities of essential oils [J]. Fitoterapia, 77 (4): 279-285.

Scott, H. W., Dehority, B. A., 1965. Vitamin requirements of several cellulolytic rumen bacteria [J]. J. Bacteriol., 89: 1169-1175.

Shabat, S. K. B., Sasson, G., Doron-Faigenboim, A., et al., 2016. Specific microbiome-dependent mechanisms underlie the energy harvest efficiency of ruminants [J]. ISME J, 10 (12): 2958-2972.

Shinkai, T., Kobayashi, Y., 2007. Localization of ruminal cellulolytic bacteria on plant fibrous materials as determined by fluorescence in situ hybridization and real-time PCR [J]. Appl. Environ. Microbiol., 73 (5): 1646-1652.

Shinkai, T., Ohji, R., Matsumoto, N., et al., 2009. Fibrolytic capabilities of ruminal bacterium *Fibrobacter succinogenes* in relation to its phylogenetic grouping [J]. FEMS Mcrobiol. Lett., 294 (2): 183-190.

Shinkai, T., Ueki, T., Kobayashi, Y., 2010. Detection and identification of rumen bacteria constituting a fibrolytic consortium dominated by *Fibrobacter succinogenes* [J]. Anim. Sci. J., 81 (1): 72-79.

Stewart, C. S., Bryant, M. P., 1988. The rumen microbial ecosystem [M]. Elsevier Science Publishers, Ltd, London, UK.

Suen, G., Stevenson, D. M., Bruce, D. C., et al., 2011a. Complete genome of the cellulolytic ruminal bacterium Ruminococcus albus 7 [J]. J. Bacteriol., 193 (19): 5574-5575.

Suen, G., Weimer, P. J., Stevenson, D. M., et al., 2011b. The complete genome sequence of *Fibrobacter succinogenes* s85 reveals a cellulolytic and metabolic specialist [J]. PLoS One, 6 (4).

Sujani, S., Seresinhe, R. T., 2015. Exogenous enzymes in ruminant nutrition: a review [J]. Asian J. Anim. Sci, 9: 85-99.

Thiele, J. H., Zeikus, J. G., 1988. Control of interspecies electron flow during anaerobic digestion: significance of formate transfer versus hydrogen transfer during syntrophic methanogenesis in flocs [J]. Appl. Environ. Microbiol., 54: 20-29.

Van Tappeiner, H., 1884. Untersuchungen iiber die garung der cellulose insbesondere iiber deren losung im darmkanale [J]. Z. Biol, 20: 52-134.

Wallace, R. J., 2008. Gut microbiology-broad genetic diversity, yet specific metabolic niches [J]. Animal, 2: 661-668.

Weimer, P. J., Stevenson, D. M., Mantovani, H. C., et al., 2010. Host specificity of the ruminal bacterial community in the dairy cow following near-total exchange of ruminal contents [J]. J. Dairy Sci., 93: 5902-5912.

Wells, J. E., Russell, J. B., Shi, Y., et al., 1995. Cellodextrin efflux by the cellulolytic ruminal bacterium *Fibrobacter succinogenes* and its potential role in the growth of nonadherent bacteria [J]. Appl. Environ. Microbiol., 61: 1757-1762.

Wright, A. D. G., Klieve, A. V., 2011. Does the complexity of the rumen microbial ecology preclude methane mitigation [J]. Anim. Feed Sci. Technol., 166-167: 248-253.

Yokoyama, M. G., Johnson, K. A., 1988. Microbiology of the rumen and intestine [M]. Pages 125-144 in The Ruminant Animal: Digestive Physiology and nutrition. D. C. Church, ed. Waveland Press, Englewood Cliffs, NJ.

Zhou, Y. W., McSweeney, C. S., Wang, J. K., et al., 2012. Effects of disodium fumarate on ruminal fermentation and microbial communities in sheep fed on high-forage diets [J]. Animal, 6 (5): 815-823.

Managing the 3 Critical Calf Periods

A. F. Kertz

ANDHIL LLC St. Louis, MO USA

Abstract

The 3 critical periods in a calf's life are around calving which includes the cow condition and colostrum management, the first 2 wk of life when most deaths occur, and the 2 wk before and after weaning. Colostrum is essential to provide antibody protection to newborn calves. The degree of this protection is dependent on how quickly after birth colostrum is fed, how much is fed, and the antibody concentration in the colostrum. Recently, lack of cleanliness of colostrum has been found to be a problem from 30% to 50% of samples checked for bacteria in several studies. Colostrum also provides more than just higher nutrient levels versus regular milk as it contains various hormones and metabolites which have anabolic growth effects. There are both short term and long term benefits to colostrum feeding. Since a calf is a non-functioning ruminant when it is born, the milk or milk replacer goes into the abomasum (true stomach) and bypasses the rumen. As dry calf starter is eaten, it empties into and leads to the functional development of the rumen. In the meantime, the liquid diet meets nutrient needs for maintenance and growth, primarily in terms of energy and protein. A transition to dry diet needs to occur, and this requires functional development of the rumen. Dependent on the amount of milk or milk replacer fed, and its energy and protein levels, this will determine how much protein and energy is available for daily gain. The weaning transition period, 2 wk before and 2 wk after full weaning, determines how well calves perform at that time and in the future. A key objective it to double calf birth weight by the end of 2 month of life. The key to rumen development is to have a good texturized (particle size) starter and to feed no hay until after 2 month of age. Water is themost essential nutrient needed in the greatest quantity for calves, the calf's birth body is 70% water, when calves have diarrhea they drink more water, it is directly related to dry matter intake (DMI) —limit it and DMI and performance are limited, water intake is 4 times DMI, and good managers achieving good results figure out how best to feed water. In cold weather, warm water should be fed. A very practical common problem is failure to separate water and starter containers to eliminate calves dribbling water into starter and starter into water. This causes dirty water and wet feed, which reduces the calf's intake of both. This can significantly reduce daily gain too. The month following the weaning transition period is also a key transition period during which group size should only be 6 to 8 calves, starter fed at 3 to 4 kg daily, and hay limited fed at 0.5 kg daily. Calf health and growth during the first two month affects heifer growth and subsequent first and later lactation milk production.

Key Words: calves, critical periods, colostrum, milk replacer, starter, hay, water, weaning transition period, rumen development, growth, milk production

Introduction

The 3 critical periods in a calf's life are around calving which includes the cow and colostrum management, the first 2 wk of life when most deaths occur (Nahms, 2016), and the 2 weeks before and after weaning. Since there is no antibody transfer from the mother to the calf during pregnancy in ruminants, newborn dairy calves are dependent on colostrum for their antibody protection. The degree of protection that calves get depends on the amount of antibodies actually absorbed. This in turn is dependent on several key factors: how soon after birth calves are administered colostrum, the antibody content of the colostrum, and the quantity of colostrum they are administered. Another major factor in determining how calves respond to colostrum is the pathogen load in their environment at birth and how clean colostrum is. In a sense, there is an equation of pathogen load plus colostrum equals degree of protection for the calf. The strength or weakness of both these factors determines how much protection the calf is provided. This degree of protection must last until about 3 wk of age (Warner and Brownstein, 1976; Hurlbert and Moisa, 2016) when the calf begins to produce its own antibodies as illustrated in Figure 1.

Immunity/Absorption

Stott et al. (1979) conducted a study with calves at a large dairy and found that doubling the amount of colostrum fed from 0.5 L to 1 L and then from 1 L to 2 L per feeding doubled the subsequent blood plasma levels of IgG (Figure 2). When colostrum was administered at birth, blood plasma IgG levels were about 25% greater compared to when initial feeding of colostrum was delayed until 4 hours (h) after birth, doubled when first feeding was delayed until 8 h after birth, and then virtually not existent (shown in figure) when delayed until 24 h after birth.

Colostrum Feeding and Management Practices

Recent studies indicate that colostrum is not being harvested, fed, or stored properly (God-

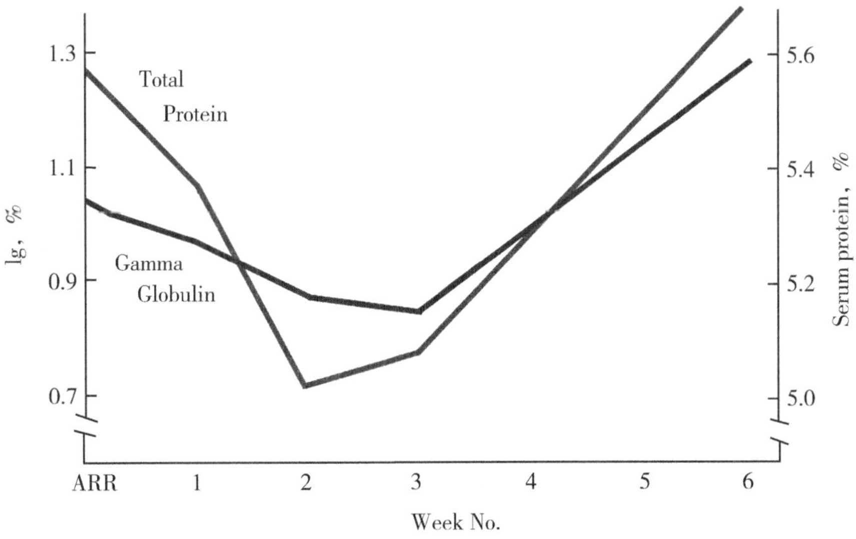

Figure 1 Age effect on levels of serum protein and gamma globulin in calves (C. U. Brownstein & Wamer, 1976)

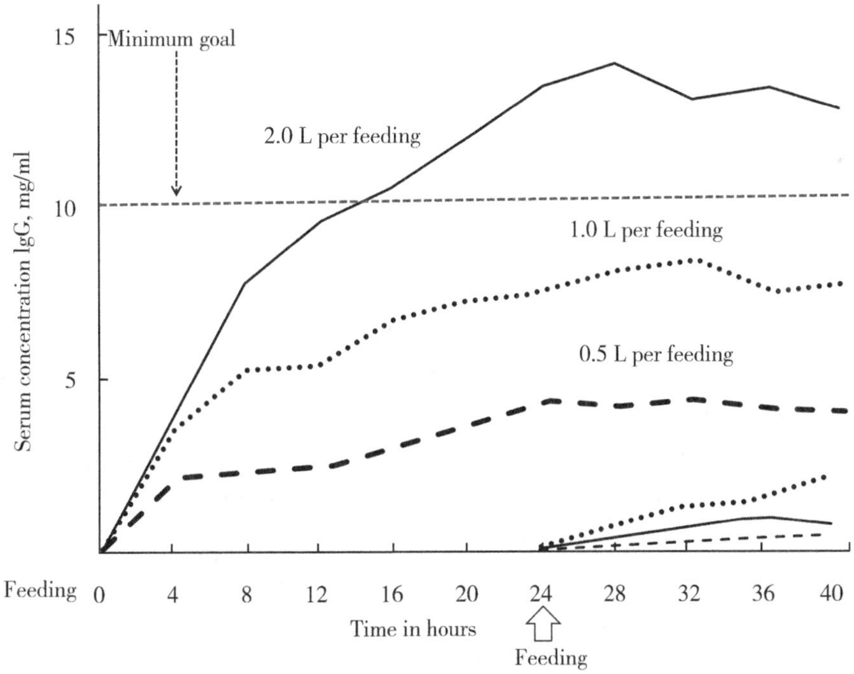

Figure 2 Serum IgG levels in calves is dependent on amount and when colostrum was fed (Stott et al., 1979)

den, 2007). As many as 30% to 50% of on-farm samples in 4 field studies in the U.S. and

Canada have exceeded upper limits for bacteria of 100, 000 CFU/mL total plate counts (TPC) and 10,000 CFU/mL total coliform counts (TCC). Harvesting of colostrum is the first route of bacterial exposure to calves. There are three primary sources of bacteria in colostrum: those shed directly from the udder, from contaminated equipmentsuch as bucket/bottleor that used for feeding, and bacterial proliferation in improperly stored colostrum. Thus, for optimal colostrum feeding, it is necessary to (1) prepare and clean the udder properly, (2) sanitize collection, storage, feeding, etc. equipment, (3) not pool sources for disease control such as Johne's (*mycobacterium avium subspecies paratuberculosis*), and (4) refrigerate by 2 hours (and use within 3 days) or freeze (Godden, 2007). These field problems are most often due to using dirty colostrum or keeping colostrum warm too long before feeding. Thus, calves areinadvertently inoculated with bacteria ("bacterial soup") when the colostrum has been kept warm too long (more than 20-30 min) before it is fed—especially in the summer time. Greater bacterial counts also reduce IgG absorption, and contribute to a false sense of security when other good practices of feeding colostrum are being followed. This contamination of a calf's system by dirty colostrum or environment often results in diarrhea 2-3 wk later, which is the major cause of illness and death in calves at this age.

Nutrient and Anabolic Contributions of Colostrum

Whiletraditionally colostrum has been thought of as being of particular value because of its antibody content, this is still true and beneficial. But compared to regular milk, it also has more than twice its energy content, much greater protein content (primarily due to antibodies), and greater content of lactoferrin, insulin, growth hormone, glucagon, prolactin, IGF-1, leptin, TGF-α, cortisol and 17 β estradiol (Van Amburgh and Soberon, 2016). Many of these components are anabolic, and promote growth. This supports the concept of epigenetics in which high levels of nutrition, IgG, and metabolic components activate more genes which facilitate growth, health, and subsequent milk production And since colostrum is only the very first milk of a cow after having given birth to its calf, transition milk from the following 5 or so milkings will have somewhat intermediate levels of these components. Thus, there is still some greater value to also feeding this milk rather than just regular milk. At one time, this was a standard practice but seems to have virtually disappeared due to larger herds not separating this transition milk and continuing to feed it to calves.

Long Term Colostrum Benefits

A study (Faber et al., 2005) conducted at a Wisconsin Brown Swiss herdfed calves either 2 L or 4 L of colostrum at the very first feeding. This was followed by feeding of transition milk for the next two days. Calves were then fed and managed the same for their growing period until completing one or two lactations. Calves fed only 2 L colostrum had twice the veterinary costs, gained 0.23 kg less daily at first breeding, had 11% less milk production in the first lactation, and had 17% less milk production in the second lactation than those calves that had been fed 4 L colostrum. These are major benefits simply from feeding more colostrum at the very first feeding of acalf.

Feeding 2 L or 4 L colostrum enhanced rate of gain (0.60 kg vs. 0.74 kg) at 80 d of age (Soberon and Van Amburgh, 2010a). Within each colostrum level feeding, calves were fed either 42.7 kg or 20.3 kg milk replacer during the first 2 months of life. This resulted in daily gains at 80 d of 0.72 kg vs. 0.61 kg and 4 times as % of body weight the amount of mammary gland parenchymal tissue developed (Soberon and Van Amburgh, 2011b). Understanding which cells may be responding to early life enhanced nutrition may help to best tailor young calf feeding and nutrition and more fully realize future milk production.

Liquid and Starter Feeding Program

Since a calf is a non-functioning ruminant when it is born, milk empties into the abomasum (true stomach) and bypasses the rumen. As dry calf starter is eaten, it empties into and leads to the functional development of the rumen. In the meantime, the liquid diet meets nutrient needs for maintenance and growth, primarily in terms of energy and protein. A transition to dry diet needs to occur, and this requires functional development of the rumen. If done properly, this leads to calves being in good shape to enter their first grouping. It is important that feeding and management changes to calves be done with transitions as the calf is the most vulnerable animal on a dairy farm.

Dependent on the amount of milk or milk replacer fed and its energy and protein levels, this will determine how much protein and energy is available for daily gain. A traditional milk replacer (MR) feeding program in the US only fed about 500 g daily of a 20% protein with 20% fat MR (Kertz and Loften, 2013). But around 2000, MR feeding programs began to change with feeding higher levels of a higher protein MR, typically 28% protein with 15% fat. A comparison of these feeding programs (Stamey et al., 2012) is illustrated inFigure 3 along with calf starter intakes. and their impact on daily gains.

In this study (Stamey et al., 2012), Holstein female and male calves were fed either a conventional 20/20 MR with 12.5% solids at 10% of birth weight daily in two feedings from wk 1 to 5 and at 5% once daily during wk 6; or 28/15 MR with 15% solids at 1.5% of BW as DM during wk 1, 2% of BW as DM during wk 2 to 5 divided into two daily feedings, and at 5% of BW during wk 6 in one daily feeding. All calves were weaned at end of 6 wk. The 28/15 MR feeding program contained two calf starter (CS) treatments of 18% or 22% CP as-fed or 19.6% and 25.5% CP DM basis; which were combined into one dataset for the graphing com-

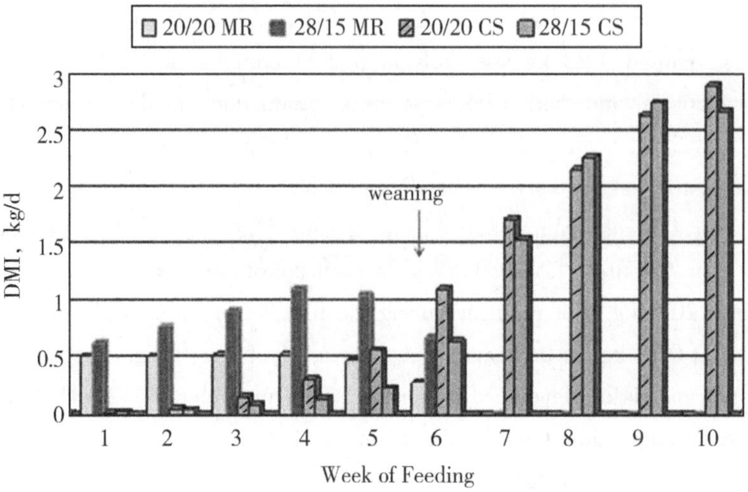

Figure 3 Milk replacer (MR) and calf starter (CS) dry matter intake (DMI) of calves fed either a 20% CP/20% fat or 28% CP/15% fat MR (Stamey et al., 2012)

parison (Figure 3) since there were no differences (P> 0.10) between the two starter treatments. Feeding less 20/20 MR (P<0.01) resulted in greater CS intake (P<0.02), but less ADG and height increase (P<0.02) than 28/15 treatments during preweaning. Total MR intakes were 19.3 kg for 20/20 and 34.7 kg for 28/15 with CS intakes before weaning of 14.0 and 7.3 kg, respectively. With greater DMI from MR, DMI of CS was reduced on 28/15 MR. This inverse relationship for MR intake and starter intake is illustrated from a compilation of intake data by Gelsinger et al. (2016). But total nutrient intake was greater on 28/15 resulting in greater ADG except for wk 7 which was just after full weaning. Figure 3 shows that loss in 28/15 MR DMI with full weaning was not equalized by CS DMI vs. 20/20 MR treatment during wk 7, but it was in wk 8-10 when ADG was similar between MR treatments. At the end of 8 wk, birth BW was approximately doubled on the 28/15 MR treatments, but was about 10 kg less on the 20/20 MR treatment. Thus, the objective of a good calf growing program is to double birth weight at the end of 2 months (mo) of age.

Rumen Development, Hay, and Texturized Starters

The key to rumen development is to have a good texturized (particle size) starter and to feed no hay until after 2 months of age. This may seem counterintuitive, but it is based on a calf starter with particle size providing the proper fermentation in the rumen to stimulate papillae development in the rumen wall without marginal rumen acidosis (pH ~ 5.0). Hay is bulky, has a low rate and extent of digestion, and produces the wrong balance of volatile fatty acids (VFAs) in rumen fermentation for papillae development. These papillae in turn absorb the VFAs as the calf's major energy source—just as occurs in adult ruminants. When a texturized starter was compared to an all pelleted starter (Porter et al., 2007), the following resulted: greater intake and daily gain, earlier rumination and more time ruminating, greater rumen papillae development, higher rumen pH (less marginal rumen acidosis), and greater dry matter and ADF/NDF digestibilities. When calves under 2 months of age were fed cottonseed hulls or hay along with a texturized starter, starter intake, daily gain, and empty body weight daily gain were all reduced (Hill et al., 2008). Unfortunately, the physical form of starters is not always indicated even in research trial reports (Kertz, 2017).

Feeding Water

The importance of feeding water to calves is often not well understood. Some of the reasons given for not feeding water are: it causes scours (diarrhea), calves don't need it, they get it through their milk replacer or milk, it freezes in the winter and they don't need it then anyway, and it is a hassle. But the reasons to feed water are: it is themost essential nutrient needed in the greatest quantity, the calf 's body is 70% water at birth, when calves scour they drink more water, it is directly related to dry matter intake (DMI) —limit it and you limit DMI and performance, water intake is 4 times DMI, and good managers achieving good results figure out how best to feed water. Not feeding any supplemental water in one study (Kertz et al., 1984) resulted in calves eating 30% less starter and gaining 40% less body weight priorto weaning. In this study, water consumption was 4 times starter intake. Another study (Quigley et al., 2006) found that the ratio of water to DMI (including milk replacer) was about double before weaning, but then increased to over 4 to 1 after

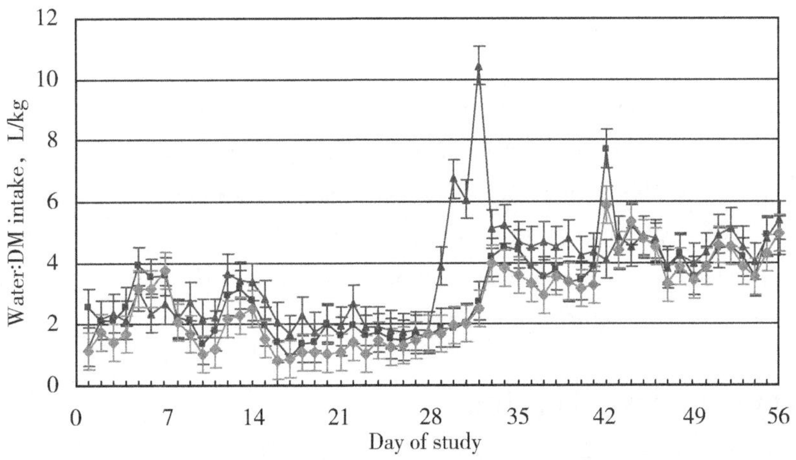

Figure 4 Ratio of water intake to dry matter intake (DMI) before and after weaning in calves (Quigley et al., 2006)

weaning (Figure 4). And in winter time, where possible, feeding calves warm water will allow them to drink it before it freezes, increases starter intake which produces additional heat during rumen fermentation to help meet their increased energy needs during cold weather, and reduces the need for calves to use energy to warm water to their rumen temperature (Dracy and Kurtenbach, 1968).

Practical Problem

A very practical common problem is failure to separate water and starter containers to eliminate calves dribbling water into starter and starter into water. This causes dirty water and wet feed, which reduces the calf's intake of both. This can significantly reduce daily gain as seen in a study of calves in the month after weaning (Figure 5).

Water/Starter Separation

2nd Month	Separated	Adjacent
BW gain, kg/d	0.84	0.72
Starter intake, kg/d	2.28	2.02
Water intake, kg/d	8.2	6.2

Ralston Purina Research Center, 1982

Figure 5 Effect of separation of water and starter containers on intake and weigh gain

Weaning Transition Starter Intake

Starter intake is the key for a good weaning transition program. If starter intake is inadequate, then rumen papillae development will be poor, and the change at and after weaning will be too great. This is stressful for calves and can result in lowered immunity resulting in outbreak of a respiratory problem. In general, starter intake should approximately double each wk from the previous week. Approximate starter intakes should be as depicted in Figure 6 for each wk of the wk 4 weaning transition period.

First Grouping Post Weaning

As part of the weaning transiton program, it is preferable to keep calves separate for 2 wk after weaning, This allows time for further increased starter intake, preferably a texturized starter, and rumen functional development. Unless calves have previously been group-fed, this will

Transition Starter Intakes

Transition Weeks	Starter Intake, kg/d
Before begin MR<intake	0.45
Week of one-half MR intake	0.91
Week after complete weaning	1.82
2nd wk after complete weaning	2.3-2.7

Figure 6 Approximate daily starter intakes for each week of the 4-wk weaning transition period

be their first time to be grouped and experience all the dynamics in group social behavior. In that first grouping, a limited number of calves should be grouped, preferably only 6 to 8 calves. There is some information that indicates calves like to pair up so even numbers of calves should be in this first group. Too often, up to 50 calves are put into the first grouping. That creates much social stress which can result in lowered immunity and respiratory problems. This then can impair a calf for life. In this first grouping, calves should continue to befedthe starter up to about 3 to 4 kg daily as this is one less change for the calf to endure. Alfalfa or a good quality hay should be limit fed to about 0.5 kg. This prevents calves from filling their gut with hay which will limit total intake and daily gain. There are multiple changes which occur in this first grouping as calves must adjust to group dynamics, new facility and feeding arrangements, change in diet, etc. If this is change not done well, calves can regress. Consequently, this month should be viewed and managed as a transition period of a month following the 4-wk weaning transition program.

Long Term Effects on Milk Production

Heinrichs and Heinrichs (2011) summarized 10 year of data from birth to 4 mo from 795 Holsteins in 21 Pennsylvania herds. They found that difficult births and days ill of calves resulted in later age at their first calving and lower milk yield in that lactation. Growth of calves was either affected negatively or positively by DMI from milk, MR, grain, and forage. First lactation milk yield was affected by weaning DMI, days treated for respiratory illness, and BW at calving. Lifetime production was similarly affected, but to lesser degree than first lactation. Thus, a variety of negative or positive effects that occurred during the first 4 month affected how these calves performed in first and later lactation.

While potential benefits of accelerated MR feeding were originally postulated as being due to reduced age at first calving, recent analyses by Soberon et al. (2012) indicated benefits may accrue to subsequent milk yield. A Test Day Model was developed utilizing inputs of preweaning

ADG, birth weight, weaning weight, calving age, birth year, birth month, and calculated energy intake over estimated maintenance requirements. For every additional 1 kg of ADG during the first 2 month (within the range of 0.10 to 1.58 kg/d), 1,244 heifers produced 850 kg more milk during their first lactation ($P<0.01$) and produced 2,280 kg more during their first 3 lactations. An additional 235 kg more milk was produced in the first lactation and 903 kg over 3 lactations for every Mcal ME intake above maintenance during the pre-weaning period. Calves born during winter mo (0.2℃) consumed an average of 1.43 Mcal/d less energy above maintenance than calves born during warmer mo (19.2℃). Preweaning ADG accounted for 22% of variation in first lactation milk yield. Age at first calving did not affect milk production within a range of 20 to 30 month. Colder weather for calves negatively affected subsequent milk production as less energy was available over increased maintenance needs for young calves resulting in their lower growth rate. Probable mechanisms for this increased milk yield are not understood, but are speculated to be related to very early mammary gland development.

A subsequent meta-analysis of this and other studies (Soberon and Van Amburgh, 2013) found that for a range in 1 kg ADG during the first 2 month of age, first-lactation milk increased by 1,550 kg; and an odds ratio of 2.09 indicated that calves fed for greater ADG were two times more likely to have greater milk yield in their first lactation.

Conclusions/Recommendations

- The 3 critical periods in a calf's life are around calving which includes the cow and colostrum management, the first 2 wks of life when most deaths occur, and the 2 wks before and after weaning.
- Colostrum is essential to provide antibody protection to newborn calves and is dependent on how quickly after birth colostrum is fed, how much is fed, the antibody concentration in the colostrum measured by a colostrometer, and how clean the colostrum is.
- Colostrum also provides more than twice the energy level of regular milk, more protein and other nutrients, and a high level of many hormones and components which are anabolic for growth. Feeding high levels of colostrum, and to a lesser extent transition milk, provides for more growth in itself and when paired with higher feeding levels of milk or milk replacer feeding.
- Dependent on the amount of milk or milk replacer fed and its energy and protein levels will determine how much protein and energy are available for daily gain.
- As dry calf starter is consumed, it empties into and leads to the functional development of the rumen.
- The key to rumen development is to have a good texturized (particle size) starter and to feed no hay until after 2 month of age.

- Water is themost essential nutrient needed in the greatest quantity, and its intake is 4 times dry matter intake (DMI) —limit it and DMI and performance are limited.
- Failure to separate water and starter containers leads to dirty water and wet feed, which reduces the calf's intake of both and daily gain as well.
- In the first grouping after the 4-wk weaning transition program, limit group size to 6 to 8 calves, and feed up to 3 to 4 kg of the texturized starter along with 0.5 kg of alfalfa or good quality grass hay,
- There are long term benefits in subsequent milk production based on how well calves grow in their first 2 month of age, but also long term detrimental effects from calves with diarrhea and respiratory problems.

References

Dracy, A. E., Kurtenbach, A. J., 1968. Temperature change within the rumen, crop area, and rectal area when liquid of various temperatures was fed to calves [J]. J. Dairy. Sci., 51: 1787-1790.

Faber, S. N., Faber, N. E., McCauley, T. C., et al., 2005. Case study: Effects of colostrum ingestion on lactation performance [J]. The Prof. Anim. Scientist., 21: 420-425.

Gelsinger, S. L., Heinrichs, A. J., Jones, C. M., 2016. A meta-analysis of the effects of preweaned calf nutrition and growth on first-lactation performance [J]. J. Dairy Sci., 99: 6206-6214.

Godden, S., 2007. Practical methods of feeding clean colostrum [R]. Proc. 11th Annual PDHGA Dairy Calf and Heifer Conf., Pre-Conf. Calf Seminar.

Heinrichs, A. J., Heinrichs, B. S., 2011. A prospective study of calf factors affecting first-lactation and lifetime milk production and age of cows when removed from the herd [J]. J. Dairy Sci., 94: 336-341.

Hill, T. M., Bateman, H. G., Aldrich, J. M., et al., 2008. Effects of the amount of chopped hay or cottonseed hulls in a textured calf starter on young calf performance [J]. J. Dairy Sci., 91: 2684-2693.

Hurlbert, L. E., Moisa, S. J., 2016. Stress, immunity, and the management of calves [J]. J. Dairy Sci., 99: 3199-3216.

Kertz, A. F., 2017. Letter to the Editor: A call for more complete reporting and evaluation of experimental methods, physical form of starters, and results in calf research [J]. J. Dairy Sci., 100: 851-852.

Kertz, A. F., Loften, J. R., 2013. A historical perspective and brief review: Holstein dairy calf milk replacer feeding programs in the U.S. [J] Prof. Anim. Scientist, 29: 321-332.

Kertz, A. F., Reutzel, L. F., Mahoney, J. H., 1984. Ad libitum water intake by neonatal calves and its relationship to calf starter intake, weight gain, feces score, and season [J]. J. Dairy Sci., 67: 2964-2969.

National Animal Health Monitoring System (NAHMS). Dairy, 2014. Dairy Cattle Management Practiced in the United States [R]. USDA, APHIS, VS, February 2016, 246 p. Fort Collins, CO. USA.

Porter, J. C., Warner, R. G., Kertz, A. F., 2007. Effect of fiber level and physical form of starter on growth and development of dairy calves fed no forage [J]. The Prof. Anim. Scientist, 23: 395-400.

Quigley, J. D., Wolfe, T. A., Elsasser, T. H., 2006. Effects of additional milk replacer feeding on calf health, growth, and selected blood metabolites in calves [J]. J. Dairy Sci., 89: 207-216.

Soberon, F., Van Amburgh, M. E., 2010a. Effects of colostrum intake and pre-weaning nutrientintake on post-weaning feed efficiency and voluntary feed intake [J]. J. Dairy Sci., 94: 69-70 Suppl. 1, abstr. #M180.

Soberon, F., Amburgh, M. E., 2010b. Effects of pre-weaning nutrient intake in the developing mammary parenchymal tissue and fat pad [J]. J. Dairy Sci., 94: 69-70 Suppl. 1, abstr. #M190.

Soberon, F., Raffrenato, E., Everett, R. W., et al., 2012. Preweaning milk replacer intake and effects on long term productivity of dairy calves [J]. J. Dairy Sci., 95: 783-793.

Soberon, F., Van Amburgh, M. E., 2013. The effect of nutrient intake from milk or milk replacer pf preweaned dairy calves on lactation milk yield as adults: A meta-analysis of current data [J]. J. Anim. Sci., 91: 706-712.

Stamey, J. A., Janovick, N. A., Kertz, A. F., et al., 2012. Influence of starter protein content on growth of dairy calves in an enhanced early nutrition program [J]. J. Dairy Sci., 95: 3327-3336.

Stott, G. H., Marx, D. B., Menefee, B. E., et al., 1979. Colostral Immunoglobulin Transfer in Calves I. Period of Absorption [J]. J. Dairy Sci., 62: 1632-1638.

Van Amburgh, M. A., Soberon, F., 2016. Developing a quality heifer: Management, economic and biological factors to consider [R]. Resource Guide of Dairy Calf and Heifer Association *Moving Forward* program, April 11-13, Madison, WI, p. 37-44.

Warner, R. G., Browstein, M. T., 1976. Factors affecting the immune response in calves and its relationship to subsequent performance [J]. Proc. Cornell Nutr. Conf., 26-33.

Interaction of Stocking Density, Cow Comfort, and Productivity: Effects on Lactating Cows

P. D. Krawczel[1], M. A. Campbell[2], J. A. Kull[1], and R. J. Grant[2]

[1]The University of Tennessee, Department of Animal Science, Knoxville, TN USA

[2]William H. Miner Agricultural Research Institute, Chazy, NY USA

Abstract

Overstocking remains a challenge management issue across commercial dairy farms. Economic analysis suggests that farm profitability may be maximized when cows are housed at 120% of feeding and/or resting resources (1.2 cows per available resting or feeding space within a pen). However, this analysis is based on published data, which has one limitation. Stocking density may serve as a subclinical stressor, which becomes problematic when presented in conjunction with another stressor. We have designed a series of studies to evaluate this concept in lactating dairy cows. First, we will present our findings that address the effects of combining stocking density with a dietary stressor. Second, we will address the concept of sleep and lying deprivation in cows. Stocking density has consistently reduced lying time, yet its influence on sleep has not been considered. Collectively, this suggests that understanding the dynamic of sleep deprivation within the context of a cumulative effect with sleep deprivation is the next critical step to establishing the negative effects of stocking density on behavioral and production responses.

Introduction

The behavior of dairy cows is dependent on the interaction between the cows and their physical environment. In the "big picture", the physical factors of the facility (stall design, flooring type, feed bunk design, environmental quality) impose baseline limitations on how the cows will interact with the housing conditions. Within these limitations, the ability of cows to engage in natural behaviors is further dictated by management routines such as grouping strategy and stocking density. The emphasis of this paper will be placed on evaluating the interactions of stocking density, cow comfort, and productivity with a special emphasis on the idea that stocking density may function as a subclinical stressor that interacts with other factors to alter productivity.

On commercial farms, overstocking is just one of many non-dietary factors affecting production (Bach et al., 2008). Previous research demonstrated that the three most important non-nutritional factors influencing cow well-being and efficiency of production are: (1) free-stall availability, (2) insufficient feed availability, and (3) heat stress (Bach et al., 2008; Bava et al., 2012). Ensuring feed availability (i.e. keeping it within reach of the cows and enough feed is delivered) was associated with 1.6 kg/d to 4.0 kg/d more milk per cow (Bach et al., 2008). Despite this, the trend toward higher feed costs encourages farmers to minimize feed refusals (i.e. feed remaining at the end of the day), even to the point of restricting access to feed. Moderate heat stress (temperature-humidity index of 73.8) reduced lying time by 3 h/d and substantially increased incidence of lameness (Cook et al., 2007).

Some degree of overstocking may be optimal, if onlyconsidering profitability. Models, built from published data, on the relationships among stocking density, lying time, and profit ($/stall/year; De Vries et al., 2016) determined that profit per stall was maximized around 120% stocking density. The profitability of overstocking was a function of revenue gained by increasing production per stall, the cost of increasing or decreasing production per cow, variable costs (i.e., costs that vary with changes in milk production), and milk price (De Vries et al., 2016).

The effects of stocking density on lying behavior have been of interest for several decades. The earliest research (Friend et al., 1977) suggested that total lying time was not affected until a stockingdensity greater than 150% was imposed. Lying time at the densities of 100%, 120%, and 150% was 14 h per day and was then reduced to 10 h and 7 h per day when stocking density was increased to 200% and 300%, respectively. The lying behavior of cows subjected to either under- (67% stocking density) or overcrowded (113%) conditions did not differ in terms of average lying times (10 h/d), time spent ruminating while lying (5 h/d), or total time within a freestall (15 h/d; Fregonesi and Leaver, 2002). This response may explain why Bach et al. (2008) found no benefit to undercrowding. However, there were fewer aggressive interactions per hour in the undercrowded pen. Conversely, increasing stocking densities incrementally from 100% to 142% or 150% resulted in reduction of lying time (Krawczel et al., 2012a; Fregonesi et al., 2007), but the extent of the reduction varied. Krawczel et al. (2012a) observed a reduction of 42 to 48 min per day for cows spending 13 h/d lying at a stocking density of 100%. Fregonesi et al. (2007) observed closer to a 2 h reduction in lying at 150% from the 13 h/d cows spent lying at 100%. Fregonesi et al. (2007) observed a reduction of latency to lie down when stocking density exceeded 120% may be misinterpreted as an increase in cow comfort, and may actually pose an increased risk of environmental mastitis. Finally, a greater number of aggressive interactions per hour occurred with each increase in stocking density in both studies. This reported reduction of lying time was consistent with the reduction in the percentage of stall usage in overcrowded cows, when stocking density was increased from 100% to 142% (Hill et al., 2007).

Though it is a highly variable relationship, overcrowding at the freestalls tends to result in overcrowding at the feed bunk. This relationship is highly dependent on the barn design (4-row versus 6-row) and severity of the freestall overcrowding. The effects of spatial allowance at the feed bunk of lactating dairy cows have been examined for the past 3 decades (Friend et al., 1977; DeVries et al., 2004; Huzzey et al., 2006). The earliest research established that reducing feed bunk space per cow to less than 10 cm per cow reduced feeding time (Friend et al., 1977). The behavioral effects of providing either 0.5 m, slightly less than the 0.6 cm commonly recommended, or 1 m of bunk space per cow were reduced the number of aggressive interactions per cow and increased the percentage of cows feeding during the 90 minutes following the delivery of fresh total mixed ration (DeVries et al., 2004). At stock densities ranging from 75% to 300%, feeding time decreased and aggression increased as stocking density increased (Huzzey et al., 2006). One potential coping strategy that was observed was the shift in feeding times, which may be problematic if the ration is sorted by the first cows to feed. Feed availability was also demonstrated to be a key management consideration related to the impacts of stocking density. A comparison of 24 h/d vs. 14 h/d of feed availability in conjunction with 100% or 200% stocking densities observed that the reduction of time that feed was available reduced DMI while stocking density did not (Collings et al., 2011).

Stocking Density May Function as a Subclinical Stressor

The concept of subclinical stressors suggests that the summation of two stressors, such as housing and feeding management, will be greater than either in isolation. A subclinical stressor depletes the animal's biological resources without generating a detectable change in function, which leaves the animal without the resources to respond to subsequent stressors (Moberg, 2000). Therefore, dairy cows may exhibit changes in behaviors that do not always result in clinical or visible outcomes, such as decreased milk yield, diminished milk quality, or increasedhealth issues. However, a subclinical stressor, i.e. stocking density, would diminish her ability to cope with further stressors resulting indetrimental outcomes. Unlike designed research trials on stocking density, additional stressors are likely to occur due to constant changes in feeding and cow management common on commercial farms. While the investigation of the interaction, or cumulative effect, of multiple stressors in lactating dairy cows is novel, the concept is not. Various studies across different species demonstrated that imposing multiple stressors have a greater effect on behavior, biological function, or stress physiology than when presented alone. The combination of stocking density and mixing altered the feeding behaviors of swine in an additive manner (Hyun et al., 1998a). Ambient temperature, stocking density, and social group collectively reduced average daily gain to a greater extent than any of the three alone (Hyun et al., 1998b). A linear relationship between the severity of the response, decreased weight gain and increased the heterophil: lymphocyte ratio, and the number of stressors, including heat stress and beak-trimming, was evident in broiler chickens (McKee and Harrison, 1995). The combination of restraint and LPS-challenge reduced growth and increased concentrations of corticosterone in mice to a greater extent that either stressor alone (Laugero and Moberg, 2000). Collectively, these studies support our hypothesis that the additive effect, or interaction, of two stressors on behavior, productivity, and the stress response will be greater than either imposed alone. Understanding the effects of stocking density with additional management stressors, such as low-fiber diets or feed restriction, are the next steps in alleviating stress and improving the well-being and long-term productive efficiency of lactating dairy cows housed in free-stall barns.

Cumulative Effect of Stocking Density and Diet Composition

In our first study, forty-eight multiparous and 20 primiparous Holstein cows were assigned to 1 of 4 pens ($n=17$ cows per pen). Pens were assigned to treatments in a 4×4 Latin square with 14-d periods using a 2×2 factorial arrangement. Two stocking densities (STKD; 100% or 142%) and 2 diets (straw, S; no straw, NS; Table 1) resulted in 4 treatments (100NS, 100S, 142NS, and 142S). Stocking density was achieved through denial of access to both headlocks and free-stalls (100%, 17 free-stalls and headlocks per pen; 142%, 12 free-stalls and headlocks per pen). Pen served as the experimental unit.

Table 1 Ingredient composition and analyzed chemical composition (dry matter basis) of TMR samples for NS (No Straw) and S (Straw) experimental diets

	NS	S	SEM[1]
Ingredient, % of DM			
Conventional corn silage	39.72	39.73	
Haycrop silage	6.91	2.33	
Wheat straw, chopped	···	3.45	
Citrus pulp, dry	4.82	4.82	
Whole cottonseed, linted	3.45	3.45	
Soybean meal, 47.5% solvent	···	1.12	
Molasses	3.20	3.20	
Concentrate mix	41.89	41.88	

(Continued)

	NS	S	SEM[1]
Chemical composition			
CP, % of DM	15.0	15.1	0.3
NDF, % of DM	30.8	30.1	0.4
Acid detergent lignin, % of DM	3.8	3.8	0.1
Starch, % of DM	25.0	25.5	0.5
Sugar, % of DM	7.4	8.1	0.4
Ether extract, % of DM	5.9	5.7	0.1
7-h starch digestibility, % of starch	73.3	74.3	0.9
Physically effective $NDF_{1.18\ mm}$, % of DM[2]	23.9	25.9	0.7
30-h uNDFom, % of DM[3]	13.1	14.9	0.3
120-h uNDFom, % of DM[3]	9.0	10.2	0.2
240-h uNDFom, % of DM[3]	8.5	9.7	0.2

[1]Standard error of the means.

[2]peNDF determined with method described by Mertens (2002).

[3]undigested NDF determined with method described by Tilley and Terry (1963) with modifications (Goering and Van Soest, 1970).

Diets were similar except that the S diet had a portion of haycrop silage replaced with chopped wheat straw and soybean meal. Each diet was formulated to meet both ME and MP requirements. The TMR was mixed and delivered once daily at approximately 06: 00 h and pushed up approximately 6 times daily.

The diets were designed to differ in a biologically meaningful way in physically effective NDF (peNDF) and undigested NDF (uNDF) measured at 30 h, 120 h, and 240 h of *in vitro* fermentation while remaining similar in analyzed chemical composition. Twelve multiparous and 4 primiparous ruminally cannulated cows were used to form 4 focal groups for ruminal fermentation data. Each focal group was balanced for DIM, milk yield, and parity. Ruminal pH was measured using an indwelling ruminal pH measurement system (Penner et al., 2006; LRCpH; Dascor, Escondido, CA) at 1-min intervals for 72 h on days 12, 13, and 14 of each period. Daily ruminal pH measurements were averaged over 10-min intervals. Measurements were then averaged across days and among cows into a pen average for each period.

As hypothesized, increasing the peNDF content of the diet reduced the time spent below pH 5.8 ($P=0.01$) as well as decreasing the severity of sub-acute ruminal acidosis (SARA) as observed through a reduction in area under the curve below pH 5.8 ($P=0.03$). Higher stocking density increased time spent below pH 5.8 ($P<0.01$) and tended to increase the severity of SARA ($P=0.06$). Furthermore, there was a trend for an interaction between stocking density and diet, indicating greater SARA when cows were housed at higher stocking density and fed the lower fiber diet. Importantly, greater stocking density had a larger effect on ruminal pH than changes to the diet, with a 1.4-h difference between 100 and 142% stocking density but only a 0.9-h difference between diets. Reductions in SARA through the addition of straw was observed at both stocking densities (0.4-h difference at 100% and 1.4-h difference at 142%), although there seemed to be greater benefit of boosting dietary peNDF or uNDF at the higher stocking density.

Cows were milked 3 times daily and milk yields were recorded electronically on d 8 to 14 of each period. Milk samples were collected across 6 consecutive milkings for each cow on d 13 and 14 of each period and analyzed for composition. Ingestive, rumination, and lying behavior as well as the location (feed bunk, stall, alley, standing or lying) of these performed behaviors were assessed on all cows using 72-h direct observation at 10-min intervals (Mitlöhner et al., 2001) on d 8, 9, and 10 of each period. Eating time (238 min/d, SEM = 4) and rumination time (493 min/d, SEM = 9) did not differ among treatments ($P > 0.10$). However, rumination within a free-stall as a percent of total rumination decreased at higher stocking density. As resting and rumination are significant contributors to buffer production (Maekawa et al., 2002b), it is possible that this shift in the location of rumination may affect the volume or rate of buffer production, partially explaining the increased risk of SARA at higher stocking densities. Ruminal pH differences between diets are likely explained by increased buffer volume produced during eating and rumination for the straw diets as evidenced by Maekawa et al. (2002a) where increases in the fiber-to-concentrate ratio resulted in increased total daily saliva production. Higher stocking density increased the latency to consume fresh feed - i.e., it took cows longer to approach the bunk and initiate eating with higher stocking density. Additionally, higher stocking density reduced lying time, but boosted the time spent lying while in a stall indicating greater stall - use efficiency. Overall, time spent standing in alleys increased markedly with overstocking. This behavior may explain the changes in ruminal pH, due to the lost of buffer capacity within the rumen.

Detrimental Effects of Lying Deprivation

Lying down is highly prioritized by dairy cows. Furthermore, other behaviors such as feeding and socializing have been given up to engage in lying down (Munksgaard et al., 2005). On average, a dairy cow spends roughly (11.9 ± 2.4) h/d lying down and (4.3 ± 1.1) h/d feeding (Gomez and Cook, 2010). There are multiple benefits to the cows achieved while they are lying; a major one relative to production is the increase in blood flow to the udder (Metcalf et al., 1992), which may help explain the relationship between increasing lying time and milk production. The high demand for lying time and the relationship with milk production suggestsit is critical to the overall welfare of dairy cows.

Effects of Lying Deprivation

With lying time being such a critical component of a dairy cow's time budget, there is growing evidence of the negative effects that result when cows are deprived of it. Cooper et al. (2007) evaluated the effects of a 2 h and 4 h lying deprivation period. During the 2-h deprivation period, cows stomped their feet and repositioned themselves more, which were interpreted as behavioral indicators of stress. Similar results were observed during the 4-h deprivation period; however, butting and continually shifting of their weight were also observed. These behaviors have been consistently seen through many lying deprivation studies suggesting cows are likely frustrated and uncomfortable (Ruckebusch, 1974; Metz, 1985, Cooper et al., 2007). One study evaluated the effects of feed deprivation, as a sole stressor, compared to the combined effects of feed and lying deprivation. When cows were deprived on feed and lying deprivation for 3 h, cows chose to lie down rather than feed (Metz, 1985). This indicated that cows prioritize lying over feeding behavior suggesting lying behavior is a basic requirement for overall well - being. However, not only can behavior be altered, but deprivation can cause physiological change as well. Munksgaard et al. (1999) reported cows deprived on lying for 14-h had a greater ACTH concentration at the beginning and end of treatment indicating an increased hypothalamic-pituitary-adrenal activity.

To date, evaluations of the effect of lying deprivation were confounded by failing to consider that the deprivation treatments also resulted in sleep deprivation. Recent work from universities in the Nordic countries has provided insight into sleep in dairy cows. Unlike some species that sleep over 15 hours a day like opossums or bats, cows only sleep 3-4 h/d (Zepelin et al., 2005, Ternman et al., 2012). More specifically, they spend roughly 3 h/d in NREM sleep, 45 min/d in REM sleep and 8 h/d drowsing (Ruckebusch, 1972). The relatively limited amount of sleep dairy cows engage in throughout a 24-h period might be indicative of their need to forage (Allison and Cicchetti, 1976). Although sleeping postures differ among species, adult cows have been characterized as sleeping when they are lying down with their head resting on their flank (Ruckebusch, 1972). However, varying behavioral postures can be used to distinguish between vigilant states. For example, to engage in REM sleep, eyelids must be completely closed and the neck muscle relaxed (Ruckebusch, 1972). This prevents cows from engaging in REM sleep while standing, however, some cows can experience NREM sleep during this time especially when lying is prevented (Ruckebusch, 1974). While cows can ruminate during drowsing, rumination is prevented during REM sleep due to the body position and reduced muscle tone required for REM sleep (Ternman et al., 2012). Although muscle tone is reduce during drowsing, it may come back in short periods due to re - positioning (Ruckebusch, 1972). Drowsing and NREM sleep are sometimes hard to differentiate because they both can display the same behavioral postures. During NREM sleep and drowsing, the cows eyelids are relaxed, but may be partially open making it difficult to distinguish between the two (Ruckebusch, 1972). Behavioral postures were established as a validated approach to assessing sleep in calves (Hänninen et al., 2008), they cannot be used applied to adult cows due to resting postures looking the same across vigilant states (Ternman et al., 2014). Because drowsing was not observed in calves, sleep estimates were more accurate for this age group compared to cows (Hänninen et al., 2008). For example, the behavioral indicator for NREM sleep was to a large extent applied to drowsing as well. Furthermore, NREM and REM sleep could not be differentiated solely based on behavioral indicators (Ternman et al., 2014). One other limiting factor with previous studies is that rumination interferes with the EEG signals during drowsing and NREM sleep. Therefore, the rumination artefacts created by the muscles from chewing make it hard to distinguish between vigilant states (Ternman et al., 2012).

Sleep deprivation may prove to be a critical stressor to dairy cows, as there is evidence of the detrimental consequences within other species. In other species, primarily humans and laboratory animals, sleep is considered the most direct measurement of rest quality and has long been considered imperative for health and well-being. Rats died after two to three weeks after complete sleep deprivation (Rechtschaffen and Bergmann, 1995). They also developed ulcerative and hyperkeratotic lesions on the tail and paw area, which are likely brought on from the deprivation. Along with physical effects, rats also experienced reduced body temperatures, high metabolic rate and decreased host defensive suggesting sleep maintains vital bodily functions. In subsequent work, bacterial infections were observed post sleep deprivation and concluded that the rats may have died from septicemia (Everson et al., 1989). However, later studies did not support this hypothesisas even antibiotics did not prevent the death of sleep-deprived rats (Bergmann et al., 1996). This suggests that sleep deprivation leads to the breakdown of the host defense system. Furthermore, even one night of sleep deprivation can impair performance such as creative and complex thinking as well as decrease attention span (Horne, 1988). All of these represent potentially detrimental outcomes for lactating dairy cows experiencing sleep deprivation.

While many studies have observed the effects of sleep deprivation in humans and rodents, few studies have looked at sleep deprivation in cattle. Ruckebusch (1974) recorded the effects of a 14 h/d, 20 h/d and a 22 h/d lying deprivation period on dairy cows. Although not directly studying the effects of sleep deprivation, REM sleep was prevented and NREM sleep was reduced during the deprivation period. Interestingly, when lying deprivation was increased to 20 h/d and when the free choice period (no deprivation) was limited to 2 h/d, cows chose to eat for that entire time rather than sleep. Even though the cows increased their rate of food intake during this time, the cows still lost weight. However, when the free choice period was 4 h/d, cows increased the rate of food intake and were able to engage in an equal amount of

REM sleep (Ruckebusch, 1974). This suggests that although cows chose to eat rather than sleep during the 22 h/d deprivation period, they do try to maintain an equilibrium between the two when possible. On average, sleep deprivation kills rats after an average of 21 days whereas food deprivation alone kills rats after 17–19 days (Everson et al., 1989; Obermeyer et al., 1991). This suggests that sleep deprivation has similar effects to deprivation of basic needs.

Our initial work to establish the varying response to the cumulative effect on the lying and sleep deprivation focused on the acute response. The objective was to determine the effectsof sleep or lying deprivation on lying behaviour and productivity of dairy cows. Data were collected from 8 multiparous and 4 primiparous cows [DIM = 199±44 (mean±SD); days pregnant=77±30; white blood cell count ≤ 12.6] between April and May 2016. Approach and brush tests were used to select cows who were not fearful of humans or stimulus around their head. Each cow experienced: (1) sleep deprivation achieved by disrupting the cow using noise or physical contact when her posture suggested the onset of sleep, and (2) lying deprivation implemented using a wooden grid within the pen. Treatments were imposed using a crossover design with 11–d data collection periods followed by 12–d washout periods. Cows were housed in individual box stalls (mattress base with no bedding) during habituation (d–2 and d–1; no data collection), baseline (d 0), and treatment (d 1) periods. After the treatment, cows returned to a sand–bedded freestall pen and lying behavior was monitored for a 7–d recovery period (d 2 to d 8). Lying time, lying bouts and bout duration were recorded using accelerometers attached to their hind legs. Milk production was recorded 2× daily automatically. Data were analyzed using a mixed model in SAS including fixed effects of treatment, day (0 to 9) and a random effect of cow. Significant main effects were separated using a PDIFF statement ($P \leqslant 0.05$). An interaction between treatment and day were evident for lying time, bouts, and bout duration ($P<0.001$). Lying time decreased during treatment (d 1) and increased on the first day of recovery (d 2) when cows were lying deprived compared to sleep deprived [d 1: (1.9±0.8) h/d vs. (8.4±0.7) h/d; $P<0.001$; d 2: (16.8±0.6) h/d vs. (13.6±0.7) h/d; $P = 0.002$]. Lying bouts were greater during treatment (d 1) for sleep–deprived cows compared to lying deprived [(7.6±0.7) h/d vs. (4.1±0.9)h/d; $P=0.01$]. No other differences were evident ($P \geqslant 0.12$) Bout duration ($P< 0.001$); decreased during lying deprivation (d 1) compared to sleep deprivation [(15.3±8.0) min/bout vs. (72.9±7.0) min/bout], but was increased on the first day of recovery [d 2; (110±6.6) min/bout vs. (89.9±7.0) min/bout]. Milk yields decreased during lying deprivation compared to sleep deprivation on the first day of recovery [d 2; (31.8±2.4) kg/d vs. (35.3±2.4) kg/d; $P = 0.002$]. Differences during treatment established lying deprivation was successfully achieved. Lying deprivation altered the behavior of cows during the recovery phase relative to sleep deprivation, as well as reduced milk yield. It needs to be established if these differences were caused by lying deprivation or the cumulative effect of lying and sleep deprivation.

Conclusions

Overstocking remains a critical issue across dairy production systems. While economic evaluations of stocking density suggest that profitability may be maximized with some degree of overstocking of housing facilities, caution must be used as there is also growing evidence that overstocking may function as a subclinical stressor. This effective was evident in our two recent trials. First, stocking density exhibited a consistent negative effect on ruminal pH and increased the risk for SARA. The presence of additional stressors in combination with stocking density exacerbated these negative effects on ruminal pH, although the magnitude varied depending on the type of stressor. However, manipulation of the feeding environment can help mitigate the negative effects of stocking density, such as increasing peNDF in the diet or reducing time without feed. Second, lying deprivation resulted in a greater behavioral change than sleep deprivation alone. Additionally, milk yield was only reduced by lying deprivation. Collectively this continues to suggest the effects of lying deprivation may actually be driven by the cumulative effects of lying and sleep deprivation. Farm management strategies should be focused mitigating overstocking. However, when it is unavoidable, it becomes critical to evaluate the system to reduce further stressors that may interact to induce a cumulative effect.

Acknowledgements

The authors would like to thank the research and farm staff at the William H. Miner Agricultural Research Institute and Dr. David Barbano at Cornell University for their assistance in completing research focused on the effects of stocking density and dietary fiber. Funding for this research was provided by both the Miner Institute Research Enhancement Fund and USDA AFRI Foundational Grant Funding (USDA#2016-67015-24733). We are also grateful to the farm staff at the East Tennessee Research and Education Center, graduate and undergraduate students at the University of Tennessee, Dr. Katy Proudfoot, Dr. Jeffrey Bewley, Dr. Bruce O'Hare, Dr. Kevin Donahue, and Dr. Gina Pighetti for their assistance in completing the work on lying and sleep deprivation. This work was funded by the USDA-NIFA-Exploratory Grant Program (USDA # 2015-67030-24295).

References

Allison, T., Cicchetti, D. V., 1976. Sleep in mammals-Ecological and constitutional correlates [J]. Science, 194: 732-734.

Bach, A., Valls, N., Solans, A., et al., 2008. Associations between nondietary factors and dairy herd performance [J]. J. Dairy Sci., 91: 3259-3267.

Bava, L., Tamburini, A., Penati, C., et al., 2012. Effects of feeding frequency and environmental conditions on dry matter intake, milk yield and behaviour of dairy cows milked in conventional or automatic milking systems [J]. Ital. J. Anim. Sci., 11: 230-235.

Bergmann, B. M., Gilliland, M. A., Penati, C., et al., 1996. Are physiological effects of sleep deprivation in the rat mediated by bacterial invasion? [J]. Sleep, 19: 554-562.

Collings, L. K. M., Weary, D. M., Chapinal, N., et al., 2011. Temporal feed restriction and overstocking increase competition for feed by dairy cattle [J]. J. Dairy Sci., 94: 5480-5486.

Cook, N. B., Mentink, R. L., Bennett, T. B., et al., 2007. The Effect of Heat Stress and Lameness on Time Budgets of Lactating Dairy Cows [J]. J. Dairy Sci., 90: 1674-1682.

Cooper, M. D., Arney, D. R., Phillips, C. J. C., 2007. Two-or Four-Hour Lying Deprivation on the Behavior of Lactating Dairy Cows [J]. J. Dairy Sci., 90: 1149-1158

De Vries, A., Dechassa, H., Hogeveen, H., 2016. Economic evaluation of stall stocking density of lactating dairy cows [J]. J. Dairy Sci., 99: 3848-3857.

De Vries, T. J., von Keyserling, M. A. G., Weary, D. M., 2004. Effects of Feeding Space on the Inter-Cow Distance, Aggression, and Feeding Behavior of Free-Stall Housed Lactating Dairy Cows [J]. J. Dairy Sci., 87: 1432-1438.

Everson, C. A., Bergmann, B. M., Rechtschaffen, A., 1989. Sleep deprivation in the rat: III. Total sleep deprivation [J]. Sleep, 12: 13-21.

Fregonesi, J. A., Leaver, J. D., 2002. Influence of space allowance and milk yield level on behaviour, performance and health of dairy cows housed in straw yard and cubicle systems [J]. Livest. Prod. Sci., 78: 245-257.

Fregonesi, J. A., Tucker, C. B., Weary, D. M., 2007. Overstocking Reduces Lying Time in Dairy Cows [J]. J. Dairy Sci., 90: 3349-3354.

Friend, T. H., Polan, C. E., McGilliard, M. L., 1977. Free stall and feed bunk requirements relative to behavior, production, and individual feed intake in dairy cows [J]. J. Dairy. Sci., 60: 108-116.

Gomez, A., Cook, N. B., 2010. Time budgets of lactating dairy cattle in commercial freestall herds [J]. J. Dairy Sci., 93: 5772-5781.

Grant, R. J., Albright, J. L., 2001. Effect of animal grouping on feeding behavior and intake of cattle [J]. J. Dairy Sci., 84 (E-Suppl.): E156-E163.

Hänninen, L., Mäkelä, J. P., Rushen, J., et al., 2008. Assessing sleep state in calves through electrophysiological and behavioural recordings: A preliminary study [J]. Appl. Anim. Behav. Sci., 111: 235-250.

Hill, C. T., Krawczel, P. D., Dann, H. M., et al., 2009. Effect of stocking density on the behavior of dairy cows with differing parity and lameness status [J]. Appl. Anim. Behav. Sci., 117: 144-149.

Horne, J. A., 1988. Sleep loss and "divergent" thinking ability [J]. Sleep, 11: 528-536.

Huzzey, J. M., DeVries, T. J., Valois, P., et al., 2006. Stocking density and feed barrier design affect the feeding and social behavior of dairy cattle [J]. J. Dairy Sci., 89: 126-133.

Hyun, Y., Ellis, M., Johnson, R. W., 1998a. Effects of feeder type, space allowance, and mixing on the growth performance and feed intake pattern of growing pigs [J]. J. Anim. Sci., 76: 2771-2778.

Hyun, Y., Ellis, M., Riskowski, G., et al., 1998b. Growth performance of pigs subjected to multiple concurrent environmental stressors [J]. J. Anim. Sci., 76: 721-727.

Krawczel, P. D., Mooney, C. S., Dann, H. M., et al., 2012. Effect of alternative models for increasing stocking density on the short-term behavior and hygiene of Holstein dairy cows [J]. J. Dairy Sci., 95: 2467-2475.

Laugero, K. D., Moberg, G. P., 2000. Effects of acute behavioral stress and LPS-induced cytokine release on growth and energetics in mice [J]. Physio. Behav., 68: 415-422.

Maekawa, M., Beauchemin, K. A., Christensen, D. A., 2002a. Effect of Concentrate Level and Feeding Management on Chewing Activities, Saliva Production, and Ruminal pH of Lactating Dairy Cows [J]. J. Dairy Sci., 85: 1165-1175.

Maekawa, M., Beauchemin, K. A., Christensen, D. A., 2002b. Chewing Activity, Saliva Production, and Ruminal pH of Primiparous and Multiparous Lactating Dairy Cows [J]. J. Dairy Sci., 85: 1176-1182.

McKee, J. S., Harrison, P. C., 1995. Effects of supplemental ascorbic-acid on the performance of broiler chickens exposed to multiple concurrent stressors [J]. Poul. Sci., 74: 1772-1785.

Metcalf, J. A., Roberts, S. J., Christensen, D. A., 1992. Variations in blood flow to and from the bovine mammary gland measured using transit time ultrasound and dye dilution [J]. Res. Vet. Sci., 53: 59-63.

Metz, J. H. M., 1985. The reaction of cows to a short-term deprivation of lying [J]. Appl. Anim. Behav. Sci., 13: 301-307.

Mitlöhner, F. M., Morrow-Tesch, J. L., Wilson, S. C., et al., 2001. Behavioral sampling techniques for feedlot cattle [J]. J. Anim. Sci., 79: 1189-1193.

Moberg, G. P., 2000. Biological response to stress: implications for animal welfare [M]. In: Moberg, G. P. and J. A. Mench (eds) The Biology of Animal Stress. CAB. International, Wallingford, UK, pp. 1-21.

Munksgaard, L., Ingvartsen, K. L., Pedersen, L. J., et al., 1999. Deprivation of Lying Down Affects Behaviour and Pituitary-Adrenal Axis Responses in Young Bulls [J]. Acta Agr. Scand. A-An., 49: 172-178.

Munksgaard, L., Jensen, M. B., Pedersen, L. J., et al., 2005. Quantifying behavioural priorities-Effects of time constraints on behavior of dairy cows, Bos Taurus [J]. Appl. Anim. Behav. Sci., 92: 3-14.

Obermeyer, W., Bergmann, B., Rechtschaffen, A., 1991. Sleep deprivation in the rat: XIV. Comparison of waking hypothalamic and peritoneal temperatures [J]. Sleep, 14: 285-293.

Rechtschaffen, A., Bergmann, B. M., 1995. Sleep deprivation in the rat by the disk-over-water method [J]. Behav Brain Res., 69: 55-63.

Ruckebusch, Y., 1972. The relevance of drowsiness in the circadian cycle of farm animals [J]. Anim. Behav., 20: 637-643.

Ruckebusch, Y., 1974. Sleep deprivation in cattle [J]. Brain Res., 78: 495-499.

Ternman, E., Hänninen, L., Pastell, M., et al., 2012. Sleep in dairy cows recorded with a non-invasive EEG technique [J]. Appl. Anim. Behav. Sci., 140: 25-32.

Ternman, E., Pastell, M., Agenäs, S., et al., 2014. Agreement between different sleep states and behaviour indicators in dairy cows [J]. Appl. Anim. Behav. Sci., 160: 12-18.

Zepelin, H., Siegel, J. M., Tobler, I., 2005. Chapter 8-Mammalian Sleep [M]. A2-Kryger, Meir H. Pages 91-100 in Principles and Practice of Sleep Medicine (Fourth Edition). T. Roth and W. C. Dement, ed. W. B. Saunders, Philadelphia.

Factors Regulating Milk Protein Synthesis and Their Implications For Feeding Dairy Cattle

Robin R. White

Department of Animal and Poultry Science, Virginia Tech, Blacksburg, VA, 24061.

Abstract

The growing global population and increasing limited resources for food production support the need to enhance food production sustainability globally. Precision feeding N is an important step toward enhancing sustainability of dairy production systems; however, precise understanding of amino acid (AA) requirements for lactation are needed before precision feeding opportunities can be realized in practice. Understanding factors that regulate milk protein synthesis provides potential to formulate dairy cattle rations that improve protein use efficiency, reduce feed costs, and limit N excretion. Previous research evaluating factors that regulate milk protein synthesis have identified the mTOR signaling pathway as a key metabolic regulator. The mTOR signaling pathway responds to growth factors, hormones, cytokines, stress, glucose, and amino acids. Both cell-culture and mathematical model experiments suggest that mTOR, and subsequently casein synthesis, are responsive to multiple amino acids. Cell culture experiments are commonly designed with an AA deficient media and several treatments where AA are additively or independently added back to that media. In such experiments, mTOR phosphorylation and the fractional rate of casein synthesis has been shown to respond simultaneously to multiple AA. This is in conflict with the single-limiting AA theory because if a single AA limited protein synthesis, one would not expect mTOR or casein synthesis rates to respond to multiple AA individually. Modeling studies have identified that variation in predicted casein synthesis rates is poorly explained by a single limiting AA theory; however, variation is explained quite well by a model that allowed multiple AA to simultaneously limit milk production. Additionally, more mechanistic models representing phosphorylation of mTOR and synthesis of casein have identified significant effects of multiple AA and energy substrate. These modeling studies suggest that amino acid supply quantitatively is more important for regulating casein synthesis than hormone status or energy availability.

Collectively, this summary of the literature suggests that milk protein synthesis is highly dependent on AA supply and that lactation appears to be simultaneously limited by multiple AA, rather than a single AA.

Introduction

Global population is expected to reach 9.4 billion by 2050 (U.S. Census Bureau, 2013) and demand for meat and milk is expected to rise substantially (Delgado, 2003). These global dynamics suggest a need to improve the sustainability of food production systems. Optimizing animal nutrition is one method of improving sustainability of ruminant production systems (White et al., 2014; White et al., 2015). In a model-based assessment of opportunities to reduce environmental impact of dairy production, White (2016) identified that enhancing efficiency of protein use dramatically reduced the cost of improving environmental impact of U.S. dairies. Before efficiency can be effectively optimized, the biological processes governing efficiency must be understood.

Efficiency of nutrient use for lactation is of key interest in U.S. agriculture because milk production provides sufficient energy and protein to meet the annual requirements of 71×10^6 and 169×10^6 people, respectively (White and Hall, 2016). Factors affecting protein synthesis have been extensively studied for several decades (Wu et al., 2014). At the signaling pathway level, we have an in-depth understanding of how substrate, hormone profiles, and energy status work together to influence milk protein synthesis (Osorio et al., 2016). For example, fairly robust mechanistic models of mammary metabolism which capture the independent effects of key essential AA, energy supply, and insulin have been constructed (Hanigan et al., 2000; Hanigan et al., 2001; Hanigan et al., 2002; Castro et al., 2016). Despite availability of these models, factors regulating milk protein synthesis have been incompletely transferred to the whole-animal level, and as a result, incompletely incorporated into ration formulation tools.

The objective of this review is to characterize factors affecting milk protein synthesis and discuss their implications in ration formulation. Despite a fairly robust understanding of what affects efficiency of milk protein synthesis, in general we lack the data necessary to construct robust and durable ration formulation tools needed to balance rations for AA. More targeted animal trials may be needed to better design ration formulation tools that mechanistically represent factors influencing milk protein synthesis.

Regulation of Milk Protein Synthesis

Previous reviews of factors regulating milk protein synthesis highlight the integral role of the Signal Transducer and Activator of Transcription 5 (STAT5) and Janus Kinase (JAK) signaling (Jak/Stat) pathway and the mechanistic target of rapamyacin (mTOR) signaling pathway in regulating casein synthesis (Bionaz et al., 2012; Osorio et al., 2016). Collectively, these pathways are stimulated by substrate supply and hormonal signals. Because mammary tissue can adapt its AA uptake capacity to meet intracellular AA demand (Bequette et al., 2000), transport of AA into the cell is also an important regulatory factor. Collectively, these factors (energy supply, AA supply, hormone status, and AA transport) generate a complex response surface.

The mTOR signaling pathway has several key elements including AMPK, AKT, and 4E-BP1. The subunit of interest for protein synthesis is the mTORC1 complex which includes regulatory-associated protein of mTOR (Raptor), mammalian lethal with SEC13 protein 8 (MLST8), among other elements. The mTORC1 complex activates translation of proteins by interacting with p70-S6 Kinase 1 (S6k1) and eukaryotic initiation factor 4E binding protein (4E-BP1). The role of the mTOR signaling pathway in regulating milk protein synthesis has been well studied and for this reason, the review will focus primarily on the mTOR signaling pathway.

A substantial body of work has also focused on the JAK2-STAT5 signaling pathway and its role in milk production. Recent studies demonstrate the interrelation between JAK2-STAT5 and the mTOR signaling pathways (Wang et al., 2014; Yang et al., 2015; Zhang et al., 2016). Whereas mTOR acts on protein translation, the JAK-STAT pathway directs transcription. Because of the roles of these pathways on different stages of protein synthesis, it is sensible that the two pathways work collaboratively to promote milk protein synthesis.

Energy Status Effects on Milk Protein Synthesis

It has long been recognized that milk protein yield is proportional to energy content of dairy cow rations (Emery, 1978). The molecular relationships defending the relationship between energy supply and milk protein synthesis are quite clear. Glucose enhances translation by reducing phosphorylation of eukaryotic initiation factor 2 α (EIF2α). Glucose also increases ATP production, which has inhibitory effects on 5' AMP-activated protein kinase. Because AMP-activated protein kinase inhibits mTOR, the reduction in this inhibition generates a stimulation of mTOR activity.

The quantity and ratio of volatile fatty acids available to mammary cells also appears to regulate mTOR expression. In an experiment with mammary epithelial cells, Sheng et al. (2015) demonstrated that mTOR expression was significantly related to the ratio of acetate to β-hydroxybutyrate. Appuhamy et al. (2011) also demonstrated that mTOR expression and casein synthesis were significantly affected by energy type (glucose versus acetate). Collectively, the effects of the quantity of energy available and the type of energy substrate highlight a key role for energy/protein interactions in regulating milk protein synthesis.

Hormone Effects on Milk Protein Synthesis

Energy status of the animal also affects signaling pathways indirectly via hormonal signaling. Insulin, prolactin, and glucocorticoids regulate activity of STAT5 through the Jak/Stat signaling pathway (Groner, 2002). In a comparison of Jak/Stat and mTOR signaling responses to a series of hormone additions to cell culture media, the Jak/Stat pathway appears to be more sensitive to hormonal regulation than the mTOR signaling pathway (Tian et al., 2016). This difference in sensitivity suggests that transcription is more sensitive to hormonal signals than translation. This is important because identifying rate - limiting steps in lactation has proved challenging. Improving our understanding of how different steps of synthesis are differentially regulated is key to identifying which of these steps are key rate limiting components under different conditions.

Despite being less sensitive than the Jak/Stat pathway, the mTOR signaling pathway is response to hormonal signals, most notably insulin. When insulin binds to cell surface insulin receptors, insulin receptor substrate 1 (IRS-1) initiates a signaling cascade that recruits protein kinase B (AKT) to the cell surface for phosphorylation (Corradetti and Guan, 2006). Additionally, insulin affects uptake of branched-chain AA and therefore indirectly simulates mTOR activity by enhancing substrate availability. In bovine MAC-T cells, the effects of insulin on casein synthesis have also been demonstrated (Appuhamy et al., 2011).

AA Transport and Milk Protein Synthesis

Predicting AA supply in dairy cattle is a major prediction challenge because AA profiles and digestibilities change across the rumen and intestine (White et al., 2016b), between intestinal flows and plasma concentrations (Patton et al., 2015), and between blood concentrations and mammary uptake (Bequette et al., 2000). Complications associated with digestion and absorption aside, mammary AA transport is a complex process. There are several types of AA transporters that each serve specific functions (Shennan and Boyd, 2014). The primary AA transport pathways include the L, A, ASC, y+, y+L, and β^{o} systems. Based on data in Shennan and Boyd (2014), these systems are summarized in Table 1.

Table 1 Transport systems of AA in mammary tissue as described by Shennan and boyd (2014)

System	Molecular components	Description	AA type	Regulation
L	LAT1 or LAT2 plus CD98hc or SLC3A2, or LAT3 or LAT4	Na^{+}-independent electroneutral transport	Neutral	Upregulated by lactogenic hormones (Sharma and Kansal, 1999)

(Continued)

System	Molecular components	Description	AA type	Regulation
A	SNAT1 through 4	Na^+ - dependent transport	Neutral	Upregulated by estradiol (López et al., 2006) and possibly AA supply
ASC	ASCT1 and ASCT2	Na^+ - dependent transport	Linear dipolar	Physiological state and milking frequency (Alemán et al., 2009)
Y+	CAT1 through 3	Na^+ -independent electrogenic transport	Cationic	Stage of lactation (Manjarin et al., 2011)
y+L	y+LAT2 and y+LAT1	Na^+ -independent (cationic) or dependent (neutral) electroneutral transport	Cationic or neutral	Physiological state (Manjarin et al., 2011)
β^o	ATB^o	NA^+ and Cl^- dependent transport	Cationic or neutral	Unclear at present

Because AA transport is regulated by factors other than supply of AA, it is presumably possible for hormonal or physiological status to generate intracellular deficiencies in AA despite extracellular surplus. As a result, the ample hormonal regulation of AA uptake provides potential molecular evidence for feeding challenges associated with periods like the transition from gestation to lactation. As a safety net to prevent excessive AA deficiency, Osorio et al. (2016) proposed that milk protein synthesis during the transition period is mediated by a general AA control/activating transcription factor 4 (GAAC-ATF4) pathway, whereby AA biosynthesis and uptake into the mammary are stimulated by ATF4 and LAT1. Safety nets such as this GAAC-ATF4 regulation may help contribute to the high levels of milk production sustained in the transition period.

In addition to hormonal or physiological stage regulation, dietary nutrient supply appears to influence AA uptake by the mammary gland. Evidence from arteriovenous difference studies conducted in dairy cattle or goats suggests that AA uptake is regulated by AA supply (Bequette et al., 2000), glucose supply (Rulquin et al., 2004), insulin (Mackle et al., 2000), and dietary fat content (Cant et al., 1993), among other factors. The large number of factors affecting AA uptake presents a major challenge when attempting to design rations that optimize AA availability for milk protein synthesis.

Amino Acid Supply Effects on Milk Protein Synthesis

Several studies have been designed to evaluate the effects of amino acid supply on milk protein synthesis machinery and milk production in cattle. For example, Nan et al. (2014) found that mTOR phosphorylation of MAC-T cells was improved by supplementing either Lys or Met or both Lys and Met to a low protein cell-culture media. This study supports the idea that AA supply affects the mTOR signaling pathway. Interestingly, the responsiveness of mTOR phosphorylation to both Lys and Met suggests that the MAC-T cells were co-limited in Lys and Met. The idea of co-limiting AA was further tested by Arriola-Apelo et al. (2014) who infused Ile, Leu, Met and Thr into cell culture media with low protein content. The casein fractional synthesis rate (CFSR) of MAC-T cells was then evaluated for each addition of AA and for multiple AA combinations. The results showed quadratic responses of CFSR to AA supplementation. Much like the results discussed in Nan et al. (2014), Arriola-Apelo et al. (2014) found that the CFSR was responsive to multiple AA suggesting the process followed co-limiting nutrient dynamics rather than single-limiting nutrient dynamics. These two studies demonstrate regulatory roles for Lys, Met, Ile, Leu and Thr within the mTOR signaling pathway.

Several potential reasons may explain why casein synthesis appears to respond to AA in a co-limiting fashion. A proposed mechanism for single-limiting nutrient theory has centered on the idea that the AA demand in protein synthesis exceeds the supply such that tRNA fail to recruit the necessary AA and protein synthesis is terminated. Given the numerous regulatory functions of AA within the protein synthesis pathway, it is unlikely that this tRNA recruitment failure would ever occur in a biological system. For example, AA act as activators of translation and influence transcription of key genes in the mTOR pathway. These compounding roles may help explain why addition of multiple AA generate production responses in cell culture systems. Interestingly, the co-limiting AA response identified in cell culture experiments also carries to the animal level. An experiment by Giallongo et al. (2016) suggests that milk production of cows fed a low protein diet also responds independently to multiple AA. Another experiment in mice found that pup growth rates were responsive to multiple AA (Liu et al., 2017). These *in vivo* studies further support that idea that milk protein synthesis follows a co-limiting nutrient model and is not limited by one nutrient at a time.

Models of Casein Synthesis

A model of casein synthesis responses to differing amino acid profiles was developed initially by Hanigan et al. (2001). This model included 19 stocks and considered 37 blood metabolites, including 22 AA. The model was fit against 38 milk protein observations from 4 studies. Although the model explained 48% of the variation in milk production when rations were fed, it was able to explain 0% of the variation when single AA were infused. As a result, the model was adapted to assume milk protein was limited by many AA at a time. This increased the model fit to explain 64% of the variation in milk production associated with fed rations and approximately 50% of the variation in milk production when individual AA were infused. The authors concluded that the remaining error was due to energy supply.

The model by Hanigan et al. (2001) was refined by Castro et al. (2016). Although the original intent was to develop a model of casein synthesis that accounted for mTOR phosphorylation and effects of AA supply, insulin, acetate, and glucose, the final model included only effects for Ile, Leu, acetate and insulin. The model suggested that casein synthesis is much more responsive to AA supply than to acetate or insulin. This result highlights the importance of AA supply and AA transport in regulating milk protein synthesis. Much like the cell culture experiments discussed previously and the model of Hanigan et al. (2001), Castro et al. (2016) found that casein synthesis was responsive to multiple AA.

Challenges Associated for Transitioning to Ration Formulation

Although the model by Castro et al. (2016) provides a mechanistic quantification of the major factors known to influence milk protein synthesis, it had high prediction error. The fractional casein synthesis rate was predicted with nearly 55% error. This highlights one major challenge in transitioning from our knowledge of the factors regulating milk protein synthesis to the design of diets with optimal AA profiles. Despite mechanistically defining several key regulatory relationships, the response in milk protein yield was poorly described, likely because of the complexity in the response surface. Although Castro et al. (2016) had over twice the amount of data that Hanigan et al. (2001) used, it was undoubtedly still insufficient to cover the full response surface of these factors. Before we can design optimally efficient rations for AA profile, we need additional data that describes the response surface of milk protein synthesis to independent and interacting effects of AA, energy, and hormones.

A nutrient requirement model accounting for co-limiting AA will have additional challenges with adoption. For example, in current dairy ration formulation tools requirements are calculated as target nutrient output (milk protein yield) divided by efficiency. However, co-limiting AA suggest that a deficiency in one AA could be made up by a surplus of one or more other AA. This means that a requirement (as traditionally defined) cannot, and perhaps should not, be calculated for individual AA. Instead, milk production should be modeled as a function of AA supply. Therefore, during ration formulation, the exercise becomes "how should I formulate the ration to achieve the most cost-effective production level" rather than "what is the lowest cost combination of ingredients required to achieve a target production level". Although related, these questions do have minor differences.

One potential way to represent the effects of AA profile on milk efficiency without doing away with nutrient requirements would be to vary metabolizable protein (MP) use efficiency based on AA supply. After updating the National Research Council (2001) model to eliminate errors in the supply portion of the model (White et al., 2017a; White et al., 2017b), it could theoretically be possible to update this model to adjust MP use efficiency based on AA profile. In a preliminary effort to test this approach, White et al. (2016a) found that adjusting MP use efficiency for AA profile helped to dramatically improve prediction accuracy of the National Research Council (2001) model. This preliminary work suggests this approach may be a promising way to incorporate AA effects on milk protein synthesis into ration formulation tools.

References

Alemán, G., López, A., Ordaz, G., et al., 2009. Changes in messenger RNA abundance of amino acid transporters in rat mammary gland during pregnancy, lactation, and weaning [J]. Metabolism, 58: 594-601.

Appuhamy, J. A. D. R. N., Bell, A. L., Nayananjalie, W. A. D., et al., 2011. Essential amino acids regulate both initiation and elongation of mRNA translation independent of insulin in MAC-T Cells and bovine mammary tissue slices [J]. J. Nutr., 141: 1209-1215.

Bequette, B., Hanigan, M., Calder, A., et al., 2000. Amino acid exchange by the mammary gland of lactating goats when histidine limits milk production [J]. J. Dairy Sci., 83: 765-775.

Bionaz, M., Hurley, W., Loor, J., 2012. Milk protein synthesis in the lactating mammary gland: Insights from transcriptomics analyses [M]. In Milk Protein. Intechopen, 285-324.

Cant, J., DePeters, E., Baldwin, R., 1993. Mammary amino acid utilization in dairy cows fed fat and its relationship to milk protein depression [J]. J. Dairy Sci., 76: 762-774.

Castro, J. J., Arriola Apelo, S. I., Hanigan, M. D., 2016. Development of a model describing regulation of casein synthesis by the mTOR signaling pathway in response to insulin, amino acids and acetate [J]. J. Dairy Sci., 99 (8): 6714-6736.

Corradetti, M., Guan, K., 2006. Upstream of the mammalian target of rapamycin: do all roads pass through Mtor? [J]. Oncogene, 25: 6347-6360.

Delgado, C. L., 2003. Rising consumption of meat and milk in developing countries has created a new food revolution [J]. J. Nutr., 188: 89075-89105.

Emery, R., 1978. Feeding For Increased Milk Protein [J]. J. Dairy Sci., 61: 825-828.

Giallongo, F., Harper, M., Oh, J., et al., 2016. Effects of rumen-protected methionine, lysine, and histidine on lactation performance of dairy cows [J]. J. Dairy Sci., 99: 4437-4452.

Groner, B., 2002. Transcription factor regulation in mammary epithelial cells [J]. Domest. Anim. Endocrin., 23: 25-32.

Hanigan, M., Crompton, L., Bequette, B., et al., 2002. Modelling mammary metabolism in the dairy cow to predict milk constituent yield, with emphasis on amino acid metabolism and milk protein production: Model evaluation [J]. J. Theor. Biol., 217: 311-330.

Hanigan, M., Crompton, L., Metcalf, J., et al., 2001. Modelling mammary metabolism in the dairy cow to predict milk constituent yield, with emphasis on amino acid metabolism and milk protein production: Model construction [J]. J. Theor. Biol., 213: 223-239.

Hanigan, M. D., France, J., Crompton, L. A., et al., 2000. Evaluation of a representation of the limiting amino acid theory for milk protein synthesis [M]. CABI, Wallingford, 127-144.

Liu, G., Hanigan, M., Lin, X., et al., 2017. Methionine, leucine, isoleucine, or threonine effects on mammary cell signaling and pup growth in lactating mice [J]. J. Dairy Sci., 100 (5): 4038-4050.

López, A., Torres, N., Ortiz, V., et al., 2006. Characterization and regulation of the gene expression of amino acid transport system A (SNAT2) in rat mammary gland [J]. Am. J. Physiol-Endoc. M., 291: E1059-E1066.

Mackle, T., Dwyer, D., Ingvartsen, K. L., et al., 2000. Effects of Insulin and Postruminal Supply of Protein on Use of Amino Acids by the Mammary Gland for Milk Protein Synthesis [J]. J. Dairy Sci., 83: 93-105.

Manjarin, R., Steibel, J., Zamora, V., et al., 2011. Transcript abundance of amino acid transporters, β-casein, and α-lactalbumin in mammary tissue of periparturient, lactating, and postweaned sows [J]. J. Dairy Sci., 94: 3467-3476.

Osorio, J. S., Lohakare, J., Bionaz, M., 2016. Biosynthesis of milk fat, protein, and lactose: roles of transcriptional and posttranscriptional regulation [J]. Physio. Genomics, 48: 231-256.

Patton, R., Hristov, A., Parys, C., et al., 2015. Relationships between circulating plasma concentrations and duodenal flows of essential amino acids in lactating dairy cows [J]. J. Dairy Sci., 98: 4707-4734.

Rulquin, H., Rigout, S., Lemosquet, S., et al., 2004. Infusion of glucose directs circulating amino acids to the mammary gland in well-fed dairy cows [J]. J. Dairy Sci., 87: 340-349.

Sharma, R., Kansal, V. K., 1999. Characteristics of transport systems of L-alanine in mouse mammary gland and their regulation by lactogenic hormones: Evidence for two broad spectrum systems [J]. J. Dairy Res., 66: 385-398.

Sheng, R., Yan, S., Qi, L., et al., 2015. Effect of the ratios of acetate and β -hydroxybutyrate on the expression of milk fat-and protein-related genes in bovine mammary epithelial cells [J]. Czech J. Anim. Sci., 60: 531-541.

Shennan, D., Boyd, C., 2014. The functional and molecular entities underlying amino acid and peptide transport by the mammary gland under different physiological and pathological conditions [J]. J. Mammary Gland Biol., 19: 19-33.

Tian, Q., Wang, H., Wang, M., et al., 2016. Lactogenic hormones regulate mammary protein synthesis in bovine mammary epithelial cells via the mTOR and JAK-STAT signal pathways [J]. Anim. Prod. Sci., 56 (11): 1803.

U.S. Census Bureau., 2013. U.S. Census Bureau International Data Base [EB/OL]. Vol. Accessed: 1/14/2013. U.S. Deparment of Commerce.

Wang, M., Xu, B., Wang, H., et al., 2014. Effects of arginine concentration on the *in vitro* expression of casein and mTOR pathway related genes in mammary epithelial cells from dairy cattle [J]. PLoS One, 9: e95985.

White, R., Roman-Garcia, Y., Firkins, J., et al., 2017a. Evaluation of the National Research Council (2001) dairy model and derivation of new prediction equations. 2. Rumen degradable and undegradable protein [J]. J. Dairy Sci., 100 (5): 3611-3627.

White, R., Roman-Garcia, Y., Firkins, J., et al., 2016a. Evaluation of the 2001 Dairy NRC and derivation of new equations [M]. In Energy and protein metabolism and nutrition. EAAP Publication, 71.

White, R., Roman-Garcia, Y., Firkins, J., et al., 2017b. Evaluation of the National Research Council (2001) dairy model and derivation of new prediction equations. 1. Digestibility of fiber, fat, protein, and nonfiber carbohydrate [J]. J. Dairy Sci., 100 (5): 3591-3610.

White, R. R., 2016. Increasing energy and protein use efficiency improves opportunities to decrease land use, water use, and greenhouse gas emissions from dairy production [J]. Agr. Syst., 146: 20-29.

White, R. R., Brady, M., Capper, J. L., et al., 2014. Optimizing diet and pasture management to improve sustainability of U.S. beef production [J]. Agr. Syst., 130: 1-12.

White, R. R., Brady, M., Capper, J. L., et al., 2015. Cow-calf reproductive, genetic, and nutritional management to improve the sustainability of whole beef production systems [J]. J. Anim. Sci., 93:

3197-3211.

White, R. R., Hall, M. B., 2016. Agriculture without animals? The environmental and economic role of livestock in food production [C]. Page 249 in 5th EAAP International Symposdium on Energy and Protein Metabolism and Nutrition. J. Skomial and H. Lapierre, ed. Wageningen Academic Publishers, Krakow, Poland.

Wu, G., Bazer, F. W., Dai, Z., et al., 2014. Amino acid nutrition in animals: protein synthesis and beyond [J]. Annu. Rev. Anim. Biosci., 2: 387-417.

Yang, J. X., Wang, C. H., Xu, Q. B., et al., 2015. Methionyl-methionine promotes α-s1 casein synthesis in bovine mammary gland explants by enhancing intracellular substrate availability and activating JAK2-STAT5 and mTOR-mediated signaling pathways [J]. J. Nutr., 145: 1748-1753.

Zhang, M., Zhao, S., Gao, H., et al., 2016. Effects of glucose and amino acids on casein synthesis via JAK2/STAT5 signaling pathway in bovine mammary epithelial cells [J]. J. Anim. Sci., 94: 404-405.

Mammary Development-Windows of Opportunity and Risk

Russell C. Hovey

Department of Animal Science 2145 Meyer Hall One Shields Avenue The University of California, Davis Davis, California 95616

Abstract

A cow's presence in the herd ultimately reflects her ability to efficiently produce milk as a function of her mammary gland's development and synthetic capacity that develop throughout the course of her lifetime. Successfully reaching this point in her life also comes at considerable economic cost, which is upwards of $ 2500 in the United States. From some angles, the processes driving mammary gland development are similar across species; from other angles the biological processes and influences in replacement heifers are distinct. Ultimately the course of mammary development is coordinately linked to various developmental and reproductive states, and is sensitive to a wide range of cues, ranging from endocrine signals to diet, the environment, the surrounding tissues and cell types, and their interaction. Given that the number of epithelial cells in the udder is proportional to milk yield potential, efforts to identify and understand the challenges and the opportunities during mammary growth stands to improve animal productivity, welfare and economic efficiency.

Species-specific Regulation of Mammary Development

When considering the growing mammary glands, particularly in dairy heifers for whom their uddersare the key determinant of their future presence in the herd, recognition must be given to what we do and don't understand about species differences in glandular development and composition. These similarities and differences stand, in turn, to inform our understanding of the processes regulating mammary growth and development.

At the gross level, all species ultimately develop a glandular epithelium that arises from the epidermis during embryogenesis (Hovey et al., 2002). However, it is the way the secretory parenchyma develops, and how it interacts with the surrounding microenvironment that differs widely (Hovey et al., 1999; Rowson et al., 2012). On the one hand, widely-studied mice develop a ductal network having a single galactophore (orifice) at the teat, similar to ruminants, whereas species such as pigs have 2 galactophores, and humans 8-15 (Rowson et al., 2012). This across-species diversity in epithelial architecture reflecting the earliest stages of organogenesis has received limited attention due to the fact that most studies to date have been in mice (Howard et al., 2013). At the same time, the anatomy of the primary duct subtending the galactophore can also vary dramatically across species, ranging from a simple enlarged lactiferous sinus in rodents, pigs, and humans, to a pronounced expansion as the gland cistern found in ruminants. Given that the size of the gland cistern is related to the ability of the udder to withstand extended milking intervals (Knight et al., 1994), our limited understanding of the factors regulating the size of this distinct structure seems ripe for further investigation.

The mammary glands also have a complex and dynamic ductal architecture that differs appreciably across species. Heifers, ewe lambs and doe kids developa network of ducts that are branched throughout their system, even as the parenchymal tissue expands. The branched apparatus in these species is best described using the same nomenclature used to define the ductal structures in the human breast, aptly defined as "terminal ductal lobular units" or TDLU (Rowson et al., 2012; Horigan et al., 2009) In essence, these TDLU provide for zonal expansion of the parenchyma through epithelial proliferation on multiple fronts. These TDLU can differ in the extent of their complexity, as described by the number of bifurcations or "ductules" within each. By contrast, the ducts in the mouse mammary gland extend as a ductal system with minimal branching that is led by concentrated zones of mitosis in pronounced "terminal end buds" (TEB) that have been studied extensively with respect to their remodeling and pluripotency potential (Hovey et al., 2002; Paine et al., 2017). Interestingly, species such as humans and pigs have a ductal architecture that develops somewhere between that found in ruminants and rodents, with both TEB and TDLU that develop concomitantly (Rowson et al., 2012).

Throughout pregnancy the parenchyma undergoes a renewed wave of proliferation, ultimately filling the organ with densely-packed milk-secreting lobules. This development is typically and loosely referred to as being of a "lobulo-alveolar" type across all species. However, the nature and origins of alveoli also vary subtly across species. In the case of mice, which are often the reference for this description, alveoli arise from primary, secondary and tertiary ducts as distinct proliferative spherical units that arise during pregnancy-specific development (Brisken et al., 2006). By contrast, ductules of the pre-established TDLUs in ruminants undergo continual bifurcation, giving rise to terminal swellings that ultimately become the single-layered alveoli in the lactating gland within the most developed TDLU, type 4 (Rowson et al., 2012).

The Mammary Gland Microenvironment

In considering the developing mammary gland, one must also take into account the stromal microenvironment in which epithelial cells grow and develop, and differences that exist across species (Hovey et al., 1999; Rowson et al., 2012). The stromal microenvironment is specialized for the mammary gland, and is absolutely essential for specification of the mammary epithelial phenotype. The stromal compartment includesvarious proportions of adipose, connective, immune, lymphatic, nervous and vascular tissues; these vary in their extent depending on species and stage of development (Hovey et al., 1999; Neville et al., 1998; Schedin et al., 2010). In mice the mammary stroma is primarily composed of adipocytes that can direct epithelial growth and function (Schedin et al., 2010); by contrast, the epithelium grows with relatively little exposure to connective tissue. In the developing mammary glands of heifers and other ruminants the stroma is much different in its composition (Hovey et al., 1999). While adipose tissue still abounds in these species, it always remains distal to the developing parenchyma. Instead, epithelial cells in ruminants grow in close apposition to a collagenous stroma that is already interwoven through the mammary fat pad. This connective tissue can be further described by its arrangement relative to the TDLU structures; the denser "intralobular" connective tissue within the TDLU structures contrasts to the less dense "interlobular" connective tissue outside each TDLU (Rowson et al., 2012). Notably, the composition of these stromal compartments also changes dramatically in response to cues ranging from endocrine signals to diet. We have also speculated that the relative abundance of connective tissue in the mammary stroma is a function of mammary gland positioning in the female body; more pendulous glands found in ruminants require additional internal suspension compared to the subcutaneous, widely-spread glands found in rodents (Hovey et al., 1999).

The Pre-and Peripubertal Period of Mammary Growth

Dairy heifers, like all females, undergo periods of mammary gland growth coincident with different stages of development. These distinct phases are linked to a changing reproductive state, ultimately to match the needs of the neonate after parturition. While a majority of epithelial development occurs after birth, growth of the future mammary glands initiates *in utero* when epithelial cells arise from the epidermis and become positioned about the future teats/nipples (Howard et al., 2014). Beyond the recognition that this period of development is important for specifying sexually-dimorphic development of the glands in males vs. females in some species such as mice (Howard et al., 2014), emerging evidence from different species also highlights this period may represent a first window of developmental sensitivity that can affect lifetime health and productivity of the mammary glands. For example, from studies of human breast cancer and mouse models comes the understanding that the developing epithelial rudiment in the mammary gland before birth is susceptible to various endocrine, environmental, and nutritional signals (Hilakivi-Clarke et al., 2006). Meanwhile, studies in sheep have highlighted that a dam's nutritional status can impact the future lactation potential of her progeny (Paten et al., 2013). While we now appreciate that adverse events such as heat stress during a heifer's fetal development can affect her lifetime milk production potential (Monteiro et al., 2016), the precise impacts on the course of mammary gland development remain to be defined, including the long-term influence on milk yield and composition.

After birth the mammary glands develop at the same rate as the rest of the body, that is, isometrically (Hovey et al., 2002). In mice, this growth phase lasts out to about 24 d of age, while in heifers it extends to approximately 2 months of age (Tucker et al., 1987). While this period has typically been considered to be relatively static in that the epithelium is not growing rapidly, such a conclusion may be oversimplified. In particular, this window may be more nutrient-sensitive than previously thought. Indeed, findings regarding the growth potential of heifers has raised important insights to suggest that pre-weaned heifers should receive a diet that is more energy-dense than would have historically been considered appropriate in order to realize significant gains in lactation performance (Soberon et al., 2013). The full effect of nutrition on the developing mammary glands during the preweaning period remain to be established; while some studies showed that an increased plane of nutrition before weaning did not increase parenchymal development in heifers (Daniels et al., 2009), others found that it did, also without dramatically increasing the deposition of fat in the surrounding stroma (Geiger et al., 2016; Brown et al., 2005). A clear conclusion from these and emerging studies is that rate of mammary growth in the isometric phase of growth is limited by undernutrition; this makes sense in that the more the body grows the more the mammary glands will grow in direct proportion, ie the definition of isometry.

As females age, epithelial cells begin to proliferate at rate faster than occurs in the rest of the body during a phase of allometric growth (Tucker et al., 1987). In mice, this growth commences around 28 d of age, while in heifers it commences around 3 months of age (~100 kg) and continues until they reach ~280 kg (Tucker et al., 1987; Meyer et al., 2006). This window of development has been a focus of considerable investigation in various circles; from a human health standpoint, the highly proliferative epithelium at this time is at increased risk for becoming cancerous, as borne out from both rodent studies as well as from data for humans exposed to high doses of radiation during puberty (Russo et al., 1996). At the same time, considerable attention has been givento the importance of ensuring that heifers do not experience a rate of gain that is too rapid for fear of them developing a "fatty udder" manifest as reduced accretion of parenchymal tissue alongside increased deposition of adipose tissue and enlargement of the extraparenchymal mammary fat pad (Lohakare et al., 2012). This presents a paradox for the management of dairy heifers-on the one hand, it is desirable for heifers entering the herd to grow at a rate that will see them reach puberty, and be mated, earlier so as to realize their income earning potential earlier (Wathes et al., 2014). At the same time, this advanced development shortens the length of time for allometric mammary growth to occur (Meyer et al., 2006), which has been proposed as an explanation for how rapid gain suppresses mammary development due to a high plane of nutrition. An alternative way of stating this interpretation is that the impaired mammary growth due to rapid growthcould be overcome werefemales not to enter puberty as early. Be that as it may, industry targets are increasingly aimed toward having heifers reach puberty and calve earlier. The proposal that this important developmental window is shortened by accelerated growth actually reinforces, in part, the alternative argument that increased rate of gain absolutely impairs mammary gland development, irrespective of advanced physiological age. Ultimately, in both scenarios, females stand to have less mammary parenchymal tissue at the time of breeding, and hence may not go on to realize their full lactation potential.

Theoverall picture of the factors that regulate allometric growth of the mammary glands is somewhatmurky. From a simple perspective, studies in rodents have led to the conclusion that increasing concentrations of circulating estrogens and growth hormone direct the increased local production of insulin-like growth factor-1 that leads to the synergistic promotion of epithelial proliferation, an axis that is also likely required for allometric growth in ruminants (Hovey et al., 1999; Meyer et al., 2006). There is a similar clear role for amphiregulin in mice (Ciarloni et al., 2007), although its role in the bovine or other domesticated livestock species has not been explored. That being said, ewe lambs underwent ovary-independent mammary gland growth (Ellis et al., 2007), implying that other factors besides estrogen and its effectors may also be at play during this period of growth.

Our recent findings in mice suggest that this period of allometric growth may not in fact be as exclusively estrogen-dependent as first thought. In particular, our laboratory has focused on dietary factors that can affect allometric mammary growth, including the effects of various "conjugated linoleic acids" (CLA) found in ruminant tissues that have also been investigated for their potential anti-cancer and weight-loss properties (McCroie et al., 2011). For these studies, we have used estrogen-deficient mice that are either ovariectomized prepubescent females, or intact males, in which the mammary glands fail to undergo allometric growth of the ductal network when the associated estrogenic signals are absent. The outcomes from these experiments challenge thestanding dogma that estrogen signaling is absolutely required for allometric growth to occur (Berryhill et al., 2012; Berryhill et al., 2017). Fascinatingly, when mice are fed a diet containing 1% of the 10, 12 isomer of CLA for as short as only a few days, the mammary ductal system initiates a pronounced burst of allometric elongation coincident with the formation of highly-proliferative TEB. This growth response is independent of estrogen or estrogen receptor signaling, and can be reversed by the insulin-sensitizing compound rosiglitazone (Berryhill et al., 2012). Intriguingly, this response is limited only to the 10, 12 isomer of CLA, and not 9, 11 CLA or other trans fatty acids as found in partially-hydrogenated vegetable oils (Berryhill et al., 2017). More recently, we have also begun investigating the question of how this seemingly benign dietary treatment can elicit such potent effects on the mammary glands. To this end we have recently completed RNA-Sequencing experiments to mine the transcriptome of this phenotype, which highlights, among other changes, that the local immune microenvironment may be a key driver of these changes. Such a finding is in keeping with several studies in mice showing that recruitment of local inflammatory mediators and immune cells into the vicinity of the growing mammary ducts is required for ductal elongation and TEB formation to occur. Given the endogenous presence of CLA isomers in ruminants, obvious questions arise as to whether this mechanism of growth stimulation is at play in dairy heifers during the allometric growth of their mammary glands, and whether it is impacted by factors such as dietary composition (leading to altered ruminal biohydrogenation) or excess energy intake.

Gestation-and Lactation-associated Mammary Development

In many cases the factors regulating growth of the mammary glands during gestation have been defined, but have not always been interrogated for their ability to impact future lactation performance (Tucker et al., 1987). Of course, gestation not only represents a period of active allometric proliferation of the mammary epithelium (Tucker et al., 1987), but also terminal differentiation of the epithelium in preparation for the synthesis of unique milk components and their secretion.

The question therefore becomes which factor (s) are most criticial in the lead up to parturition, and can they be manipulated positively or negatively to impact lactation success? These questions are best answered byremoval and replacement experiments. Of the various effectors of mammary function during gestation, prolactin isperhaps among the most importantregulators during this window; indeed, cows that have their prolactin secretion blocked prior to calving fail to initiate lactation (Akers et al., 1981), highlighting the importance of this hormone for both mammary growth and lactogenic differentiation. A similar situation exists for sows, where the suppression of serum prolactin just prior to farrowing reduces mammary gland development and abolishes milk production during the subsequent lactation (Farmer et al., 2003). Similar outcomes also occur when cows and maresare exposed to alkaloids produced from ergot-infected pastures such as fescue, leading to the suppression of both serum prolactin and lactation, as well as reproductive competence (Klotz et al., 2015).

Most recently we posed the converse question-canserum prolactin levels be elevated during late pregnancy to increase milk production or shift composition? To this end we treated primiparous gilts with the dopamine D2 receptor antagonist domperidone to elicit a transient increase in serum prolactin around d 90 of gestation during a prolactin-sensitive window when the mammary epithelium is both proliferating and differentiating (Vanklompenberg et al., 2013). Changes in mammary gland gene expression and development were evaluated from mammary biopsies collected throughout the course of the treatment period and into lactation. Gilts went on to farrow and nurse normally, and milk production and piglet growth was monitored weekly. Sows exposed to a transient increase in serum prolactin produced ~ 22% more milk without any change in milk solids. One hypothesis was that elevated circulating prolactin levels during gestation increased growth of the epithelium either during the treatment window, or into early lactation. This proved to not be the case, raising the alternative question that one or more aspects of milk synthetic capacity had been hyperstimulated by gestational hyperprolactinemia.Indeed, there were clear and pronounced carryover positive effects of brief elevated prolactin levels on the expression of all the various milk protein genes within the mammary glands throughout the rest of lactation (Vanklompenberg et al., 2013). These findings emphasize the importance of the late gestational window forthe preparation of epithelial cells prior to parturition, and also suggest that brief exposures during this period can lead to lasting epigenetic modifications in the lactational transcriptome. Several other examples also support the existence of such a mechanism. For example, the amount of light, or photoperiod, during the dry period affects subsequent lactation performance, where shorter days increase milk yield, which has been attributed to altered circulating prolactin levels (Crawford et al., 2015). In a similar way, heat stress during the dry period has a pronounced negative carryover effect on the next lactation (Wolfenson et al., 1988). Alternatively, ewes that carry twins or triplets produce increasingly more milk proportional to the level of placental lactogen produced by the placenta (Akers et al., 1985). Along the same lines, the results from a recent study suggest that dams carrying heifer calves go on to produce more milk than those carrying bulls (Hinde et al., 2014), although this may merely be a function of sex-specific lactation length (Hess et al., 2016).

It is worth noting that a similar lasting impact on milk production potential is manifest during early lactation by increased milking frequency and/or milk removal (Wall et al., 2012). While epithelial proliferation also continues into early lactation, increased milking frequency does not increase the amount of epithelial growth, leading to the conclusion that this carryover effect of frequent milking is due to epigenetic changes in the mammary epithelium. Interestingly, this window of sensitivity corresponds to the recently-described postpartum round of epithelial cell division that yields a distinct population of binucleated cells in the lactating mammary glands (Rios et al., 2016). The fact that this phenomenon occurs in of several species suggests that this window is a conserved requirement for milk production, perhaps to accommodate the massive increase in the transcriptional and synthetic capacity of epithelial cells at the start of lactation. The implications of this finding remain to be established in dairy cows (Table 1).

Table 1 Key characteristics of the developing mammary glands across various key species

Property	Mouse	Bovine	Pig	Human
Galactophores/gland	1	1	2	8-15
Gland sinus/cistern	Sinus	Cistern	Sinus	Sinus
Pregestational growth	Simple ductal	Complex ductal	Simple/Complex ductal	Simple/Complex ductal
Proliferative structures (pregestational)	Terminal end bud	Ductules of TDLU[1]	Terminal end bud/ Ductules of TDLU	Terminal end bud/ Ductules of TDLU
Stromal composition	Adipose>> connective	Inter/Intralobular connective>>adipose	Inter/Intralobular connective>>adipose	Inter/Intralobular connective>>adipose

[1]TDLU = Terminal Ductal Lobular Unit.

Conclusions

Given that development of the mammary glands spans much of a female's life, it is not surprising that each stage is sensitive to influences that can positively or negative impact growth and function of the glandular epithelium. In some cases, the mammary glands are plastic in their ability to compensate for adverse events, while in other cases they are rigid. While negative effects on mammary function have been identified (ie rapid growth), the extent of these effects on future lactation performance have not been fully quantified, which prevents their translation and integration into management decisions. On the other hand, opportunities to improve production (ie increased milking frequency) are already being implemented and used, despite the need to further understand how these responses are realized. Lastly, while studies in rodents stand to inform our understanding of the most basic biological mechanisms, the mammary glands of heifers are distinct and require independent study for many of these processes.

References

Akers, R. M., Bauman, D. E., Capuco, A. V., et al., 1981. Prolactin regulation of milk secretion and biochemical differentiation of mammary epithelial cells in periparturient cows [J]. Endocrinology, 109 (1): 23-30.

Akers, R. M., 1985. Lactogenic hormones: binding sites, mammary growth, secretory cell differentiation, and milk biosynthesis in ruminants [J]. J. Dairy Sci., 68 (2): 501-519.

Berryhill, G. E., Gloviczki, J. M., Trott, J. F., et al., 2012. Diet-induced metabolic change induces estrogen-independent allometric mammary growth [J]. Proc. Natl. Acad. Sci. U S A., 109 (40):

16294-16299.

Berryhill, G. E., Miszewski, S. G., Trott, J. F., et al., 2017. Trans-Fatty Acid-Stimulated Mammary Gland Growth in Ovariectomized Mice is Fatty Acid Type and Isomer Specific [J]. Lipids, 52 (3): 223-233.

Brisken, C., Rajaram, R. D., 2006. Alveolar and lactogenic differentiation [J]. J. Mammary Gland Biol. Neoplasia, 11 (3-4): 239-248.

Brown, E. G., Vandehaar, M. J., Daniels, K. M., et al., 2005. Effect of increasing energy and protein intake on mammary development in heifer calves [J]. J. Dairy Sci., 88 (2): 595-603.

Ciarloni, L., Mallepell, S., Brisken, C., 2007. Amphiregulin is an essential mediator of estrogen receptor alpha function in mammary gland development [J]. Proc. Natl. Acad. Sci. U S A, 104 (13): 5455-5460.

Crawford, H. M., Morin, D. E., Wall, E. H., et al., 2015. Evidence for a Role of Prolactin in Mediating Effects of Photoperiod during the Dry Period [J]. Animals (Basel), 5 (3): 803-820.

Daniels, K. M., Capuco, A. V., McGilliard, M. L., et al., 2009. Effects of milk replacer formulation on measures of mammary growth and composition in Holstein heifers [J]. J. Dairy Sci., 92 (12): 5937-5950.

Ellis, S., McFadden, T. B., Akers R M., 1998. Prepuberal ovine mammary development is unaffected by ovariectomy [J]. Domest. Anim. Endocrinol., 15 (4): 217-225.

Farmer, C., Petitclerc, D., 2003. Specific window of prolactin inhibition in late gestation decreases mammary parenchymal tissue development in gilts [J]. J. Anim Sci., 81 (7): 1823-1829.

Geiger, A. J., Parsons, C. L., Akers, R. M., 2016. Feeding a higher plane of nutrition and providing exogenous estrogen increases mammary gland development in Holstein heifer calves [J]. J. Dairy Sci., 99 (9): 7642-7653.

Hess, M. K., Hess, A. S., Garrick, D. J., 2016. The Effect of Calf Gender on Milk Production in Seasonal Calving Cows and Its Impact on Genetic Evaluations [J]. PLoS One, 11 (3): e0151236.

Hilakivi-Clarke, L., de Assis, S., 2006. Fetal origins of breast cancer [J]. Trends Endocrinol. Metab., 17 (9): 340-348.

Hinde, K., Carpenter A. J., Clay J. S., et al., 2014. Holsteins favor heifers, not bulls: biased milk production programmed during pregnancy as a function of fetal sex [J]. PLoS One, 9 (2): e86169.

Horigan, K. C., Trott, J. F., Barridollar, A. S., et al., 2009. Hormone interactions confer specific proliferative and histomorphogenic responses in the porcine mammary gland [J]. Domest. Anim. Endocrinol., 37 (2): 124-138.

Hovey, R. C., McFadden, T. B., Akers, R. M., 1999. Regulation of mammary gland growth and morphogenesis by the mammary fat pad: a species comparison [J]. J. Mammary Gland Biol. Neoplasia, 4 (1): 53-68.

Hovey, R. C., Trott, J. F., Vonderhaar, B. K., 2002. Establishing a framework for the functional mammary gland: from endocrinology to morphology [J]. J. Mammary Gland Biol. Neoplasia, 7 (1): 17-38.

Howard, B. A., Lu, P., 2014. Stromal regulation of embryonic and postnatal mammary epithelial development and differentiation [J]. Semin. Cell Dev. Biol., 25-26: 43-51.

Howard, B. A., Veltmaat, J. M., 2013. Embryonic mammary gland development; a domain of fundamental research with high relevance for breast cancer research. Preface [J]. J. Mammary Gland Biol. Neoplasia, 18 (2): 89-91.

Klotz, J. L., 2015. Activities and Effects of Ergot Alkaloids on Livestock Physiology and Production [J]. Toxins (Basel)., 7 (8): 2801-2821.

Knight, C. H., Dewhurst, R. J., 1994. Once daily milking of dairy cows: relationship between yield loss and cisternal milk storage [J]. J. Dairy Res, 61 (4): 441-449.

Lohakare, J. D., Sudekum, K. H., Pattanaik, A. K., 2012. Nutrition-induced Changes of Growth from Birth to First Calving and Its Impact on Mammary Development and First-lactation Milk Yield in Dairy Heifers: A Review [J]. Asian-Australas J. Anim Sci., 25 (9): 1338-1350.

McCrorie, T. A., Keaveney, E. M., Wallace, J. M. W., et al., 2011. Human health effects of conjugated linoleic acid from milk and supplements [J]. Nutr. Res. Rev., 24 (2): 206-227.

Meyer, M. J., Capuco, A. V., Ross, D. A., et al., 2006. Developmental and nutritional regulation of the prepubertal bovine mammary gland: II. Epithelial cell proliferation, parenchymal accretion rate, and allometric growth [J]. J. Dairy Sci., 89 (11): 4298-4304.

Meyer, M. J., Rhoads, R. P., Capuco, A. V., et al., 2007. Ontogenic and nutritional regulation of steroid receptor and IGF - I transcript abundance in the prepubertal heifer mammary gland [J]. J. Endocrinol., 195 (1): 59-66.

Monteiro, A. P. A., Tao, S., Thompson, I. M. T., et al., 2016. In utero heat stress decreases calf survival and performance through the first lactation [J]. J. Dairy Sci., 99 (10): 8443-8450.

Neville, M. C., Medina D, Monks J et al., 1998. The mammary fat pad [J]. J. Mammary Gland Biol. Neoplasia, 3 (2): 109-116.

Paine, I. S., Lewis, M. T., 2017. The Terminal End Bud: the Little Engine that Could [J]. J. Mammary Gland Biol. Neoplasia, 22 (2): 93-108.

Paten, A. M., Kenyon, P. R., Lopez-Villalobos, N., et al., 2013. Lactation Biology Symposium: maternal nutrition during early and mid-to-late pregnancy: Comparative effects on milk production of twin-born ewe progeny during their first lactation [J]. J. Anim Sci., 91 (2): 676-684.

Rios, A. C., Fu, N. Y., Jamieson, P. R. et al., 2016. Essential role for a novel population of binucleated mammary epithelial cells in lactation [J]. Nat. Commun., 7: 11400.

Rowson, A. R., Daniels, K. M., Ellis, S. E., et al., 2012. Growth and development of the mammary glands of livestock: a veritable barnyard of opportunities [J]. Semin. Cell Dev. Biol., 23 (5): p. 557-566.

Russo, J., Russo, I. H., 1996. Experimentally induced mammary tumors in rats [J]. Breast Cancer Res. Treat, 39 (1): 7-20.

Schedin, P., Hovey, R. C., 2010. Editorial: The mammary stroma in normal development and function [J]. J. Mammary Gland Biol. Neoplasia, 15 (3): 275-277.

Soberon, F., Van Amburgh, M. E., 2013. Lactation Biology Symposium: The effect of nutrient intake from milk or milk replacer of preweaned dairy calves on lactation milk yield as adults: a meta-analysis of current data [J]. J. Anim Sci., 91 (2): 706-712.

Tucker, H. A., 1987. Quantitative estimates of mammary growth during various physiological states: a review [J]. J. Dairy Sci., 70 (9): 1958-1966.

Vanklompenberg, M. K., Manjarin, R., Trott, J. F., et al., 2013. Late gestational hyperprolactinemia accelerates mammary epithelial cell differentiation that leads to increased milk yield [J]. J. Anim Sci., 91 (3): 1102-1111.

Wall, E. H., McFadden, T. B., 2012. Triennial Lactation Symposium: A local affair: How the mammary gland adapts to changes in milking frequency [J]. J. Anim Sci., 90 (5): 1695-1707.

Wathes, D. C., Pollott, G. E., Johnson, K. F., et al., 2014. Heifer fertility and carry over consequences for life time production in dairy and beef cattle [J]. Animal, 8 (Suppl 1): 91-104.

Wolfenson, D., Flamenbaum, I., Berman, A., 1988. Dry period heat stress relief effects on prepartum progesterone, calf birth weight, and milk production [J]. J. Dairy Sci., 71 (3): 809-818.

Physiologic and Molecular Implications of Amino Acid Balancing During the Periparturient Period in Dairy Cows

J. J. Loor, Z. Zhou and M. Vailati-Riboni

Department of Animal Sciences University of Illinois at Urbana-Champaign

Immunonutrition: a Future Frontier

Perhaps because of a lack of formal training outside classical concepts in nutrition, ruminant nutritionists primarily focus on meeting production requirements of dairy cows, without truly dissecting what components of maintenance requirement might affect performance. There is substantial evidence indicating that the immune system is intimately involved with other mechanisms that allow cows to adjust quickly to the onset of lactation without suffering chronic disorders. In fact, cows that lag behind the rest of the herd in terms of production outcomes (including fertility) often display a greater inflammatory status and compromised liver function (Bionaz et al., 2007; Bertoni et al., 2008; Trevisi et al., 2012). The lower dry matter intake (DMI) of the health-impaired animals (Trevisi et al., 2012) is not surprising because inflammatory molecules often have anorexogenic effects (Plata-Salaman, 1998, 2001; Wong and Pinkney, 2004). Because it is now generally-agreed that the postpartum negative energy balance is mainly caused by the reduction in DMI, rather than the increase demand of the mammary gland (Grummer et al., 2010), health and immunity has to be a focus of nutritionists aiming to improve the adaptation of the cow to lactation (LeBlanc, 2010).

Several studies reported that both the innate and adaptive immune systems in peripartal cows are often compromised; for example, cytokine production is impaired (Sordillo and Babiuk, 1991; Ishikawa et al., 1994), oxidative burst activity is reduced (Dosogne et al., 1999), and consequently phagocytic activity by leukocytes is often (Ingvartsen et al., 2003), but not always (Graugnard et al., 2012) reduced. Initial studies implicated neutrophils as the main target of parturition-induced changes in function driven in part by alterations in gene expression (Madsen et al., 2000). Currently, most of the research efforts are focused on this specific cell population.

Pharmaceutical treatment through the use of different drugs (e.g. non-steroidal anti-inflammatory compounds) have been used in different scenarios to neutralize the inflammatory state displayed, at different degrees, by almost every dairy cow, albeit with different results (Bertoni et al., 2004; Farney et al., 2013; Meier et al., 2014). Nutritional strategies, even if sometimes ineffective, such as level of dietary energy in the dry-period diet (Graugnard et al., 2012; Zhou et al., 2015), dietary amino acid balance (Yuan et al., 2014), or natural additives (Garcia et al., 2015) have resulted in some positive effects on the immune function. From a consumer standpoint, much more attentive now than before to the use of drug in food animals, nutritional approaches to boost the immune system and reduce incidence of disorders are seen in a positive light.

As the metabolic and immune networks are deeply connected (Mathis and Shoelson, 2011), a systems approach clearly is beneficial when attempting to understand the underlying mechanisms elicited by different immunonutrition strategies. Because immune cell function seems to be dysfunctional during early lactation, the feed additive industry has placed some emphasis on developing "immunostimulants" as dietary supplements. For example, recent results indicated a positive role of a commercially-available immunostimulant (OmiGen-AF, Phibro Animal Health Corporation, USA) in enhancing leukocyte function which would provide added antibacterial capacity during the peripartum (Nace et al., 2014). Molecular analyses revealed how these improvements might be due to changes in expression that might alter neutrophil apoptosis, signaling, sensitivity, and response (Wang et al., 2007; Wang et al., 2009).

Specific dietary treatments that have been successful in improving leucocyte activity include methionine and other rumen-protected amino acids (AA) (Osorio et al., 2013a; Yuan et al., 2014), dietary nitrogen level (Raboisson et al., 2014), alpha1-acid glycoprotein (Rinaldi et al., 2008), and orange oil (Garcia et al., 2015). However, the molecular technologies (i.e. holistic approach) are mostly helping us understand some of the "mistakes" in what is considered standard in today's dairy management systems. The most resounding case regards the management of dry cows, and its effect, among others, on cow health status and immunity. Traditional management provides "far-off" dry cows with a high-fiber/low-energy density diet, while in the last month of gestation ("close-up" dry period) the diet increases in energy density with a lower fiber content. However, studies from different research groups have demonstrated that prepartum overfeeding of energy often results in prepartum hyperglycemia and hyperinsulinemia and marked postpartum adipose tissue mobilization (i.e., greater blood NEFA concentration) (Holtenius et al., 2003; Janovick et al., 2011; Ji et al., 2012; Ji et al., 2014; Khan et al., 2014). In addition, higher-energy close-up diets also have been associated with negative effects on postpartum health indices, underscoring possible detrimental effects of this management approach (Dann et al., 2006; Soliman et al., 2007; Graugnard et al., 2013; Shahzad et al., 2014).

Transcriptome profiling of neutrophils (i.e. measuring changes in expression of several genes simultaneously) revealed that allowing cows free access to higher-energy diets during late-pregnancy resulted in the alteration of genes encompassing pathways associated with the immune response (Moyes et al., 2014; Zhou et al., 2015). Furthermore, phagocytosis activity of these cells was impaired, and early prepartal activation of inflammatory genes suggested a chronic state of compromised health (Moyes et al., 2014; Zhou et al., 2015). High-trhoughput transcriptomic studies (via microarrays) further highlighted how this practice not only impairs immune function, but affects the whole system of the cow as indicated by alterations in endoplasmatic reticulum stress in hepatocytes, probably as a consequences of a higher activation of inflammatory-related functions (Shahzad et al., 2014). As indicated by pro-inflammatory gene expression profiles, these 'omics' data also revealed a predisposition of cows to fatty liver while compromising overall liver health during the periparturient period (Loor et al., 2006).

Key Amino Acids and Choline for Transition Cows

Due to extensive microbial degradation in the rumen, dietary availability of key methyl donors [(e.g., Methionine (MET) and choline (CHOL)] to mammary and liver is limited (Sharma and Erdman, 1989; Girard and Matte, 2005). Consequently, the mobilization of body protein in dairy cows close to calving compensates in part for this shortfall (Komaragiri and Erdman, 1997). Supplementing rumen-protected methyl donors may help fulfill the daily methyl group requirement, and possibly improve the overall production and health of dairy cows during the transition period (Zom et al., 2011; Osorio et al., 2013; Osorio et al., 2014).

The availability of MET and CHOL is important for various biological functions. For instance, MET together with Lys are the most-limiting AA for milk synthesis (NRC, 2001). Being the only essential sulfur-containing AA, MET acts as the precursor for other sulfur-containing AA such as cysteine (Cys), homocysteine and taurine (Brosnan and Brosnan, 2006). It has been estimated in lactating goats that as much as 28% of absorbed MET could be used for CHOL synthesis (Emmanuel and Kennelly, 1984). Hence, it is thought that rumen-protected CHOL supplementation could spare MET to help cows achieve better overall performance (Hartwell et al., 2000; Pinotti, 2012). Current recommendations for duodenal supply of Lys and MET to maximize milk protein content and yield in established lactation are 7.2% and 2.4% of MP, respectively (NRC, 2001). In terms of production performance, a Lys : Met ratio close to 2.8 : 1 of MP during the periparturient period by supplementing rumen-protected MET was beneficial (Osorio et al., 2013).

As a lipotropic agent, MET is directly involved in very low density lipoprotein (VLDL) synthesis via the generation of S-adenosylmethionine (SAM), the most important methyl donor (Martinov et al., 2010). In turn, SAM can be used to methylate phosphatidylethanolamine (PE) to generate PC, which is essential for VLDL synthesis (Auboiron et al., 1995). In the context of VLDL synthesis and liver lipid metabolism, CHOL-containing nutrients (mainly in the form of PC) are indispensable for the synthesis and release of chylomicrons and VLDL (Pinotti et al., 2002). Thus, supplementation of rumen-protected MET and/or CHOL (Zom et al., 2011) may increase hepatic triacylglycerol (TAG) export and consequently decrease lipidosis.

The immune system benefits greatly from proper nutrition, which in turn prepares the cow for periods of stress, reducing adverse effects and enhancing recovery. These concepts become of central importance when applied to the transition cow, as a successful transition to lactation sets the stage for a profitable lactation, with optimal production, reproduction, and health, avoiding premature culling. Metabolic disorders are common during this time and can easily erase the entire profit potential for dairy cow farms (Drackley, 1999). The immune dysfunction during transition (Kehrli et al., 1989; Waller, 2000) and the state of oxidative stress (Abuelo et al., 2015) can lead to a cow that might be hyposensitive and hyporesponsive to antigens, hence, more susceptible to infectious diseases such as mastitis (Mallard et al., 1998).

Rumen-protected Methyl Donors and Production Performance

To date, the reported effects of rumen-protected MET and/or CHOL supplementation on dairy cow production performance have been inconsistent. Although previous studies from our group and others have observed beneficial effects from MET (Chen et al., 2011; Osorio et al., 2013) or CHOL (Pinotti et al., 2003; Zom et al., 2011) supplementation, other studies did not detect significant improvements on peripartal production performance with MET (Socha et al., 2005; Ordway et al., 2009; Preynat et al., 2009) or CHOL (Guretzky et al., 2006; Leiva et al., 2015) supplementation. Similarly, studies evaluating whether CHOL alone or in combination with MET provide equal or different benefits in terms of production performance also yielded different results. For instance, MET and CHOL supplementation both led to greater DMI in previous studies (Ardalan et al., 2010; Sun et al., 2016). In contrast, only a MET effect was observed in the most recent transition cow study from our group (Zhou et al., 2016c). Production performance results from peripartal MET and CHOL supplementation studies are summarized in Table 1.

Table 1 Summary of production performance results from transition cow studies supplementing rumen-protected MET or CHOL. Responses to feeding rumen-protected methionine or choline were positive (+), modest (/), or did not change (-)

Study	DMI	Milk	Protein	Fat	Dosage	Product	Duration	Cows
Methionine studies								

(Continued)

Study	DMI	Milk	Protein	Fat	Dosage	Product	Duration	Cows
Overton et al., 1996	–	+	–	–	17 g	RPM (Degussa)	-7 to 126	24
Piepenbrink et al., 2003	–	+	–	–	2.34% or 2.7% MP pre 2.36 or 2.63% MP post	Alimet	-21 to 84	48
Socha et al., 2005	–	–	–	–	10.5 g	SM	-14 to 105	48
Ordway et al., 2009	–	–	–	–	Met : Lys =3.0 : 1	MS and SM	-21 to 140	60
Preynat et al., 2009	–	–	–	–	15.3 g	Mepron-85	-21 to 112	60
Ardalan et al., 2010	+	–	/	/	14.4 g	SM	-28 to 70	40
Osorio et al., 2013	+	+	+	+	0.19% MS or 0.07% SM DM	MS and SM	-21 to 30	56
Zhou et al., 2016	+	+	+	+	0.08% DM	SM	-21 to 30	81
Sun et al., 2016	+	+	+	+	15 g	Mepron-85	-21 to 21	48
Choline studies								
Hartwell et al., 2000	–	–	–	–	0, 6 g, 12 g	Capshure	-21 to 120	48
Pinotti et al., 2003	–	+	–	+	20 g	Overcholine	-14 to 30	26
Piepenbrink et al., 2003	–	+	–	+	11 g, 15 g, 19 g	ReaShure	-21 to 63	48
Guretzky et al., 2006	–	–	–	–	15 g	ReaShure	-21 to 21	42
Lima et al., 2007	+	+	+	+	15 g	ReaShure	-25 to 80	369
Elek et al., 2008	–	+	+	+	25/50 g pre/post	Norcol-25	-21 to 60	32
Ardalan et al., 2010	+	+	/	/	14 g	Col 24	-28 to 70	40
Zom et al., 2011	+	–	+	–	15 g	ReaShure	-21 to 42	38

(Continued)

Study	DMI	Milk	Protein	Fat	Dosage	Product	Duration	Cows
Leiva et al., 2015	-	-	+	-	9.4/18.8 g pre/post	CholiPearl	-21 to 45	23
Zhou et al., 2016	-	-	-	-	15 g	ReaShure	-21 to 30	81
Sun et al., 2016	+	+	+	+	15 g	ReaShure	-21 to 21	48

Production performance benefits in response to MET supplementation during the periparturient period are likely associated with an enriched AA and sulfur - containing antioxidant pool. Inadequate MET availability could limit the utilization of other circulating AA according to von Liebig's hypothesis which is commonly described with the analogy of the water barrel with broken staves (Mitchell and Block, 1946). The fact that circulating MET concentration decreased markedly through parturition and was not restored to prepartum levels until 28 d postpartum (Zhou et al., 2016b) suggest increasing MET availability during this period could potentially benefit production performance. Additionally, enhancing MET availability (Graulet et al., 2005) is likely to increase its entry into the 1-carbon metabolism cycle in liver which consequently increases the production of downstream compounds such as Cys. Glutathione (GSH) is another downstream compound arising in the MET cycle that can supply AA such as Cys to the mammary gland for milk synthesis (Pocius et al., 1981).

Apart from providing MET, GSH (as a potent intracellular antioxidant) may contribute to better overall performance by alleviating oxidative stress and subsequent inflammation. Previous work has demonstrated a positive effect of MET supplementation on intrahepatic GSH concentration during the peripartal period (Osorio et al., 2014; Zhou et al., 2016a). Such effect may be directly associated with MET supplementation considering that it can be incorporated upstream in the *de novo* synthesis pathway for GSH (Halsted, 2013). Both *in vitro* (Hartman et al., 2002) and *in vivo* (Tabachnick and Tarver, 1955) studies using radioactive-labelled MET demonstrated hepatic incorporation of [^{35}S] into GSH. Therefore, the higher hepatic GSH concentration observed in MET - supplemented cows helps alleviate oxidative stress and contributes to greater DMI through an overall alleviation of the inflammatory status (Zhou et al., 2016a).

Although CHOL does not contain sulfur, MET can be generated in tissues like the liver from CHOL when homocysteine accepts a methyl group from CHOL through betaine (Wong and Thompson, 1972; Li and Vance, 2008). Hence, if comparable MET can be generated in response to CHOL supplementation, similar production performance benefits would be expected. Despite the fact that milk yield, DMI, and milk composition benefits were not observed in CHOL-supplemented cows in a recent transition cow study from our group (Zhou et al., 2016c), lactation performance benefits were detected in previous studies, indicating that CHOL may exert its lactation benefits through various means. For instance, the increase in hepatic mRNA expression of carnitine transporter suggested an increase in fatty acid uptake capacity and intracellular transport in CHOL cows, which was associated with reduced liver TAG accumulation (Goselink et al., 2013). Additionally, the CHOL supplementation from precalving through early lactation led to increased glycogen in liver tissue, implying a benefit to liver metabolism (Piepenbrink and Overton, 2003). Furthermore, CHOL can be used to generate PC, which is essential for VLDL synthesis and help reduce liver lipidosis by promoting TAG export (Pinotti et al., 2002).

Rumen-protected Methyl Donors and Metabolism

During the transition period cows will normally experience an increase in adipose tissue lipolysis due to changes in hormones such as insulin (decrease) and growth hormone (increase), and consequently blood non-esterified fatty acid (NEFA) concentrations increase. Once NEFA reach the liver these can be oxidized to provide energy, partially oxidized to produce ketone bodies, or esterified to triglyceride (TAG). A major organelle within hepatocytes where NEFA oxidation takes place is the mitochondria, and carnitine is essential for transport of NEFA from cytosol into mitochondria for subsequent β-oxidation (Drackley, 1999). Methionine is essential for carnitine synthesis (Carlson et al., 2007), thus, the greater hepatic concentration of carnitine (82.1 vs. 37.5 nmol/g of tissue) that was detected in MET-supplemented cows indicates a greater bioavailability of MET to methylate Lys (Osorio et al., 2014).

Previous studies reported no significant effect of CHOL on blood glucose or BHBA concentrations (Guretzky et al., 2006; Zahra et al., 2006; Zom et al., 2011). In a recent study from our group (Zhou et al., 2016c), the tendency for lower BHBA in response to CHOL supplementation agreed with the greater glucose concentration. Although speculative, the pattern of BHBA and glucose detected in CHOL-supplemented cows was associated with numerically lower negative energy balance as a result of lower milk production. The exact mechanisms for the lower milk yield in these cows that maintained greater blood glucose is not known.

Immunonutritional Role of Methyl Donors

In addition to being considered one of the most-limiting AA for milk production, MET and several of its metabolites display an immunonutritional role, i.e. they help support and boost certain activities of the immune system in humans (Grimble, 2006; Li et al., 2007). Since these properties have been tested on immune-suppressed human subjects with positive outcomes (Van Brummelen and du Toit, 2007), we hypothesize that enhancing

MET supply would have a positive effect on immune function in the transition period, where cows seems to be in an immuno-compromised state. A study with mid-lactation cows reported that supplementation with 30 g/d of rumen-protected MET compared with 0 or 15 g/d led to greater T lymphocyte proliferation *in vitro* in response to various mitogens (Soder and Holden, 1999). Since human lymphocytes seem to have an absolute requirement for MET to proliferate (Hall et al., 1986), these results were not unexpected.

In one of our previous studies (Osorio et al., 2013) where transition cows were fed either Smartamine (0.07%DM) or Metasmart (0.19%DM) from-21 to +30 days relative to parturition, we observed increased phagocytosis (pathogen killing ability) in neutrophils, the cells that make up the first line of defense in the animal immunity. In a follow-up study feeding Smartamine (0.08%DM) between 21 and 30 days relative to parturition (Zhou et al., 2016a) we detected both greater *in vitro* blood neutrophil phagocytosis and oxidative burst (another pathogen - killing mechanism) from day 1 post - calving through day 28 postpartum. Furthermore, supplementation with rumen - protected MET optimized the response to lipopolysaccharide (LPS) (a component of bacteria cell walls), by controlling the inflammatory ability of the immune cells (Vailati-Riboni et al., 2017). This is very relevant during the transition period, as cows might mount an excessive inflammatory response to pathogens, creating more damage than benefit (Jahan et al., 2015). One possible mechanism could be related to the ability of MET to influence the oxidative status of periparturient cows as it is a precursor for glutathione and taurine (Atmaca, 2004). Through its chloramine metabolites, taurine has a well - known immuno - modulatory capacity (Schuller-Levis and Park, 2004).

Despite the interconnection between MET and CHOL (via the one - carbon metabolism), there is a paucity of data on CHOL and the bovine immune response. Immune cells lack the ability to convert CHOL into MET through the betaine pathway, which in bovine is present only in liver and kidney (Lambert et al., 2002). In a recent study (Zhou et al., 2016a), compared with feeding MET feeding 60 g/d of Reashure (rumen protected CHOL) from 21 to 30 days in milk had no effect on immune cell killing capacity of neutrophils and monocytes (another cell type of the animal immune response). However, there are data generated using other animal models. For example, supplementation of choline in the diet improved immune indices in both fish (Wu et al., 2013) and suckling rats (Lewis et al., 2016). Authors did not speculate on the mode of action, but most probably CHOL efficacy is mediated by betaine (a choline derivate). Data from broilers revealed that dietary betaine supplementation improved intestinal health, and induced a boost in the intestinal immune response to a coccidiosis challenge (Klasing et al., 2002).

Methionine and Liver Function: Balance Between Inflammation and Oxidative Stress

The periparturient inflammatory response is characterized by an increase in the hepatic production of positive acute-phase proteins (posApp), such as haptoglobin and serum amyloid A (SAA), and a concomitant decrease in the production of negative App (negApp) such as albumin (Bertoni et al., 2008). At the level of liver, the well-established triggers of these responses are the pro-inflammatory cytokines IL-6, IL-1β, and TNF-α (Kindt et al., 2007). In contrast, oxidative stress is driven by the imbalance between the production of reactive oxygen metabolites (ROM) and the neutralizing capacity of antioxidant mechanisms in tissues and in blood. Pro-inflammatory cytokines have also been identified as a cause of oxidative stress, hence, linking the two conditions (Sordillo and Mavangira, 2014).

Both inflammation and oxidative stress reduce liver function in periparturient dairy cows (Bionaz et al., 2007; Trevisi et al., 2012). Using changes in plasma concentrations of albumin, cholesterol, and bilirubin, Bertoni and Trevisi (2013) developed the liver functionality index (LFI), which characterizes the extent of the inflammatory response and helps predict consequences on health and well-being of the cow. For instance, a low LFI value is indicative of a pronounced inflammatory response, suggestive of a more difficult transition from gestation to lactation, whereas a high LFI is suggestive of a smooth transition. In our experiments, supplementation of MET consistently increased blood albumin (Osorio et al., 2014; Zhou et al., 2016a) well above the concentration range in cows with adequate postpartum liver function (Bionaz et al., 2007). Furthermore, supplementation with rumen-protected MET decreased the concentrations of inflammation-related biomarkers such as ceruloplasmin, haptoglobin, IL-1β, and SAA (Osorio et al., 2014; Zhou et al., 2016a). MET supplementation increased liver glutathione and antioxidant capacity (Osorio et al., 2014; Zhou et al., 2016a). Using the cow data from the Zhou et al. experiment, compared with 35% of the cows without MET supplementation ending-up in the Low LFI group, only 10% of the MET-supplemented cows ended-up in the Low LFI group, hence, supporting the existence of a positve effect of MET supplementation on liver function.

The Link Between Intake and Health

The transient inflammatory-like status around parturition appears to be a "normal" aspect of the adaptations to lactation (Bradford et al., 2015), with its positive or negative impact depending on its degree. Cows that approach parturition with a greater (but still subclinical) level of circulating cytokines have greater inflammation and oxidative stress, and lower liver function often through 30 days in milk together with lower milk yield and lower postpartum DMI (Bertoni et al., 2008; Trevisi et al., 2015). In addition to their fundamental function in immunity, cytokines (ILs), interferons (IFNs) and TNF-α also elicit pathophysiological effects. This leads to what is commonly known as "sickness behavior", whose primarymanifestation is satiety. Similar to how cows react during an inflammatory state around parturition, the reduction in DMI around calving is an example of this behavior. In mice, these cytokines have been shown to reduce meal size and duration, as well as decrease meal frequency and prolong inter-meal intervals (Plata-Salaman, 1995). Furthermore, cytokines directly affect the hypothalamus; IL-1β and IFN act directly and specifically on the glucose-sensitive neurons in the brain "satiety" and "hunger" sites (Plata-Salaman, 1995). Thus, the increased DMI observed when feeding rumen-protected MET (Smartamine or Metasmart) can be partly explained by a reduction in inflammation, as it directly (at the hepatic level and by dampening the immune cell overresponse) and indirectly (reducing oxidative stress) decreases circulating pro-inflammatory cytokines. Overall, results available to date indicate that certain feed additives or nutrients have a positive immunomodulatory effect in dairy cows particularly during periods of stress such as the transition into lactation. It is our view that the feed industry should place more emphasis on understanding better mechanisms of actions whereby feed additives and supplements elicit biological effects on the animal.

References

Abuelo, A., Hernandez, J., Benedito, J. L., et al., 2015. The importance of the oxidative status of dairy cattle in the periparturient period: revisiting antioxidant supplementation [J]. J. Anim. Physiol. Anim. Nutr. (Berl), 99: 1003-1016.

Ardalan, M., Rezayazdi, K., Dehghan-Banadaky, M., 2010. Effect of rumen-protected choline and methionine on physiological and metabolic disorders and reproductive indices of dairy cows [J]. J. Anim. Physiol. Anim. Nutr. (Berl), 94: e259-265.

Atmaca, G., 2004. Antioxidant effects of sulfur-containing amino acids [J]. Yonsei. Med. J., 45: 776-788.

Auboiron, S., Durand, D., Robert, J. C., et al., 1995. Effects of dietary fat and L-methionine on the hepatic metabolism of very low density lipoproteins in the preruminant calf, Bos spp [J]. Reprod. Nutr. Dev., 35: 167-178.

Bertoni, G., Trevisi, E., 2013. Use of the liver activity index and other metabolic variables in the assessment of metabolic health in dairy herds [J]. Vet. Clin. North Am. Food Anim. Pract., 29: 413-431.

Bertoni, G., Trevisi, E., Han, X., et al., 2008. Effects of inflammatory conditions on liver activity in puerperium period and consequences for performance in dairy cows [J]. J. Dairy Sci., 91: 3300-3310.

Bertoni, G., Trevisi, E., Piccioli-Cappelli, F., et al., 2004. Effects of acetyl-salicylate used in post-calving of dairy cows [J]. Vet. Res. Commun., 28 (Suppl 1): 217-219.

Bionaz, M., Trevisi, E., Calamari, L., et al., 2007. Plasma paraoxonase, health, inflammatory conditions, and liver function in transition dairy cows [J]. J. Dairy Sci., 90: 1740-1750.

Bradford, B. J., Yuan, K., Farney, J. K., et al., 2015. Invited review: Inflammation during the transition to lactation: New adventures with an old flame [J]. J. Dairy Sci., 98: 6631-6650.

Brosnan, J. T., Brosnan, M. E., 2006. The sulfur-containing amino acids: an overview [J]. J. Nutr., 136: 1636S-1640S.

Carlson, D. B., Woodworth, J. C., Drackley, J. K., 2007. Effect of L-carnitine infusion and feed restriction on carnitine status in lactating Holstein cows [J]. J. Dairy Sci., 90: 2367-2376.

Chen, Z. H., Broderick, G. A., Luchini, N. D., et al., 2011. Effect of feeding different sources of rumen-protected methionine on milk production and N-utilization in lactating dairy cows [J]. J. Dairy Sci., 94: 1978-1988.

Dann, H. M., Litherland, N. B., Underwood, J. P., et al., 2006. Diets during far-off and close-up dry periods affect periparturient metabolism and lactation in multiparous cows [J]. J. Dairy Sci., 89 (9): 3563-3577.

Dosogne, H., Burvenich, C., Freeman, A. E., et al., 1999. Pregnancy-associated glycoprotein and decreased polymorphonuclear leukocyte function in early post-partum dairy cows [J]. Vet. Immunol. Immunopathol., 67 (1): 47-54.

Drackley, J. K., 1999. ADSA Foundation Scholar Award. Biology of dairy cows during the transition period: the final frontier? [J]. J. Dairy Sci., 82: 2259-2273.

Emmanuel, B., Kennelly, J. J., 1984. Kinetics of methionine and choline and their incorporation into plasma lipids and milk components in lactating goats [J]. J. Dairy Sci., 67: 1912-1918.

Farney, J. K., Mamedova, L. K., Coetzee, J. F., et al., 2013. Anti-inflammatory salicylate treatment alters the metabolic adaptations to lactation in dairy cattle [J]. Am. J. Physiol. Regul. Integr. Comp. Physiol., 305 (2): R110-117.

Garcia, M., Elsasser, T. H., Biswas, D., et al., 2015. The effect of citrus-derived oil on bovine blood neutrophil function and gene expression *in vitro* [J]. J. Dairy Sci., 98 (2): 918-926.

Girard, C. L., Matte, J. J., 2005. Folic acid and vitamin B_{12} requirements of dairy cows: A concept to be revised [J]. Livest. Prod. Sci., 98: 123-133.

Goselink, R. M. A. Baal, J., van Widjaja, H. C. A., et al., 2013. Effect of rumen-protected choline supplementation on liver and adipose gene expression during the transition period in dairy cattle [J]. J. Dairy Sci., 96: 1102-1116.

Graugnard, D. E., Bionaz, M., Trevisi, E., et al., 2012. Blood immunometabolic indices and polymorphonuclear neutrophil function in peripartum dairy cows are altered by level of dietary energy prepartum [J]. J. Dairy Sci., 95 (4): 1749-1758.

Graugnard, D. E., Moyes, K. M., Trevisi, E., et al., 2013. Liver lipid content and inflammometabolic indices in peripartal dairy cows are altered in response to prepartal energy intake and postpartal intramammary inflammatory challenge [J]. J. Dairy Sci., 96 (2): 918-935.

Graulet, B., Richard, C., Robert, J. C., 2005. Methionine availability in plasma of dairy cows supplemented with methionine hydroxy analog isopropyl ester [J]. J. Dairy Sci., 88: 3640-3649.

Grimble, R. F., 2006. The effects of sulfur amino acid intake on immune function in humans [J]. The J. Nutr., 136: 1660S-1665S.

Grummer, R. R., Wiltbank, M. C., Fricke, P. A., et al., 2010. Management of dry and transition cows to improve energy balance and reproduction [J]. J. Reprod. Dev., 56 (Suppl): S22-28.

Guretzky, N. A., Carlson, D. B., Garrett, J. E., et al., 2006. Lipid metabolite profiles and milk production for Holstein and Jersey cows fed rumen-protected choline during the periparturient period [J]. J. Dairy Sci., 89: 188-200.

Hall, C. A., Begley, J. A., Chu, R. C., 1986. Methionine dependency of cultured human lymphocytes [J]. Proc. Soc. Exp. Biol. Med., 182: 215-220.

Halsted, C. H., 2013. B-Vitamin dependent methionine metabolism and alcoholic liver disease [J]. Clinical chemistry and laboratory medicine: CCLM / FESCC, 51: 457-465.

Hartman, N. R., Cysyk R. L., Bruneau, W. C., et al., 2002. Production of intracellular 35S - glutathione by rat and human hepatocytes for the quantification of xenobiotic reactive intermediates [J]. ChemBiol. Interact., 142: 43-55.

Hartwell, J. R., Cecava, M. J., Donkin, S. S., 2000. Impact of dietary rumen undegradable protein and rumen-protected choline on intake, peripartum liver triacylglyceride, plasma metabolites and milk production in transition dairy cows [J]. J. Dairy Sci., 83: 2907-2917.

Holtenius, K., Agenas, S., Delavaud, C., et al., 2003. Effects of feeding intensity during the dry period. 2. Metabolic and hormonal responses [J]. J. Dairy Sci., 86 (3): 883-891.

Ingvartsen, K. L., Dewhurst, R. J., Friggens, N. C. et al., 2003. On the relationship between lactational performance and health: is it yield or metabolic imbalance that cause production diseases in dairy cattle? A position paper [J]. Livest. Prod. Sci., 83 (2-3): 277-308.

Ishikawa, H., Shirahata, T., Hasegawa, K., 1994. Interferon-gamma production of mitogen stimulated peripheral lymphocytes in perinatal cows [J]. J. Vet. Med. Sci., 56 (4): 735-738.

Jahan, N., Minuti, A., Trevisi, E., 2015. Assessment of immune response in periparturient dairy cows using ex vivo whole blood stimulation assay with lipopolysaccharides and carrageenan skin test [J]. Vet. Immunol. Immunopathol., 165: 119-126.

Janovick, N. A., Boisclair, Y. R., Drackley, J. K., 2011. Prepartum dietary energy intake affects metabolism and health during the periparturient period in primiparous and multiparous Holstein cows [J]. J. Dairy Sci., 94 (3): 1385-1400.

Ji, P., Drackley, J. K., Khan, M. J., et al., 2014. Overfeeding energy upregulates peroxisome proliferator-activated receptor (PPAR) gamma-controlled adipogenic and lipolytic gene networks but does not affect proinflammatory markers in visceral and subcutaneous adipose depots of Holstein cows [J]. J. Dairy Sci., 97 (6): 3431-3440.

Ji, P., Osorio, J. S., Drackley, J. K., et al., 2012. Overfeeding a moderate energy diet prepartum does not impair bovine subcutaneous adipose tissue insulin signal transduction and induces marked changes in peripartal gene network expression [J]. J. Dairy Sci., 95 (8): 4333-4351.

Kehrli, M. E., Jr., Nonnecke, B. J., Roth, J. A., 1989. Alterations in bovine neutrophil function during the periparturient period [J]. Am. J. Vet. Res., 50: 207-214.

Khan, M. J., 2014. Overfeeding Dairy Cattle During Late-Pregnancy Alters Hepatic PPARalpha-Regulated Pathways Including Hepatokines: Impact on Metabolism and Peripheral Insulin Sensitivity [J]. Gene Regul. Syst. Bio., 8: 97-111.

Kindt, T. J. et al., 2007. Kuby immunology [M]. 6th ed. W. H. Freeman, New York.

Klasing, K. C., Adler K. L., Remus, J. C., et al., 2002. Dietary betaine increases intraepithelial lymphocytes in the duodenum of coccidia-infected chicks and increases functional properties of phagocytes [J]. J. Nutr., 132: 2274-2282.

Komaragiri, M. V., Erdman, R. A., 1997. Factors affecting body tissue mobilization in early lactation dairy cows. 1. Effect of dietary protein on mobilization of body fat and protein [J]. J. Dairy Sci., 80: 929-937.

Lambert, B. D., Titgemeyer, E. C., Stokka, G. L., et al., 2002. Methionine supply to growing steers affects hepatic activities of methionine synthase and betaine-homocysteine methyltransferase, but not cystathionine synthase [J]. J. Nutr., 132: 2004-2009.

LeBlanc, S., 2010. Monitoring metabolic health of dairy cattle in the transition period [J]. J. Reprod. Dev., 56 (Suppl): S29-35.

Leiva, T., Cooke, R. F., Brandão, A. P., et al., 2015. Effects of rumen-protected choline supplementation on metabolic and performance responses of transition dairy cows [J]. J. Anim. Sci., 93: 1896-1904.

Lewis, E. D., Richard, C., Goruk, S., et al., 2016. The Form of Choline in the Maternal Diet Affects

Immune Development in Suckled Rat Offspring [J]. J. Nutr., 146: 823-830.

Li, P., Yin, Y. L., Li, D. F., et al., 2007. Amino acids and immune function [J]. Br. J. Nutr, 98: 237-252.

Li, Z. Y., Vance, D. E., 2008. Phosphatidylcholine and choline homeostasis [J]. J. Lipid Res., 49: 1187-1194.

Loor, J. J., Dann H. M., Guretzky, N. A. J., et al., 2006. Plane of nutrition prepartum alters hepatic gene expression and function in dairy cows as assessed by longitudinal transcript and metabolic profiling [J]. Physiol. Genomics., 27 (1): 29-41.

Madsen, S. A., Chang, L. C., Hickey, M. C., et al., 2004. Microarray analysis of gene expression in blood neutrophils of parturient cows [J]. Physiol. Genomics., 16 (2): 212-221.

Mallard, B. A., Dekkers, J. C., Ireland, M. J., et al., 1998. Alteration in immune responsiveness during the peripartum period and its ramification on dairy cow and calf health [J]. J. Dairy Sci., 81: 585-595.

Martinov, M. V., Vitvitsky, V. M., Banerjee, R., et al., 2010. The logic of the hepatic methionine metabolic cycle [J]. Biochim. Biophys. Acta., 1804: 89-96.

Mathis, D., Shoelson, S. E., 2011. Immunometabolism: an emerging frontier [J]. Nat. Rev. Immunol., 11 (2): 81.

Meier, S., Priest, N. V., Burke, C. R., et al., 2014. Treatment with a nonsteroidal antiinflammatory drug after calving did not improve milk production, health, or reproduction parameters in pasture-grazed dairy cows [J]. J. Dairy Sci., 97 (5): 2932-2943.

Mitchell, H. H., Block, R. J., 1946. Some relationships between the amino acid contents of proteins and their nutritive values for the rat [J]. J. Biol. Chem., 163: 599-620.

Moyes, K. M., Graugnard, D. E., Khan, M. J., et al., 2014. Postpartal immunometabolic gene network expression and function in blood neutrophils are altered in response to prepartal energy intake and postpartal intramammary inflammatory challenge [J]. J. Dairy Sci., 97 (4): 2165-2177.

Nace, E. L., Nickerson, S. C., Kautz, F. M., et al., 2014. Modulation of innate immune function and phenotype in bred dairy heifers during the periparturient period induced by feeding an immunostimulant for 60 days prior to delivery [J]. Vet. Immunol. Immunopathol., 161 (3-4): 240-250.

NRC, 2001. Nutrient Requirements of Dairy Cattle: Seventh Revised Edition, 2001 [M]. USA: National Academies Press.

Ordway, R. S., Boucher, S. E., Whitehouse, N. L., et al., 2009. Effects of providing two forms of supplemental methionine to periparturient Holstein dairy cows on feed intake and lactational performance [J]. J. Dairy Sci., 92: 5154-5166.

Osorio, J. S., Ji, P., Drackley, J. K., et al., 2013. Supplemental Smartamine M or MetaSmart during the transition period benefits postpartal cow performance and blood neutrophil function [J]. J. Dairy Sci., 96: 6248-6263.

Osorio, J. S., Trevisi, E., Ji, P., et al., 2014. Biomarkers of inflammation, metabolism, and oxidative stress in blood, liver, and milk reveal a better immunometabolic status in peripartal cows supplemented with Smartamine M or MetaSmart [J]. J. Dairy Sci., 97: 7437-7450.

Piepenbrink, M. S., Overton, T. R., 2003. Liver metabolism and production of cows fed increasing amounts of rumen-protected choline during the periparturient period [J]. J. Dairy Sci., 86: 1722-1733.

Pinotti, L., Baldi, A., Dell'Orto, V., 2002. Comparative mammalian choline metabolism with emphasis on the high-yielding dairy cow [J]. Nutr. Res. Rev., 15: 315-332.

Pinotti, L., Baldi, A., Politis, I., et al., 2003. Rumen - protected choline administration to transition cows: effects on milk production and vitamin E status [J]. J. Vet. Med. A Physiol. Pathol. Clin. Med., 50: 18-21.

Plata - Salaman, C. R., 1995. Cytokines and feeding suppression: an integrative view from neurologic to molecular levels [J]. Nutrition, 11: 674-677.

Plata-Salaman, C. R., 1998. Cytokines and Feeding [J]. News Physiol. Sci., 13: 298-304.

Plata-Salaman, C. R., 2001. Cytokines and feeding [J]. Int. J. Obes. Relat. Metab. Disord., 25 (Suppl 5): S48-52.

Pocius, P. A., Clark, J. H., Baumrucker, C. R., 1981. Glutathione in bovine blood: possible source of amino acids for milk protein synthesis [J]. J. Dairy Sci., 64: 1551-1554.

Preynat, A., Lapierre, H., Thivierge, M. C., et al., 2009. Influence of methionine supply on the response of lactational performance of dairy cows to supplementary folic acid and vitamin B12 [J]. J. Dairy Sci., 92: 1685-1695.

Raboisson, D., Caubet, C., Tasca, C., et al., 2014. Effect of acute and chronic excesses of dietary nitrogen on blood neutrophil functions in cattle [J]. J. Dairy Sci., 97: 7575-7585.

Rinaldi, M., Ceciliani, F., Lecchi, C., et al., 2008. Differential effects of alpha1-acid glycoprotein on bovine neutrophil respiratory burst activity and IL-8 production [J]. Vet. Immunol. Immunopathol., 126 (3-4): 199-210.

Schuller-Levis, G. B., Park, E., 2004. Taurine and its chloramine: modulators of immunity [J]. Neurochem. Res., 29: 117-126.

Shahzad, K., Bionaz, M., Trevisi, E., et al., 2014. Integrative analyses of hepatic differentially expressed genes and blood biomarkers during the peripartal period between dairy cows overfed or restricted-fed energy prepartum [J]. PLoS One, 9 (6): e99757.

Sharma, B. K., Erdman, R. A., 1989. Effects of dietary and abomasally infused choline on milk production responses of lactating dairy cows [J]. J. Nutr., 119: 248-254.

Socha, M. T., Putnam, D. E., Garthwaite, B. D., et al., 2005. Improving intestinal amino acid supply of pre-and postpartum dairy cows with rumen-protected methionine and lysine [J]. J. Dairy Sci., 88: 1113-1126.

Soder, K. J., Holden, L. A., 1999. Lymphocyte proliferation response of lactating dairy cows fed varying concentrations of rumen-protected methionine [J]. J. Dairy Sci., 82: 1935-1942.

Soliman, M., Kimura, K., Ahmed, M., et al., 2007. Inverse regulation of leptin mRNA expression by short-and long-chain fatty acids in cultured bovine adipocytes [J]. Domest. Anim. Endocrinol., 33 (4): 400-409.

Sordillo, L. M., Babiuk, L. A., 1991. Modulation of bovine mammary neutrophil function during the periparturient period following *in vitro* exposure to recombinant bovine interferon gamma [J]. Vet. Immunol. Immunopathol., 27 (4): 393-402.

Sordillo, L. M., Mavangira, V., 2014. The nexus between nutrient metabolism, oxidative stress and inflammation in transition cows [J]. Anim. Prod. Sci., 54: 1204-1214.

Sun, F., Cao, Y. C., Cai, C. J., et al., 2016. Regulation of Nutritional Metabolism in Transition Dairy Cows: Energy Homeostasis and Health in Response to Post-Ruminal Choline and Methionine [J]. PLoS One, 11: e0160659.

Tabachnick, M., Tarver, H., 1955. The conversion of methionine-S35 to cystathionine-S35 and taurine-S35 in the rat [J]. Arch. Biochem. Biophys., 56: 115-122.

Trevisi, E., Amadori, M., Cogrossi, S., et al., 2012. Metabolic stress and inflammatory response in high-yielding, periparturient dairy cows [J]. Res. Vet. Sci., 93: 695-704.

Trevisi, E., Jahan, N., Bertoni, G., et al., 2015. Pro-Inflammatory Cytokine Profile in Dairy Cows: Consequences for New Lactation [J]. Italian J. Anim. Sci., 14: 3862.

Vailati-Riboni, M., Zhou, Z., Jacometo, C. B., et al., 2017. Supplementation with rumen-protected methionine or choline during the transition period influences whole-blood immune response in periparturient dairy cows [J]. J. Dairy Sci., 100 (5): 3958-3968.

Van Brummelen, R., du Toit, D., 2007. L-methionine as immune supportive supplement: a clinical evaluation [J]. Amino Acids, 33: 157-163.

Waller, K. P., 2000. Mammary gland immunology around parturition. Influence of stress, nutrition and genetics [J]. Adv. Exp. Med. Biol., 480: 231-245.

Wang, Y., Puntenney, S. B., Burton, J. L., et al., 2007. Ability of a commercial feed additive to modulate expression of innate immunity in sheep immunosuppressed with dexamethasone [J]. Animal, 1 (7): 945-951.

Wang, Y. Q., Puntenney, S. B., Burton, J. L., et al., 2009. Use of gene profiling to evaluate the effects of a feed additive on immune function in periparturient dairy cattle [J]. J. Anim. Physiol. Anim. Nutr. (Berl), 93 (1): 66-75.

Wong, S., Pinkney, J., 2004. Role of cytokines in regulating feeding behaviour [J]. Curr. Drug. Targets, 5 (3): 251-263.

Wong, E. R., Thompson, W., 1972. Choline Oxidation and Labile Methyl Groups in Normal and Choline-Deficient Rat-Liver [J]. Biochim. Biophys. Acta., 260: 259-271.

Wu, P., Jiang, J., Liu, Y., et al., 2013. Dietary choline modulates immune responses, and gene expressions of TOR and eIF4E-binding protein 2 in immune organs of juvenile Jian carp (Cyprinus carpio var. Jian) [J]. Fish Shellfish Immunol., 35: 697-706.

Yuan, K., Vargas-Rodriguez, C. F., Mamedova, L. K., et al., 2014. Effects of supplemental chromium propionate and rumen-protected amino acids on nutrient metabolism, neutrophil activation, and adipocyte size in dairy cows during peak lactation [J]. J. Dairy Sci., 97 (6): 3822-3831.

Zahra, L. C., Duffield, T. F., Leslie, K. E., et al., 2006. Effects of rumen-protected choline and monensin on milk production and metabolism of periparturient dairy cows [J]. J. Dairy Sci., 89: 4808-4818.

Zhou, Z., Bu, D. P., Riboni, M. V., et al., 2015. Prepartal dietary energy level affects peripartal bovine blood neutrophil metabolic, antioxidant, and inflammatory gene expression [J]. J. Dairy Sci., 98 (8): 5492-5505.

Zhou, Z., Bulgari, O., Vailati-Riboni, M., et al., 2016a. Rumen-protected methionine compared with rumen-protected choline improves immunometabolic status in dairy cows during the peripartal period [J]. J. Dairy Sci., 99: 8956-8969.

Zhou, Z., Loor, J. J., Piccioli-Cappelli, F., et al., 2016b. Circulating amino acids during the peripartal period in cows with different liver functionality index [J]. J. Dairy Sci., 99: 2257-2267.

Zhou, Z., Vailati-Riboni, M., Trevisi, E., et al., 2016c. Better postpartal performance in dairy cows supplemented with rumen-protected methionine than choline during the peripartal period [J]. J. Dairy Sci., 99: 8716-8732.

Zom, R. L., Baal, J. van, Goselink, R. M. A., et al., 2011. Effect of rumen-protected choline on performance, blood metabolites, and hepatic triacylglycerols of periparturient dairy cattle [J]. J. Dairy Sci., 94: 4016-4027.

New (and Old) Technologies to Improve Feed Efficiency of the Dairy Industry

Michael J. VandeHaar

Department of Animal Science, Michigan State University, East Lansing, MI 48827

Summary

Dairy feed efficiency, as defined by the fraction of feed energy and protein captured in products, has more than doubled in many countries over the past 100 years. This increase occurred mostly because of a focus on improving production per cow, but, for the future, we must focus more directly on efficiency. Efficiency can be improved further by using genomic technologies to select for more efficient cattle and by feeding cows to more optimally meet nutrient requirements during different stages of the lactation cycle.

Introduction

Dairy feed efficiency in North America has doubled in the past 100 years (VandeHaar and St-Pierre, 2006; Capper et al., 2009), largely as a byproduct of selecting and managing cows for increased milk production. Milk synthesis is an efficient process, and increasing milk per cow results in a greater percentage of total feed intake being used for milk instead of cow maintenance. The current average milk production of dairy cattle in North America is ~ 10,000 kg/(cow · year), and elite dairy cattle currently partition >3 times more feed energy toward milk than toward maintenance over their lifetime (VandeHaar and St-Pierre, 2006). Much of the gain in feed efficiency from increasing productivity has already occurred in these cattle. Future increases in feed efficiency will still rely on increasing productivity, but we must specifically target feed efficiency as a farm goal. More detailed reviews of dairy feed efficiency have recently been published (Berry and Crowley, 2013; Connor, 2015; Pryce et al., 2014b; VandeHaar et al., 2016). In this paper, I will briefly discuss methods to further improve feed efficiency in the modern dairy cow. These include (1) selection for efficient genetics using genomics and milk output relative to body size, and (2) management to take advantage of the genetic potential of superior cattle.

Defining Feed Efficiency

Feed efficiency is a complex trait for which no single definition is adequate (Figure 1). Feed efficiency should be considered over the lifetime of a cow and include all feed used as a calf, growing heifer, and dry cow and all products including milk, meat, and calves. At the farm level, feed efficiency also should account for feed that is wasted and for products that are not suitable for human-consumption. We should also consider inputs of human-consumable vs. other foods, fossil fuels, water, and land, and outputs of greenhouse gasses, pollutants, fertilizers and other products not used for human consumption. How we feed dairy cattle also impacts ecosystem services (such as availability of wildlife habitat), rural aesthetics and sociology, soil conservation, food quality and healthfulness, food security, animal well-being, the need for imported oil, and how many beef cows are needed to produce calves. Developing a metric that includes all relevant factors for feed efficiency would be difficult. Thus, we use simple metrics such as milk to feed ratio, feed cost per unit milk, and income over feed cost. Whereas protein could be considered the most important component of milk, energy intake generally limits milk production, and feed energy includes the energy of protein. Thus, this paper focuses on energetic efficiency.

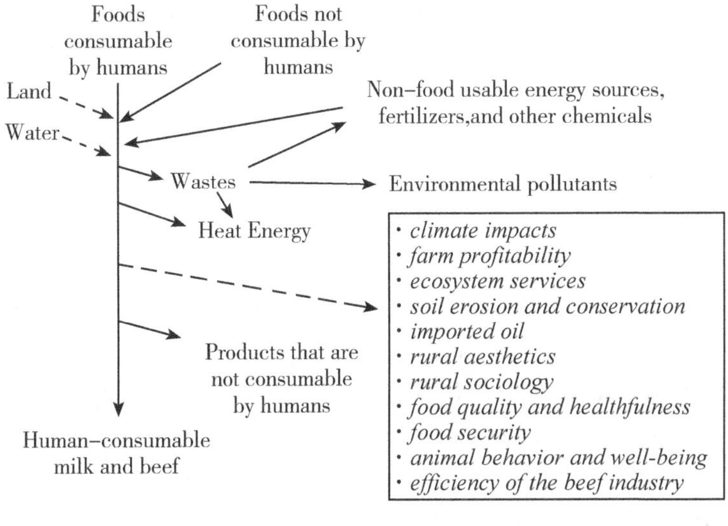

Figure 1 Factors to consider in defining feed efficiency. solid lines indicate direct energy transfers, whereas dashed lines indicate other factors that should be considered in a holistic view of efficiency

Gross energy (GE) is the total chemical energy of a feed but some of it is lost as the chemical energy in feces, gasses, and urine, and some is lost as the heat associated with the metabolic work of fermenting, digesting, and processing nutrients (Figure 2). The remaining chemical energy is known as net energy (NE). Some NE is used to support maintenance functions and is subsequently lost as heat. Some NE is the chemical energy of secreted milk or accreted body tissue and conceptus. For this paper, Gross Feed Efficiency (GEff) is defined as the energy captured in milk and body tissue divided by the GE consumed by a cow in her lifetime, and it is highly correlated to milk energy output per unit body weight (BW). GEffis a useful way to examine how well the dairy industry is stewarding resources.

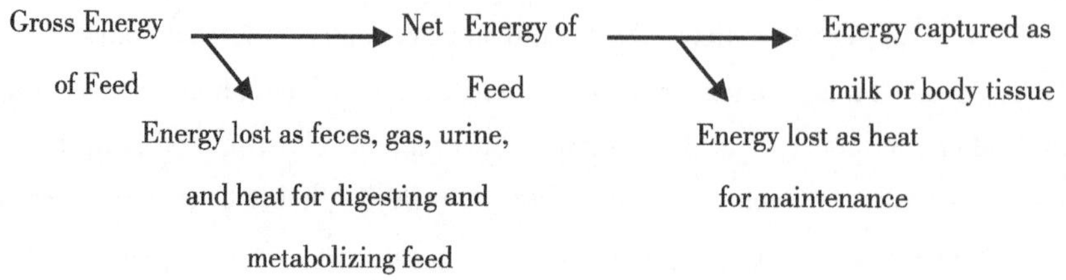

Figure 2 Energy flow in a cow

The major components affecting feed efficiency can be divided into (1) those that alter maintenance and the dilution of maintenance, or the portion of NE that is captured in milk or body tissues instead of used for maintenance, and (2) those that alter the conversion of GE to NE, which include diet and cow effects.

Selecting for Cows that Capture NE More Efficiently

The typical Holstein cow has a maintenance requirement of ~ 10 Mcal of NE/day (equivalent to ~25 Mcal of GE and ~6 kg of feed). This feed is used for basal life-sustaining functions even if animal is not producing milk, growing, working, or pregnant. Any extra feed consumed above that needed for maintenance can be converted to milk or body tissues. If the cow eats twice as much feed as she needs for maintenance, so 2X maintenance, then only half of her feed NE intake is used for maintenance with the remainder used for milk. As she eats even more feed, the portion used for maintenance becomes a smaller fraction of total feed intake. This "dilution of maintenance" increases efficiency and has been known for a long time (Freeman, 1975; VandeHaar et al., 2016). However, the marginal increase in efficiency from diluting maintenance diminishes with each successive increase in feed intake and, efficiency likely plateaus at ~5X maintenance intake (Figure 3).

Production relative to maintenance can be increased by increasing production or by decreasing maintenance. Maintenance is highly correlated to a cow's body weight, and, over the past 50 years, the body size of dairy cattle has increased. Because of this, the US genetic base for body size traits in all dairy breeds is continually being adjusted up. However, our latest analysis on 5,000 Holsteins in mid-lactation (Tempelman et al., 2015) demonstrated no genetic correlation between BW and milk energy output (VandeHaar et al., 2014); moreover,

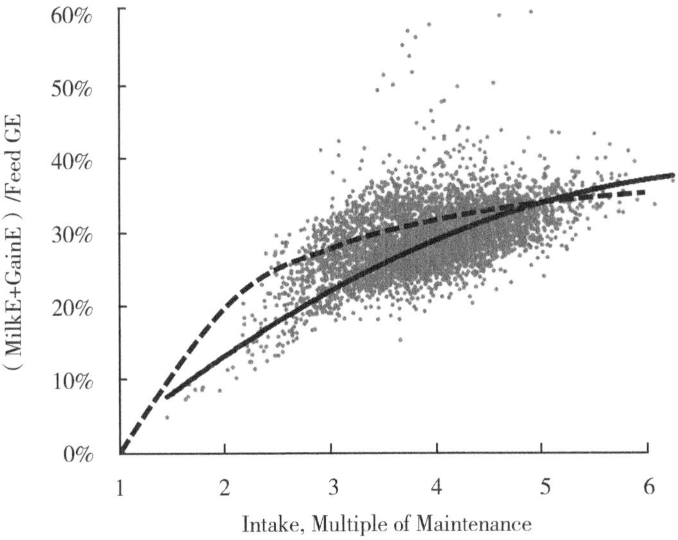

Figure 3 Change in gross feed efficiency (GEff) as intake increases. each intake multiple of maintenance will be about 10 mcal of NEL and 5 to 6 kg of feed DM. the dashed line depicts the expected GEff based on the NRC (2001) maintenance requirement of 0.08 mcal per unit BW^ 0.75 and demonstrates that GEff diminishes with increasing intake. this expected response assumes no depression in digestibility as intake increases. thus, the actual response should plateau even sooner and at a lower level. the symbols depict the measured geff in ~5,000 mid-lactation holstein cows assuming a feed GE value of 4.5 mcal/kg (unpublished data from the study of tempelman et al., 2015). the solid line is the trend line for GEff and demonstrates that the diminishing response is $GEff = -0.10 + 0.13 \times MM - 0.0094 \times MM^2$, where MM = multiple of maintenance

BW was genetically correlated negatively with GEff. In a smaller subset of that data, Manzanilla-Pech et al. (2015) showed that milk energy output had zero or negative genetic correlations with BW and that stature was genetically correlated negatively with GEff. The fact that cows have gotten larger over the past 50 years is counter to the goal of increasing efficiency. This is especially true when considering that the maintenance requirement per unit BW of dairy cows has increased in the past 50 years (Moraes et al., 2016). Thus, we can continue to improve efficiency by selectingfor cows with greater milk production but also we should consider selecting for smaller BW at the same time.

Using the projected curve for the dilution of maintenance of Figure 3, efficiency should begin to plateau as cows achieve about 5X maintenance. However, this projection is overly optimistic because as cows eat more, the percentage of feed that is digested is depressed (NRC, 2001; Huhtanen et al., 2009). According to the equations for digestibility used in the NRC (2001), efficiency peaks at ~4X maintenance intake, which is ~45 kg milk (3.5% fat) per day for a 680 kg cow. The NRC (2001) model likely discounts digestibility too much at high intakes, as shown in Huhtanen et al. (2009) and supported by the actual data in Figure 3, so efficiency might not be close to maximal until cows produce at least 60 kg if they weight ~680 kg BW. Importantly, however, feed efficiency should be considered on a lifetime basis, so we must account for body tissue gain and the feed consumed as a heifer and dry cow. Based on the theoretical model of VandeHaar (1998), lifetime GEff is ~20% for 1,600 kg cows producing 10,000 kg milk/year (the current US average), 25% for cows at 20,000 kg/yr, and would likely never exceed 30%. Certainly, substantial gains can still be made in lifetime feed efficiency from increasing production relative to body size. However, top dairy farms are at a point where the return in efficiency from further gains in productivity will be smaller than they have been in the past. Thus, along with continuing to breed for more milk per unit BW, we should also develop new methods to select for feed efficiency directly and focus on ways to save on feed inputs through better feed management and nutritional grouping.

Selecting for Cows that Convert GE to NE More Efficiently

One way to select for feed efficiency as a breeding goal, independently of production level, is to select for lowResidual Feed Intake (RFI). RFI is a measure of actual versus predicted intake for an individual and is essentially "unjustified feed intake" and has been previously described (Berry and Crowley, 2013; Pryce et al., 2015; Connor, 2015; Tempelman et al., 2015; VandeHaar et al., 2016). Usually, RFI is determined statistically as the deviation of actual dry matter intake (DMI) of a cow from the average DMI of other cows that are fed and managed the same (Cohort) after adjusting for the major energy sinks of BW (related to maintenance), milk energy output, and body energy change, where the residual error termis RFI (Figure 4). Thus, RFI includes error that is true variation amongst cows due to genetics, true variation that is due to permanent environmental effects, and variation from measurement error. Cows that eat less than expected have negative RFI, and thus are desirable when comparing animals for selection purposes as long as RFI is only seen as one factor to use in selecting for efficiency. Selecting for high milk production relative to BW also remains an important selection criteria.

Based on our data examining the GEff of cows compared to their level of production (Figure 2), efficiency varies considerably among cows within a production level. This variation can also be examined in intake units, or RFI. Whereas part of the variationin RFI is error in measurements, some RFI is biological with a heritability of 0.17 based on 4,900 cows (Tempelman et al., 2015). During the 20th century, selection of superior genetics relied heavily on quantification of the phenotype in daughters of young sires; sires with outstanding daughters were deemed genetically superior. Because DMI cannot be measured easily and routinely on individual cows in commercial farms, direct selection for feed efficiency was impossible. Genomic se-

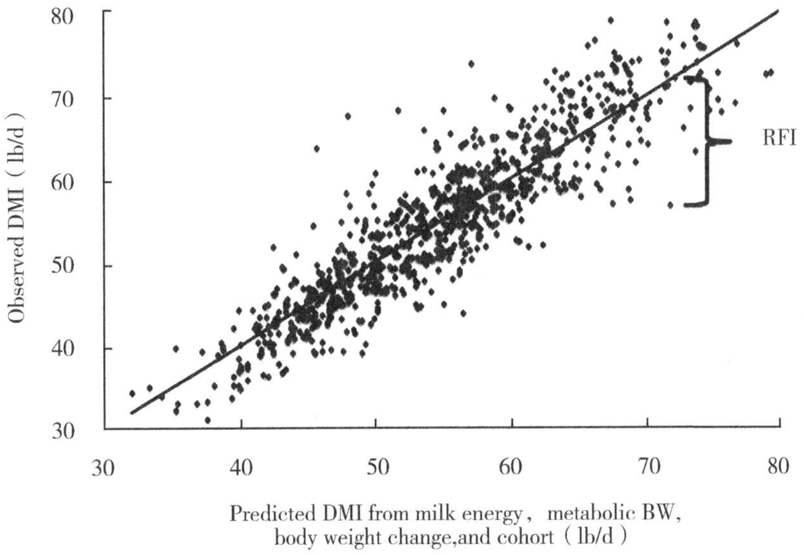

Figure 4 Residual feed intake (RFI) as a measure of feed efficiency. cows that truly eat less than predicted are more efficient at converting feed GE to NE or require less NE for maintenance than expected based on their BW

lection enables selection for traits like feed efficiency for which daughter phenotypes are unknown. Excellent reviews on the general methodology of genomic selection are Eggen (2012) and Hayes et al. (2010).

The biological basis for variation in RFI is not clear. This variation is associated mostly with the conversion of GE to NE, and thus is due to differences in digestibility, methane production, urinary energy losses, and metabolic pathways involved in processing nutrients. Variation among cows in the actual maintenance requirement relative to MBW also contributes to RFI; for example, a maintenance requirement of only 0.07 Mcal per unit MBW would contribute toward a negative RFI for a cow compared to a maintenance requirement of 0.12 Mcal per unit MBW. Most of the key regulatory hormones do not seem related to RFI, although those related to regulation of appetite are correlated with RFI, as might be expected (Xi et al., 2016). Digestibility may account for up to one-third of the variation in RFI in cows fed high fiber diets, but very little of the variation in cows fed high starch diets (Potts et al., 2017). Additional thoughts on the variation amongst cows can be found in VandeHaar et al., (2016) and Herd and Arthur (2009). However, regardless of the reasons that some genotypes are more efficient, genomic selection can and will be used.

Evidence that genomic selection for RFI can work in the dairy industry has been demonstrated by Davis et al. (2014), who calculated genomic breeding values (BV) for RFI based on studies in heifers and then genotyped an independent set of lactating cows. 100 cows with highest genomic BV for RFI and 100 with the lowest genomic BV for RFI were brought to a common location. Cows with genomic BV for low RFI required 0.6 kg/d less feed to produce milk than cows with genomic BV for high RFI, even though milk production and BW were similar. This savings in feed is similar to the expected savings in feed for maintenance in a cow weighing 80 kg less. The use of genomics in selection against RFI or DMI is already beginning in Australia (Pryce et al., 2015) and the Netherlands (Veerkamp et al., 2014) and will likely occur in North America in the near future. If selection for efficiency is to be realized by selection for RFI, RFI should be a repeatable trait across diets, climate conditions, lactations, stages within a lactation, and even stages of life. Data to date suggest that it is (Tempelman et al., 2015; Potts et al., 2015; Connor et al., 2013; MacDonald et al., 2014). It is important to note that RFI is only part of feed efficiency. Selection for efficiency must also consider the optimal levels of milk production relative to body weight. The approach used by Pryce et al. (2015) seems reasonable, with an index to select against body size and against RFI while also selecting for milk yield and composition. Improvements in feed efficiency must not occur at the expense of health and fertility of dairy cows. Many traits must be optimized as we consider the ideal cow of the future to promote profitability of farms and sustainability of the dairy industry (Table 1).

Table 1 Breeding goals for the cow of the future to enhance efficiency and sustainability

Efficiency goals	Efficiently captures (partitions) lifetime NE to product because maintenance represents a small portion of required feed NE
	Has a negative RFI, indicating greater efficiency at converting GE to NE or lower maintenance than expected based on BW
	Is profitable (high production dilutes out farm fixed costs)
	Has minimal negative environmental impacts
	Can efficiently use human-inedible foods, pasture, and high fiber feeds
	Requires less protein and phosphorus per unit of milk
	Produces milk and meat of high quality and salability
Other goals	Is healthy, long-lived, and thrives through the transition period
	Is fertile and produces high-value offspring
	Is adaptable to different climates and diets
	Has a good disposition

Managing for Feed Efficiency

Using the model described in Vande Haar (1998), the impacts of various management changes on efficiency were predicted. Increasing average daily milk production by 10% or increasing cow longevity from 3 to 4 lactations is expected to increase lifetime energetic efficiency ~0.7%. Reducing feed use by 2% with no change in milk production, by selecting against RFI or for smaller cows, or decreasing feed wastage, would improve energy efficiency ~ 0.5%.Reducing the age at first calving by 2 months, or reducing calving interval by 1 month would increase lifetime efficiency ~0.3%. How cows are fed and managed at each stage of life can alter milk yield per day of life and thereby dilute maintenance and increase efficiency. These management changes promote similar improvements in the efficiency of converting feed protein to milk or body protein. However, the single biggest impact farms could make on efficiency of protein use is to simply quit overfeeding protein, as is often done in late lactation. Feeding cows past 150 days postpartum a diet with 2 percent less protein (15% vs. 17% CP) would increase efficiency of lifetime protein use by ~1.3%.

Nutrient requirements vary as lactation progresses, and the optimal diet for maximum efficiency and profitability changes as well (Figure 5; NRC, 2001; Allen and Piantoni, 2014). The widespread adoption of totally mixed ration (TMR) feeding in North America has improved productivity and efficiency because cows eat a consistent diet, but cows are less likely to receive a diet that matches their individual requirements. This is especially true if all lactating cows are fed the same TMR; feeding a single TMR across lactation can never maximize production and efficiency. A single TMR is usually formulated for the higher producing cows and is more nutrient-dense than optimal for cows in later lactation, resulting in inefficient use of most nutrients for these cows. In addition, although a single TMR is formulated for the high producers, it will not maximize milk income over feed costs for the herd because forages, grains, and expensive supplements cannot be allocated optimally.

Contreras-Govea et al. (2015) found the two major constraints to nutritional grouping on commercial farms were that "It makes things too complicated" and "Low diets decrease milk yield". Thus, in my opinion, the job of a nutritionist is to (1) develop diets that consistently meet needs optimally for fresh, peak, and maintenance groups and demonstrate their benefits, (2) use supplements, metabolic modifiers, feed additives, and low cost alternative feeds to improve efficiency within groups, (3) help farms make rules based on milk and BCS for moving cows and design systems to track cows, and (4) develop protocols for feeding an extra diet.

The optimal number of rations on any farm depends on many factors, but we recommend at least three based on feeding goals and cow biology. The regulation of voluntary feed intake must be considered in diet formulation. Intake is likely limited by hepatic oxidation of fuels in fresh cows and perhaps late lactation cows, but by rumen fill throughout much of the duration of lactation (Allen and Piantoni, 2014). The goal in feeding cows around parturition is optimal health, so expensive supplements are warranted. The goal in feeding cows for the first half of lactation, which includes peak lactation, is maximal milk, rebreeding, and health; these cows should be fed minimum fiber diets with plenty of digestible

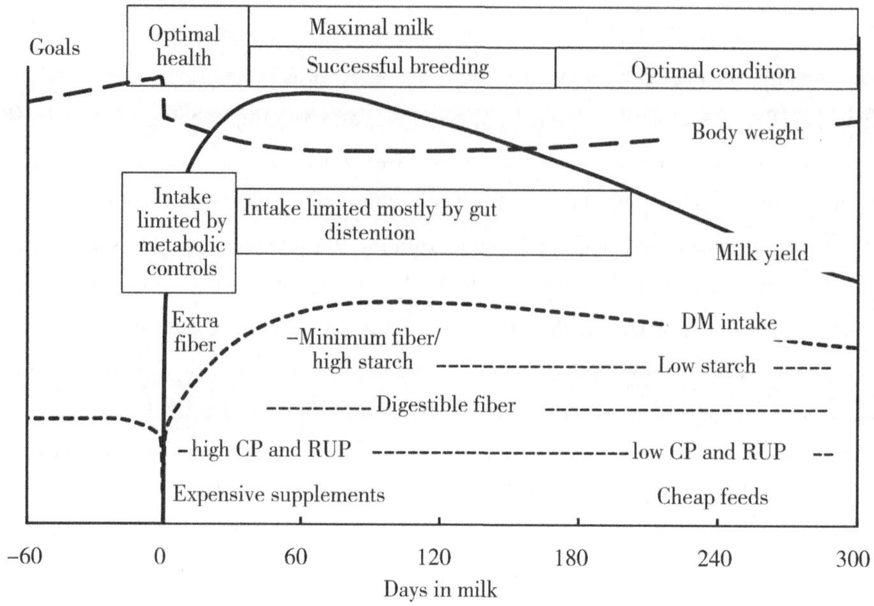

Figure 5 Considerations in nutritional grouping. Optimal formation of nutritional groups should consider goals for the cow, the primary drivers of appetite, the potential value of expensive supplements, and the effects of diet on nutrient partitioning between milk and body tissues

starch to maximize energy intakes, and again expensive supplements, including amino acids, might be warranted. Cows in later lactation, after replenishing their body stores to a body condition score of 3, should be fed to optimize milk and maintain body condition; they should be fed less fermentable starch and more fermentable fiber to promote partitioning of nutrients toward milk instead of body tissues (Allen and Piantoni, 2014; Boerman et al., 2015). Late lactation cows also should be fed lower protein diets to maximize efficiency of protein use (NRC, 2001); expensive supplements are almost never needed. Nutritional grouping and multiple TMR undoubtedly do increase capital, management, and labor costs; however, feeding cows according to requirements enhances production, efficiency, profitability, and sustainability of the industry (Vande Haar and St-Pierre, 2006).

Should We Choose Diets that Enhance the Conversion of GE to NE?

Diet can significantly and directly alter feed efficiency. Fats are energy-dense, so can increase the ratio of milk to feed. More importantly, some feeds are simply more digestible than others; feeds high in starch are almost always more digestible than feeds high in fiber. Unless rumen health is compromised by inadequate forage fiber, a greater percentage of the GE of high starch diets will be converted to milk or body tissue, compared to high fiber diets. If attaining the highest GEff was the goal of the dairy industry, all dairy cattle would be fed diets high in grains and fats with minimal forage and byproduct feeds. However, for purposes of efficiency, sustainability, and profitability, feeding to maximize GEff is illogical. Diet composition can alter a cow's voluntary feed intake (Allen and Piantoni, 2014), and therefore, because feed intake alters the dilution of maintenance, the effect of diet on GEff is not always easy to predict. Moreover, energy-dense feeds are often expensive, even on a per unit of energy basis, so energy diets might be less profitable even if they increase efficiency. Finally, one of the values of the ruminant system is its ability to obtain energy from fibrous feeds, such as forages and high-fiber byproducts. From a global perspective, forages and byproduct feeds high in fiber should be fed to dairy cattle when possible and grains and fats should be fed only as needed to optimize production and health. For making decisions about feeding and management, GEff is almost never useful; instead metrics such as income over feed costs on a farm basis, the efficiency of using human-consumable foods, or milk per acre, seem more reasonable. For making decisions about animal selection, however, GEff and IOFC would be highly correlated.

Efficiency and Stewardship

Life Cycle Analysis (LCA) are used to best assess the environmental impact of management decisions. Thomassen et al. (2008) compared conventional and organic Dutch dairy farms. Conventional farms used 60% more energy and caused 50% more eutrophication per unit of milk produced. However, with no difference in climate change gasses. However, organic farms required 40% more land to get the same amount of milk. Climate change gasses were similar per unit of milk, but in our view, the higher productivity of the conventional farms would spare land for biofuel production or carbon sequestration. Thus, the conventional dairies would have had less negative climate change impact. This is consistent with an LCA study by Capper et al. (2009) showing that the US dairy industry has decreased greenhouse gas emissions 60% per unit of milk in the last 60 years, mostly because of the enhanced feed efficiency from higher productivity. Thus, increased lifetime productivity increases efficiency, and increased efficiency generally is good for the environment—more people can be fed with less resources and less negative environmental impact. A recent FAO report (FAO, 2010) shows that even scientists who are not part of the US dairy science community agree with this view. Improving efficiency of milk production by using new technologies seems the responsible thing to do for the environment, at least in the foreseeable future.

Conclusion

We have made major gains in feed efficiency in the past 50 years as a byproduct of selecting, feeding, and managing cows for increased productivity, which dilutes maintenance. Average production is currently ~10,000 kg/yr in North America, and most cows have the genetics for even higher production. We must harness the genetics of the current dairy cattle population to improve feed efficiency even further and help feed people sustainably. Better feeding and management may be especially helpful for many lower producing herds to help them achieve at least 15,000 kg/yr. To enhance efficiency further, we should take advantage of new genomic tools that will enable us to select for cows that require less feed per unit of milk by using a selection index that favors greater milk production and components, smaller cow size, and negative RFI.

References

Allen, M. S., Piantoni, P., 2014. Carbohydrate nutrition: managing energy intake and partitioning through lactation [J]. Vet. Clin. North Am. Food Anim. Pract., 30: 577-597.

Berry, D. P., Crowley, J. J., 2013. Genetics of feed efficiency in dairy and beef cattle [J]. J. Anim. Sci., 91: 1594-1613.

Boerman, J. P., Potts, S. B., VandeHaar, M. J., et al., 2015. Milk production responses to a change in dietary starch concentration vary by production level in dairy cattle [J]. J. Dairy Sci., 98: 4698-4706.

Capper, J. L., Cady, R. A., Bauman, D. E., 2009. The environmental impact of dairy production: 1944 compared with 2007 [J]. J Animal Sci., 87: 2160-2167.

Connor, E. E., Hutchison, J. L., Norman, H. D., et al., 2013. Using residual feed intake in Holsteins during early lactation shows potential to improve feed efficiency through genetic selection [J]. J. Anim. Sci., 90: 1687-1694.

Connor, E. E., 2015. Invited review: improving feed efficiency in dairy production: challenges and possibilities [J]. Animal, 9: 395-408.

Contreras-Govea, F. E., Cabrera, V. E., Armentano, L. E., et al., 2015. Constraints for nutritional grouping in Wisconsin and Michigan dairy farms [J]. J. Dairy Sci., 98: 1336-1344.

Davis, S. R., Macdonald, K. A., Waghorn, G. C., et al., 2014. Residual feed intake of lactating Holstein-Friesian cows predicted from high-density genotypes and phenotyping of growing heifers [J]. J. Dairy Sci., 97: 1436-1445.

Eggen, A., 2012. The development and application of genomic selection as a new breeding paradigm [J]. Anim. Frontiers, 2: 10-15.

FAO., 2010. Greenhouse gas emissions from the dairy sector-A life cycle assessment [R]. P. Gerber, T. Vellinga, P., C. Opio, B. Henderson, and H. Steinfeld. Food and Agriculture Organization of the United Nations: Rome.

Freeman, A. E., 1975. Genetic variation in nutrition of dairy cattle [M]. In: The Effect of Genetic Variation on Nutrition of Animals. National Academy of Science, Washington, DC. pp. 19-46.

Hayes, B. J., Pryce, J., Chamberlain, A. J., et al., 2010. Genetic architecture of complex traits and accuracy of genomic prediction: coat colour, milk-fat percentage, and type in Holstein cattle as contrasting model traits [J]. PLoS Genet., 6: e1001139.

Herd, R. M., Arthur, P. F., 2009. Physiological basis for residual feed intake [J]. J. Anim. Sci., 87 (E. Suppl.): E64-E71.

Huhtanen, P., Rinne, M., Nousiainen, J., 2009. A meta-analysis of feed digestion in dairy cows.

2. The effects of feeding level and diet composition on digestibility [J]. J. Dairy Sci., 92: 5031-5042.

Mac Donald, K. A., Pryce, J. E., Spelman, R. J., et al., 2014. Holstein-Friesian calves selected for divergence in residual feed intake during growth exhibited significant but reduced residual feed intake divergence in their first lactation [J]. J. Dairy Sci., 97: 1427-1435.

Manzanilla-Pech, C., Veerkamp, R. F., Tempelman, R. J., et al., 2016. Genetic parameters between feed-intake-related traits and conformation in 2 separate dairy populations-the Netherlands and United States [J]. J. Dairy Sci., 99: 443-57.

Moraes, L. E., Kebreab, E., Strathe, A. B., et al., 2015. Multivariate and univariate analysis of energy balance data from lactating dairy cows [J]. J. Dairy Sci., 98: 4012-4029.

National Research Council., 2001 Nutrient Requirements of Dairy Cattle [M]. 7th revised edition. National Academy Press, Washington, D. C.

Potts, S. B., Boerman, J. P., Lock, A. L., et al., 2015. Residual feed intake is repeatable for lactating Holstein dairy cows fed high and low starch diets [J]. J. Dairy Sci., 98: 4735-4747.

Potts, S. B., Boerman, J. P., Lock, A. L., et al., 2017. Relationship between residual feed intake and digestibility for lactating Holstein cows fed high and low starch diets [J]. J. Dairy Sci., 100: 265-278.

Pryce, J. E., Wales, W. J., de Haas, Y., et al., 2014b. Genomic selection for feed efficiency in dairy cattle [J]. Animal, 8: 1-10.

Pryce, J. E., Gonzalez-Recio, O., Nieuwhof, G., et al., 2015. Hot Topic: Definition and implementation of a breeding value for feed efficiency in dairy cows [J]. J. Dairy Sci., 98: 7340-7350.

Tempelman, R. J., Spurlock, D. M., Coffey, M., et al., 2015. Heterogeneity in genetic and nongenetic variation and energy sink relationships for residual feed intake across research stations and countries [J]. J. Dairy Sci., 98: 2013-2026.

Thomassen, M. A., Dalgaard, R., Heijungs, R., et al., 2008. Attributional and consequential LCA of milk production [J]. Internatl. J. Life Cycle Assessment, 13: 339-349.

Vande Haar, M. J., 1998. Efficiency of nutrient use and relationship to profitability on dairy farms [J]. J. Dairy Sci., 81: 272-282.

Vande Haar, M. J., St-Pierre, N., 2006. Major Advances in Nutrition: Relevance to the sustainability of the dairy industry [J]. J. Dairy Sci., 89: 1280-1291.

Vande Haar, M. J., Lu, Y., Spurlock, D. M., et al., 2014. Phenotypic and genetic correlations among milk energy output, body weight, and feed intake, and their effects on feed efficiency in lactating dairy cattle [J]. J. Dairy Sci., 97 (E-Suppl): 80.

Vande Haar, M. J., Armentano, L. E., Weigel, K. A., et al., 2016. Harnessing the genetics of the modern dairy cow to continue improvements in feed efficiency [J]. J. Dairy Sci., 99: 4941-4954.

Veerkamp, R. F., Calus, M. P. L., de Jong, G., et al., 2014. Breeding Value for Dry Matter Intake for Dutch Bulls based on DGV for DMI and BV for Predictors [EB/OL]. Proc. Of 10th World Congress of Genetics Applied to Livestock Production. https: //asas. org/docs/default-source/ wcgalp-proceedings-oral/115_paper_8665_manuscript_206_0. pdf? sfvrsn=2.

Xi, Y. M., Wu, F., Zhao, D. Q., et al., 2016. Biological mechanisms related to differences in residual feed intake in dairy cows [J]. Animal, 10: 1311-1318.

Urea Metabolism and Regulation by Rumen Bacterial Urease in Ruminants-A Review

Di Jin[1,2], Shengguo Zhao[1,3], Nan Zheng[1,3], Yves Beckers[2], Jiaqi Wang[1,3]

[1]State Key Laboratory of Animal Nutrition, Institute of Animal Science, Chinese Academy of Agricultural Sciences, Beijing, 100193, China;[2]University of Liège, Gembloux Agro-Bio Tech, Precision Livestock and Nutrition Unit, Passage des Déportés 2, B-5030 Gembloux, Belgium;[3]Ministry of Agriculture-Milk Risk Assessment Laboratory, Institute of Animal Science, Chinese Academy of Agricultural Sciences, Beijing 100193, China

Abstract

Urea is used as non-protein nitrogen in the rations of ruminants as an economical replacement for feed proteins. Urea transferred from the blood to the rumen is also an important source of nitrogen for rumen microbial growth. It is rapidly hydrolyzed by rumen bacterial urease to ammonia (NH_3) and the NH_3 is utilized for the synthesis of microbial proteins required to satisfy the protein requirements of ruminants. Urea has commonly become an accepted ingredient in the diets of ruminants. In recent decades, urea utilization in ruminants has been investigated by using traditional research methods. Recently, molecular biotechnologies have also been applied to analyze urea-degrading bacteria or urea nitrogen metabolism in ruminants. Combining traditional and molecular approaches, we can retrieve better information and understanding related to the mechanisms of urea metabolism in ruminants. This review focuses on urea utilization in ruminants and its regulation by rumen bacterial urease in the host. The accumulated research provides foundations for proposing further new strategies to improve the efficiency of urea utilization in ruminants.

Key words: rumen, urea utilization, ureolytic bacteria, urease, regulation

Introduction

Urea has been used as non-protein nitrogen (NPN) in ruminant rations for some time. Kertz (2010) wrote that more than one hundred years ago, German workers suggested that urea could be used to replace a portion of dietary protein in ruminants. Thereafter, some studies were conducted on the use of NPN in ruminant diets. During the 1970s and 1980s, multiple studies were conducted on the utilization of urea as a replacement for protein in ruminant diets, especially its effect on dry matter intake (Wilson et al., 1975; Polan et al., 1976), rumen fermentation (Pisulewski et al., 1981; Kertz et al., 1983), milk yield and reproduction-related parameters (Ryder et al., 1972; Erb et al., 1976). Since then, research attempting to understand the mechanisms of urea utilization in ruminants has been conducted (Balcells et al., 1993; Huntington and Archibeque, 2000; Stewart and Smith, 2005).

Studies for improving urea utilization in ruminants are ongoing. It is known that the performance and metabolism of dairy cows depend upon the amount of urea they are fed (Sinclair et al., 2012; Giallongo et al., 2015). For example, ruminal nitrogen metabolism and urea kinetics of Holstein steers fed diets containing either rapidly degrading or slowly degrading urea at various levels of degradable intake protein (DIP) were estimated by Holder et al. (2015). They found that the rapidly degrading urea group had higher dry matter digestibility than the slow-release urea group, and gastrointestinal entry of urea nitrogen (urea-N), urea-N lost to feces and urea-N apparently used for anabolism were not different between treatments, while plasma urea concentrations were greater in higher DIP diets and higher for the rapidly degrading urea group than the slow release urea group. When 2% of urea was fed to lactating dairy cows as a replacement for soybean meal, both the milk protein content and milk yield decreased, while plasma urea-N increased (Imaizumi et al., 2015). Urea supplementation could also increase nitrogen availability for ruminal microorganisms. A study by Wanapat et al. (2016) showed that when swamp buffaloes were fed rice straw supplemented with urea, the feed intake, nutrient digestibility, and microbial protein synthesis increased. More importantly, the authors also tried to determine the effect of urea supplementation on rumen microbes and they found that fungal zoospores, total bacteria and the three predominant cellulolytic bacteria (*Ruminococcus albus*, *Fibrobacter succinogenes*, and *Ruminococcus flavefaciens*) were increased by urea supplementation.

Following extensive research on urea utilization in rumens, interests began to focus on urea-degrading microbes and urea utilization mechanisms in dairy cows. Research studying the regulation of bacterial urease for improving urea utilization has also been conducted. Advanced molecular biotechnologies provide new strategies to reveal the mechanisms of urea hydrolysis, transportation, and utilization in ruminants, and provide more knowledge for the improvement of nitrogen utilization efficiency in practical ruminant production systems. This review focuses on urea recycling in ruminants, urea hydrolysis, utilization and its regulation by rumen bacterial urease in recent research.

Urea Nitrogen Recycling in Ruminants

In ruminants, ammonia arises in the rumen from the diet and recycled urea. Urea in the rumen is rapidly hydrolyzed to ammonia and CO_2 by the bacterial enzyme urease. Ammonia from urea or from degraded dietary protein is used by the ruminal microbiota for the synthesis of microbial proteins, which are subsequently digested in the intestine. The excess ammonia is transported to the liver for endogenous urea synthesis, and urea recycling via the ruminal wall, and salivary secretion. Urea recycling to the rumen is an evolutionary advantage for ruminants because it provides part of the N required for microbial protein synthesis and enhances survival (Reynolds and Kristensen, 2008).

Reutilization of endogenous urea

Ruminants fed on diets with high NPN had higher portal blood flow, greater hepatic uptake of excess NH_3 and increased rates of urea synthesis (Symonds et al., 1981; De Visser et al., 1997; Holder et al., 2015). Redundant NH_3 transported to the liver is likely to enter the ornithine cycle (Zhou et al., 2015). Therefore, ammonia detoxification in the liver is likely to be one of the reasons for increased plasma urea concentration (Law et al., 2009). Blood urea-N concentrations are influenced by many parameters, especially dietary nitrogen intake (Puppel and Kuczynska, 2016), and it also has been used to predict nitrogen excretion and efficient nitrogen utilization in cattle and several different species of farm animals (Kohn et al., 2005).

Ruminants recycle substantial amounts of nitrogen as urea by the transfer of urea across the ruminal wall, and salivary secretion (Huntington and Archibeque, 2000). In ruminants, urea that is recycled to the rumen is an important source of nitrogen for microbial growth and reported data indicate that 40 to 80% of endogenously produced urea-N is returned to the gastrointestinal tract (Harmeyer and Martens, 1980; Lapierre and Lobley, 2001). There is a reciprocal change between urea recycling and excretion in urine depending on the crude protein intake (Reynolds and Kristensen, 2008). When growing cattle were fed prairie hay with very low protein concentrations, virtually almost all urea entering the blood pool was returned to the gut, and little was excreted in urine. In addition, ruminal fermentation products such as short-chain fatty acids and CO_2 acutely stimulate urea transport across the ruminal epithelium, and the effects are pH-dependent (Abdoun et al., 2010). The presence of ammonia has a negative impact on urea transport rates and is concentration dependent, with saturation at 5 mmol/l (Lu et al., 2014). At physiological pH, uptake of $NH4^+$ into the cytosol may be a key signaling event regulating ruminal urea transport. Therefore, in ruminants, urea-N recycling is affected by a number of factors including plasma urea-N concentration and fermentable carbohydrates in the gastrointestinal tract.

Urea kinetics has been obtained by the infusion of labeled urea to provide an estimate of urea entry rate. Wickersham et al. (2008) evaluated the effect of increasing the amount of rumen-DIP on urea kinetics in steers consuming prairie hay with jugular infusions of $^{15}N^{15}N$-urea. They found that the transfer of urea from the blood to the rumen contributes between one-fourth and one-third of the N utilized by ruminal microbes for the synthesis of microbial protein. Provision of supplemental DIP increased forage utilization and N retention in cattle consuming low-quality forage. Zhou et al. (2015) also used $^{15}N^{15}N$-urea to detect urea kinetics and nitrogen balance in Tibetan sheep when fed oat hay. Urea-N entry rate, gastrointestinal tract entry rate, return to ornithine cycle and fecal urea-N excretion all increased linearly with an increase in dry matter intake. The Tibetan sheep demonstrated low N requirements for maintenance compared with other ruminants. The estimated N requirements for maintenance were 0.50 g/kg $BW^{0.75}$ per day, that is, only 66% of the amount recommended by NRC for growing sheep of its size. However, it should be noted that the NRC values were for sheep without marked feed restriction. Therefore, for different ruminants, there are a number of differences in N metabolism and recycling except for in some common responses.

Currently, meta-analytical approaches have been used to evaluate the efficiency of urea utilization inruminants. In the study of Marini et al. (2008), by utilizing a statistical approach and data obtained from studies reporting duodenal, ileal, and fecal N flows in cattle, the endogenous N losses and true digestibility of N were estimated for different segments of the gastrointestinal tract. The N transactions for the reference diet (24.2 g of N/kg of organic matter [OM], 32% neutral detergent fiber [NDF] and carbohydrates of medium fermentation rate) were estimated as shown in Figure 1. The results showed that the minimal contribution of endogenous N to the N available in the rumen was 39%. In addition, Batista et al. (2017) also estimated urea kinetics and microbial usage of recycled urea-N in ruminants by combining data from studies with ruminants (beef cattle, dairy cows, and sheep), which were published from 2001 to 2016 and analyzed according to meta-analysis techniques using linear or non-linear mixed models. They concluded that urea-N synthesized in the liver and urea-N recycled to the gut linearly increased as N intake (g/body $weight^{0.75}$) increased, with increases corresponding to 71.5% and 35.2% of N intake, respectively. However, increasing dietary crude protein intake led to decreases in the fractions of urea-N recycled to the gastrointestinal tract and of recycled urea-N incorporated into microbial N. Therefore, a better understanding of the factors involved in endogenous urea losses will allow for a more accurate estimation of both N supply and N requirements. Since urea-N recycling to the gut is influenced by many dietary and ruminal factors, some modulation could be made in the rations of ruminants in order to improve the efficiency of utilizing endogenous urea.

Urea Transport Across the Rumen Epithelium

Urea produced in the liver is transferred across the rumen wall from the blood and then it is hy-

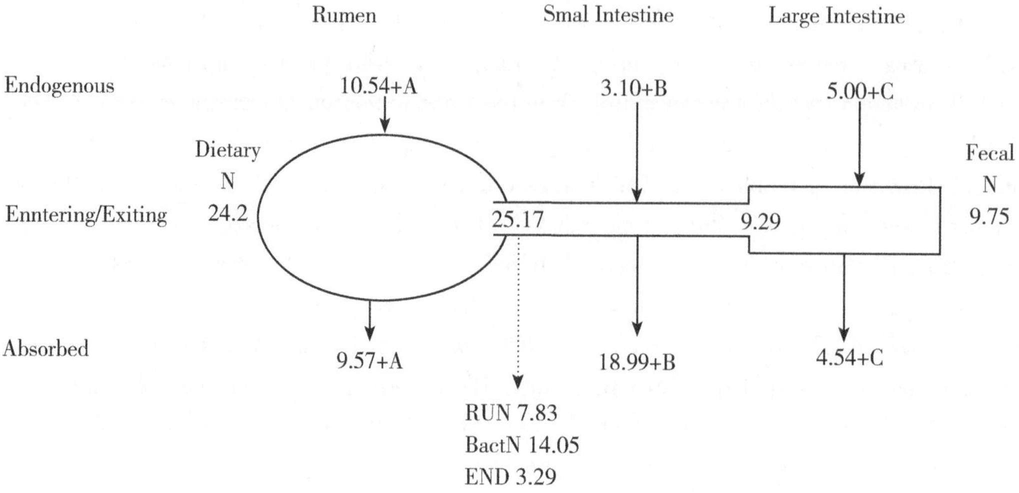

Figure 1 Nitrogen transactions along the gastrointestinal tract of cattle in an example diet (32% neutral detergent fiber [NDF], carbohydrates of medium degradation rate, consumed at 2% of bodyweight daily, on an OM basis) containing 24.2 g of N/kg of OM. All fluxes expressed in g of N/kg of OM intake. nitrogen entering the small intestine is composed of ruminal undegraded feed N (RUN), bacterial N (BactN), and free endogenous N reaching the duodenum (END).24.2 is N from the diet entering the rumen; 25.17 is the total duodenal N; 9.29 is the total ileal N; 9.75 is the total fecal N; 10.54, 3.10 and 5.00 are the endogenous N entering the rumen, the small intestine and the large intestine, respectively; 9.57, 18.99 and 4.54 are N absorbed by the foregut, in the small intestine and in the large intestine, respectively; xyz is absorbed and xyz is secreted etc. for the other parts of gastrointestinal tract. The estimates for the endogenous N entering each compartment are minimal estimates, which is indicated by the letters A, B, and C (Marini et al., 2008)

drolyzed to ammonia by resident bacteria (Lapierre and Lobley, 2001). As is already known, urea transport across the ruminant wall is mediated via urea transporters in the epithelium membrane (Abdoun et al., 2006). These transporters allow the passage of urea across cell membranes, down a concentration gradient (Smith and Rousselet, 2001). Facilitative urea transporters are derived from the UT-A and UT-B genes (Bankir et al., 2004). UT-B mRNA or protein expressions have been characterized in the rumen epithelium (Stewart et al., 2005; Simmons et al., 2009; Lu et al., 2015). In the study of Coyle et al. (2016), UT-B transporters were identified to be specifically localized to certain regions of tissue in the bovine gastrointestinal tract. UT-B2 was the predominant UT-B mRNA transcript expressed in dorsal, ventral and cranial ruminal sacs, while alternative UT-B transcripts were present in other gastrointestinal tissues (Figure 2).

In addition to the UT-B transporters, some alternative transport mechanisms are also involved in urea transport across the epithelium. The aquaporins (AQP) are a family of mem-

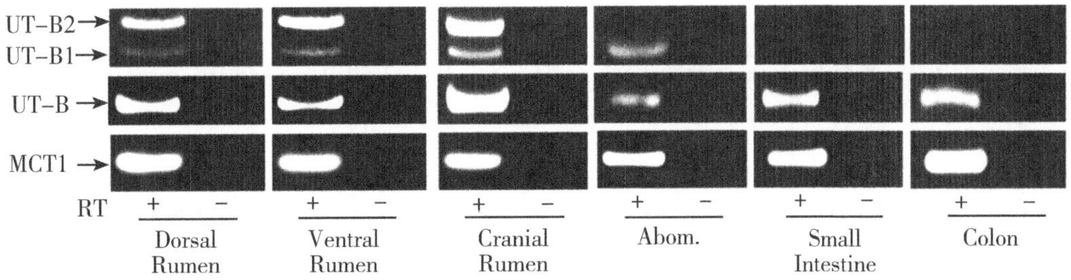

Figure 2 RT-PCR experiments investigating cDNA derived from total RNA samples from bovine gastrointestinal tissues. Analysis with BSF/BODR UT-B primers confirmed strong UT-B2 (900 bp) and weak UT-B1 (750 bp) expression in the dorsal, cranial and ventral rumen. Using these isoform-specific BSF/BODR primers, only UT-B1 was detected in the abomasum and no signals at all in either the small intestine or colon. In contrast, using BODF/BODR primers, general UT-B transcripts were detected in all six tissues tested. In addition, strong MCT1 signals were also detected in ruminal sac, abomasum, small intestine, and colon samples. MCT 1, monocarboxylate transporter 1; RT, reverse transcriptase; abom., abomasum (Coyle et al., 2016)

brane-spanning proteins predominantly involved in water movement, and some of them also play a role in urea movement. AQP-3, -7, -9 and -10 have been proven to be involved in urea uptake or transport, while AQP-8 is permeable to ammonia (Rojek et al., 2008; Litman et al., 2009). Rojen et al. (2011) showed that messenger RNA expression of AQP3, AQP7, and AQP10 and the abundance of AQP8 increased with increasing nitrogen intake, but their findings do not point to these proteins as the cause of increased rumen epithelial urea permeability in dairy cows fed a low N diet. Walpole et al. (2015) investigated the roles of UT-B and AQP in the serosal-to-mucosal urea flux across rumen epithelium using Ussing chambers. The urea flux markedly decreased when Phloretin and $NiCl_2$ were added to inhibit UT-B- or AQP-mediated urea transport, respectively, which proved that both AQP and UT-B play significant functional roles in urea transport. Gene transcript abundance for UT-B and AQP was observed to be significantly correlated with the ruminal serosal to mucosal urea fluxes. However, the mechanism by which the increased gene expression occurred is unclear. Transcriptome analysis has been used to analyze the rumen epithelium metabolic pathway changes under various conditions (Baldwin et al., 2012; Dionissopoulos et al., 2014; Naeem et al., 2014), and this approach may provide a better means to understand the regulation of these urea transport mechanisms across the rumen wall.

Rumen Ureolytic Bacteria

Rumen ureolytic bacteria play an important role in dietary urea hydrolysis, for they produce ureases that catalyze the breakdown of urea to ammonia (NH_3) and carbon dioxide (Owens et al., 1980). In the rumen, the ammonia can be assimilated by many rumen bacteria for the synthesis of microbial proteins (Owens et al., 1980; Milton et al., 1997). However, the efficiency of urea N utilization in ruminants is low and this is attributed to the rapid hydrolysis of urea to NH_3, which occurs at a higher rate than NH_3 utilization by rumen bacteria (Patra, 2015). Due to the difficulty in cultivating rumen bacteria, only a small number of bacteria have been isolated (Kim et al., 2011). The lack of sufficient understanding of the ruminal microbiome is one of the major knowledge gaps that hinder effective enhancement of rumen functions (Firkins and Yu, 2006). In addition, limited information about rumen urea-degrading bacteria makes regulation of the urea hydrolysis rate by targeting predominant ureolytic bacteria difficult.

Ureolytic bacteria isolated using culture-dependent methods

Early studies have isolated some ureolytic bacteria from the rumen (Cook, 1976; On et al., 1998). Wozny et al. (1977) described a rapid qualitative procedure to detect urease in strains isolated from the bovine rumen, and found that many species including *Succinivibrio dextrinosolvens*, *Treponema* sp., *Ruminococcus bromii*, *Butyrivibrio* sp., *Bifidobacterium* sp., *Bacteroides ruminicola*, and *Peptostreptococcus productus* had urease activity and most *P. productus* strains contain urease. Kakimoto et al. (1989) assayed about 16, 000 isolates from animal feces and intestines for the production of acid urease and found that most of the selected strains belonged to the genera *Streptococcus* and *Lactobacillus*. In a similar study by Lauková and Koniarová (1994), 909 strains from the rumen of 104 domestic and wild ruminants were tested for urease activity, and their results showed that some *Selenomonas ruminantium* strains and *Lactobacilli* manifested medium urease activity and most of the *Enterococcus faecium* and all of the *E. faecalis* isolates expressed urease activity. In addition, *Howardella ureilytica*, a Gram-positive bacterium that has been isolated from the rumen fluid of sheep; was found to be strongly ureolytic and generated ATP through the hydrolysis of urea (Cook et al., 2007). All these above studies were conducted using culture-based methods; however, due to the difficulty in cultivating rumen bacteria, those that have been isolated represent only 6.5% of the community (Kim et al., 2011), and, therefore, only very limited information is known about rumen ureolytic bacteria. These previous studies exploring ureolytic bacteria only identified the urease activity of isolated bacteria and did not consider the information about the urease genes that express the urease. With the help of modern molecular technologies, we can acquire more information of the ureolytic bacteria at the DNA level.

Culture-independent methods of studying ureolytic bacteria

In order to get further information about the function of rumen microbes, sequencing and phylogenetic analysis of 16S rRNA and functional geneshave been extensively carried out in studies focused on members of uncultivable bacteria (Chaucheyras-Durand and Ossa, 2014). For ureolytic bacteria, the *ureC* gene encodes the largest urease functional subunit and contains several highly conserved regions that are suitable as PCR priming sites (Mobley et al., 1995). Previously, Reed (2001) successfully designed urease PCR primers that can amplify a 340 bp fragment of the *ureC* gene from a variety of urease producing bacteria. Primers for *ureC* genes have been developed and applied to the analysis of urea-degrading microorganisms in various environments (Collier et al., 2009; Singh et al., 2009; Su et al., 2013).

Because rumen ureolytic bacteria are the key organisms that produce urease for the breakdown of urea, further insights into the abundant ureolytic bacteria or urease functional genes could provide the basis for designing strategies to efficiently manipulate the rumen bacteria and improve urea utilization in ruminants. Zhao et al. (2015) attempted to examine rumen ureolytic bacterial diversity by cloning and sequencing *ureC* genes and found that among the total 317 *ureC* sequences from the rumen digesta, some were about 84% identical (based on amino acid sequence) to the *ureC* gene of *Helicobacter pylori*. They also developed a vaccine based on *ureC* of *H. pylori*, vaccinated cows had significantly reduced urease activity in the rumen compared to control cows that were mock immunized. Therefore, a vaccine based on *ureC* of *H. pylori* could be a useful approach to decrease bacterial ureolysis in the rumen. A vaccine prepared from a combination of representatives of different rumen *ureC* clusters may be more effective than *ureC* of *H. pylori* or a single rumen bacterial *ureC*.

In order to get moreaccurate information about the rumen ureolytic bacteria, Jin et al. (2016) investigated abundant ureolytic bacterial communities by high-throughput sequencing when treated with an activator (urea) or inhibitor (acetohydroxamic acid, AHA) of ureolytic bacteria *in vitro*. Results from 16S rRNA gene sequencing showed that rumen ureolytic bacteria were abundant in the genera of *Pseudomonas*, *Haemophilus*, *Neisseria*, *Streptococcus*, *Actinomyces*, *Bacillus*, and unclassified Succinivibrionaceae. Recently, Jin et al. (2017) studied the differences in ureolytic bacterial composition between the rumen digesta and rumen wall based on *ureC* gene classification, and found that more than 55% of the *ureC* sequences did not affiliate with any known taxonomically assigned urease genes, and the most abundant *ureC* genes were affiliated with the families of Methylococcaceae, Clostridiaceae, Paenibacillaceae, Helicobacteraceae, and Methylophilaceae (Figure 3).

Studies which target the*ureC* genes provide a basis for obtaining the full-length urease functional gene information (Yuan et al., 2012). This survey has expanded our knowledge of information relating to the predominant *ureC* gene in the rumen ureolytic microbial community, and provides a basis for obtaining vaccine targets of urease for regulating rumen bacterial urease activities, and moderating urea hydrolysis and utilization in the rumen.

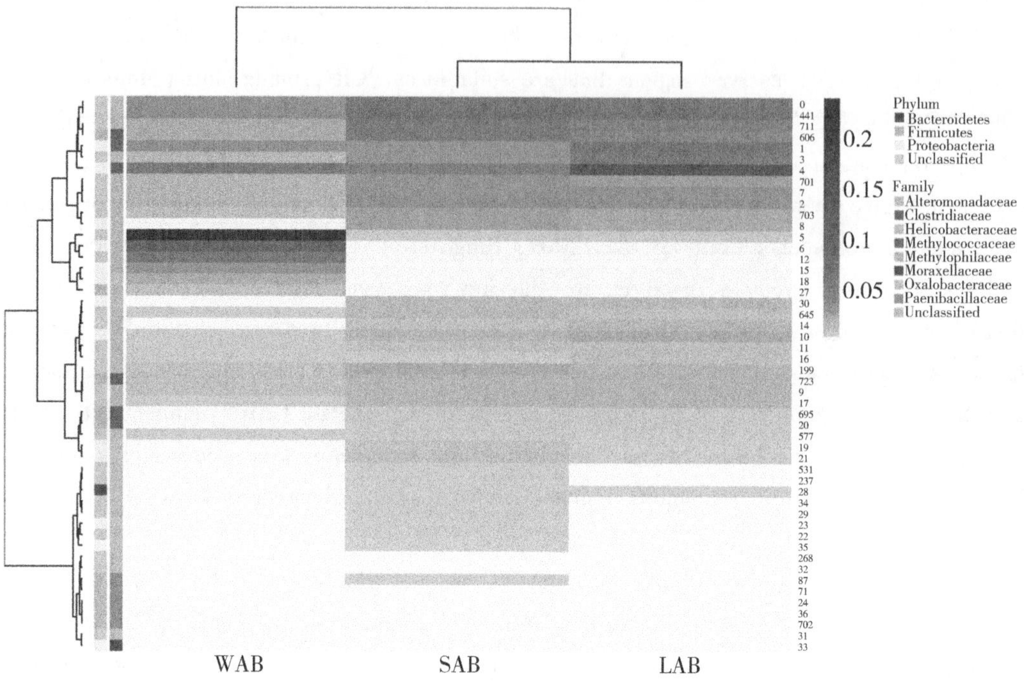

Figure 3 Rumen *ureC* gene community heatmaps and clustering of the most abundant 50 operational taxonomic units (OTUs) from different rumen fractions. Ward' s minimum variance method was used for hierarchical clustering of the computed distance matrix for samples based on the Jaccard dissimilarity indices of the OTU data in the vegan package. LAB, liquid-associated bacteria; SAB, solid-adherent bacteria; WAB, wall-adherent bacteria (Jin et al., 2017)

Bacterial Urease

Characterization and activation of bacterial ureases

Ureases (urea amidohydrolases, EC 3.5.1.5) are nickel-dependent enzymes, found in plants, fungi, and bacteria, which are commonly composed of two or three subunits (encoded by genes *ureA*, *ureB*, and *ureC*), and require up to several accessory proteins for activation (Mobley et al., 1995). For example, the urease of *Klebsiella aerogenes* has three subunits $(UreABC)_3$ (Jabri et al., 1995). The urease of *H. pylori* consists of two subunits $((ureAB)_3)_4$, and *ureB* in the *Helicobacter* species is equivalent to *ureC* in the organisms possessing a three-subunit enzyme (Hu and Mobley, 1990). Urease accessory genes (such as *ureD*, *ureE*, *ureF*, *ureG*, *ureH*, and *ureI*) are required for synthesis of catalytically active urease when the gene clusters are expressed in a recombinant bacterial host. Some of the accessory genes have been shown to play a role in the activation of the apoenzyme, and these genes are known to be required for assembly of the nickel metallocenter within the active site of the enzyme (Mehta et al., 2003; Witte et al., 2005; Boer and Hausinger, 2012). Taking the urease activation of *K. aerogenes* as an example, the UreD, UreF, UreG, and UreE are sequentially complexed to UreABC as required for its activation (Farrugia et al., 2013).

Urease inhibitors are targeted to the functional area of active urease. Therefore, investigation of the bacterial urease structure and activation of urease are important for finding the binding sites between urease inhibitor and urease, and for the regulation of the activation process of urease. Some studies have been done to explore the structures for the activation complex of urease (Biagi et al., 2013; Fong et al., 2013). Ligabue-Braun et al. (2013) provided an atomic-level model for the $(UreABC - UreDFG)_3$ complex from *K. aerogenes* by employing comparative modeling associated to sequential macromolecular dockings, validated through small-angle X-ray scattering profiles. The resulting model included a putative orientation for UreG at the $(UreABC - UreDFG)_3$ oligomer. Fong et al. (2013) proposed a mechanism on how urease accessory proteins facilitate the maturation of urease. They reported the crystal structure of the UreG/UreF/UreH complex in *H. pylori*, which illustrates how UreF and UreH facilitate dimerization of UreG and assembles its metal binding site by juxtaposing two invariant Cys66-Pro67-His68 metal binding motifs at the interface to form the $(UreG/UreF/UreH)_2$ complex. Furthermore, Zambelli et al. (2014) identified the nickel binding properties of *H. pylori* UreF in the nickel-based activation of urease (Figure 4). UreF binds two Ni^{2+} ions per dimer, with a micromolar dissociation constant. Two nearly identical and symmetric tunnels were found, going from the central cavity in the UreG/UreF/UreH complex, and UreF was involved in the metal ion transport through these tunnels during urease activation. Currently, many aspects of the urease metallocenter assembly still remain obscure. The activation mechanism and roles of each accessory protein in urease maturation still need to be answered.

Regulation of bacterial urease synthesis

The regulation of urease synthesis in ureolytic bacteria is complex. In some organisms such as *Bacillus pasteurii*, and *Morganella morganii* isolated from soil, urease synthesis is constitutive (Mörsdorf and Kaltwasser, 1989; Burbank et al., 2012). However, urease synthesis in some bacteria is regulated by environmental conditions, such as the concentration of urea and nitrogen or pH (Weeks and Sachs, 2001; Dyhrman and Anderson, 2003; Belzer et al., 2005; Liu et al., 2008). Urease activity of *Providencia stuartii*, for example, is induced by the presence of urea (Armbruster et al., 2014), while *Klebsiella pneumoniae*, a facultative anaerobic organism, can use urea as the sole source of nitrogen, and the urease expression is regulated by the supply of nitrogen in the growth medium (Liu and Bender,

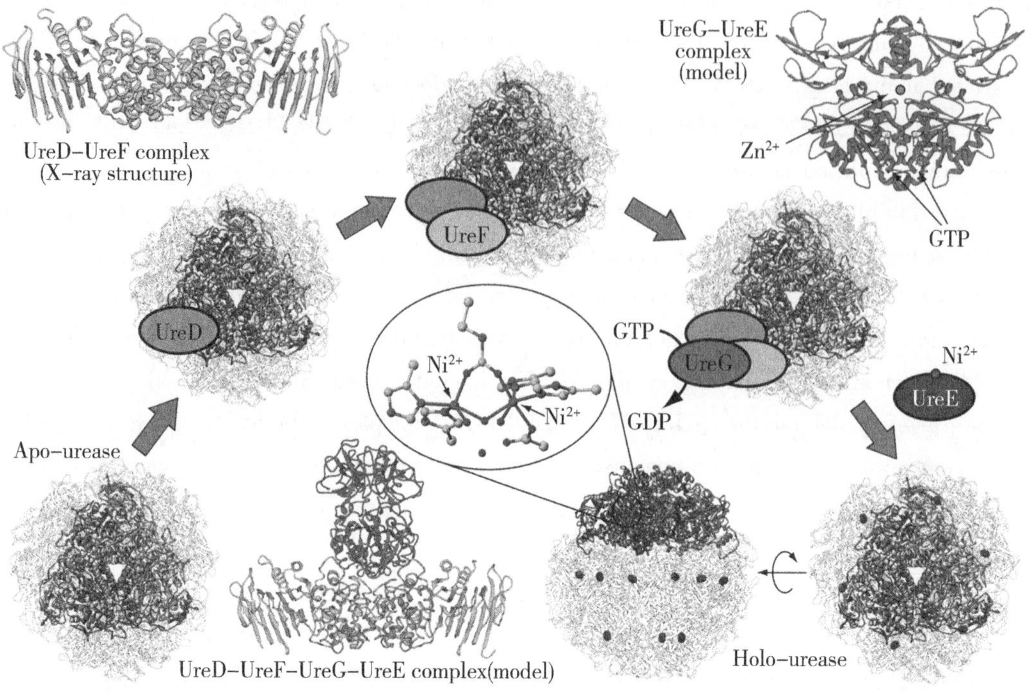

Figure 4 The *helicobacter pylori* urease activation process starting from the apoenzyme and leading to holo–urease. The ribbon diagrams show the structure of *H. pylori* urease in its $[(ab)_3]_4$ quaternary structure; each blue chain, gold chain, and green chain represents one (ab) heterodimer, and together they reveal the similarity of the $(ab)_3$ moiety in this urease with the $(abc)_3$ quaternary structure of other bacterial ureases, such as those of *sporosarcina pasteurii* and *klebsiella aerogenes*. the details of the coordination environment of the Ni^{2+} ions in the active site are shown in the central inset. the crystal structures or models of the various protein complexes involved in the process are also shown as ribbon diagrams: *H. pylori* UreD (HpUreD) in light green, *H. pylori* UreF (HpUreF) in orange, *H. pylori* UreG (HpUreG) in red, and *H. pylori* UreE in dark green. GDP guanosine 5′–diphosphate, GTP guanosine 5′–triphosphate (Zambelli et al., 2014)

2007). The regulation of urease gene expression of *Actinomyces naeslundii* under different environmental conditions has been investigated by Liu et al. (2008). *A. naeslundii* is considered anaerobic or microaerophilic, the conditions of neutral pH, fast dilution rate, increased carbohydrate supply or low nitrogen supply in the medium all resulted in the enhancement of urease activity in *A. naeslundii*.

Helicobacter are Gram–negative, microaerophilic bacteria. In research comparing the regulation of urease activity in *Helicobacter hepaticus* and *H. pylori*, the urease activity of *H. hepaticus* was found to be acid–independent, which contrasts with the acid–induced urease system of *H. pylori* (Belzer et al., 2005).

When the model rumen Firmicutes organism *Ruminococcus albus* 8 were supplied with different nitrogen sources (urea, ammonia, and peptides), the urease activity was higher in the presence of urea than in the presence of ammonia and peptides (Kim et al., 2014). However, urease transcript abundance in *R. albus* 8 is not predicated by the presence of urea in the medium. This urease activity may demonstrate that *R. albus* 8 expresses urease to acquire urea as an alternative nitrogen source when the ammonia concentration in the medium is limited. Because the regulation of urease activity is complex and the rumen harbors a large diversity of ureolytic bacteria, the mechanisms controlling urease synthesis in the complicated rumen environment need further research.

Improved urea utilization in ruminants with urease inhibitors

In ruminants, reducing the rate of urea hydrolysis in the rumen is of great importance for improving urea utilization and minimizing ammonia wastage. Urease inhibitors are one available option found to be effective in control urea hydrolysis. Several urease inhibitors, including AHA (Brent et al., 1971; Jones and Milligan, 1975), phenylphosphorodiamidate (Voigt et al., 1980a; Voigt et al., 1980b; Whitelaw et al., 1991), and N- (n-butyl) thiophosphoric triamide (NBPT) have been investigated (Ludden et al., 2000). Zhang et al. (2001) also studied the effect of hydroquinone on ruminal urease activity and found that concentrations of 0.01 ppm to 10 ppm inhibited urease activity of intact rumen microbes *in vitro* by 25% to 64%. Urease inhibitors also provide an insight into understanding the mechanism of enzyme catalytic activity present at the active site of the enzyme and the importance of nickel to urease, the metalloenzyme (Upadhyay, 2012).

The mechanism of *B. pasteurii* urease inhibition with acetohydroxamic acid has been resolved. The inhibitor anion symmetrically bridges the two Ni ions in the active site through the hydroxamate oxygen and chelating one Ni ion through the carbonyl oxygen (Benini et al., 2000). Although recent studies have already evaluated the function of different urease inhibitors in improving urea utilization efficiency (Ludden et al., 2000; Giallongo et al., 2015), further research is needed to investigate the response of the rumen bacteria community, especially ureolytic bacteria, to these inhibitors.

Summary

Urea is one of the major non-protein nitrogen feeds for ruminants and the optimal utilization of urea in feed can alleviate to some extent the cost of dietary protein. Urea is hydrolyzed quickly by ureolytic bacteria in the rumen. Because about 90% of rumen microbes have not been pure-cultured to date, only limited information about active ureolytic bacteria communities is known, which limits the regulation and efficient application of urea in ruminant production. Increased knowledge about ureolytic microbiomes will permit the development of mitigation strategies, such as urease inhibitors and vaccines, to target the dominant ureolytic bacteria species or urease successfully. There are breakthroughs in molecular strategies, the rapid advancement of "~omics" technologies, including metagenomics, metatranscriptomics, metabolomics, and bioinformatics could give a better understanding of the microbial and molecular mechanisms of ruminal urea hydrolyzation and utilization, and will provide knowledge for manipulating urea utilization efficiency in ruminants.

References

Abdoun, K., Stumpff, F., Martens, H., 2006. Ammonia and urea transport across the rumen epithelium: a review [J]. Anim. Health Res. Rev., 7: 43-59.

Abdoun, K., Stumpff, F., Rabbani, I., et al., 2010. Modulation of urea transport across sheep rumen epithelium *in vitro* by SCFA and CO_2 [J]. American Journal of Physiology-Gastrointestinal and Liver Physiology, 298: G190-G202.

Armbruster, C. E., Smith, S. N., Yep, A., et al., 2014. Increased incidence of urolithiasis and bacteremia during Proteus mirabilis and Providencia stuartii coinfection due to synergistic induction of urease activity [J]. J. Infect Dis., 209: 1524-1532.

Balcells, J., Guada, J., Castrillo, C., et al., 1993. Rumen digestion and urinary excretion of purine derivatives in response to urea supplementation of sodium-treated straw fed to sheep [J]. Brit. J. Nutr., 69: 721-732.

Baldwin, R. L. t., Wu, S., Li, W., et al., 2012. Quantification of transcriptome responses of the rumen epithelium to butyrate infusion using RNA-seq technology [J]. Gene Regul. Syst. Bio., 6: 67-80.

Bankir, L., Chen, K., Yang, B., 2004. Lack of UT-B in vasa recta and red blood cells prevents urea-induced improvement of urinary concentrating ability [J]. Am. J. Physiol-Renal., 286: F144-F151.

Batista, E. D., Detmann, E., Valadares Filho, S. C., et al., 2017. The effect of CP concentration in the diet on urea kinetics and microbial usage of recycled urea in cattle: a meta-analysis [J]. Animal, 11 (8): 1303-1311.

Belzer, C., Stoof, J., Beckwith, C. S., et al., 2005. Differential regulation of urease activity in *Helicobacter hepaticus* and *Helicobacter pylori* [J]. Microbiology, 151: 3989-3995.

Benini, S., Rypniewski, W. R., Wilson, K. S., et al., 2000. The complex of Bacillus pasteurii urease with acetohydroxamate anion from X-ray data at 1. 55 Å resolution [J]. J. Biol. Inorg. Chem., 5: 110-118.

Biagi, F., Musiani, F., Ciurli, S., 2013. Structure of the UreD-UreF-UreG-UreE complex in Helicobacter pylori: a model study [J]. J. Biol. Inorg. Chem., 18: 571-577.

Boer, J. L., Hausinger, R. P., 2012. Klebsiella aerogenes UreF: identification of the UreG binding site and role in enhancing the fidelity of urease activation [J]. Biochemistry, 51: 2298-2308.

Brent, B., Adepoju, A., Portela, F., 1971. Inhibition of rumen urease with acetohydroxamic acid [J]. J. Anim. Sci., 32: 794-798.

Burbank, M. B., Weaver, T. J., Williams, B. C., et al., 2012. Urease activity of ureolytic bacteria isolated from six soils in which calcite was precipitated by indigenous bacteria [J]. Geomicrobiol. J., 29: 389-395.

Chaucheyras-Durand, F., Ossa, F., 2014. Review: The rumen microbiome: Composition, abundance, diversity, and new investigative tools [J]. The Professional Animal Scientist, 30: 1-12.

Collier, J. L., Baker, K. M., Bell, S. L., 2009. Diversity of urea - degrading microorganisms in open-ocean and estuarine planktonic communities [J]. Environ. Microbiol., 11: 3118-3131.

Cook, A., 1976. Urease activity in the rumen of sheep and the isolation of ureolytic bacteria [J]. J. Gen. Microbiol., 92: 32-48.

Cook, A. R., Riley, P. W., Murdoch, H., et al., 2007. Howardella ureilytica gen. nov., sp. nov., a Gram-positive, coccoid-shaped bacterium from a sheep rumen [J]. Int. J. Syst. Evol. Microbiol., 57: 2940-2945.

Coyle, J., McDaid, S., Walpole, C., et al., 2016. UT-B Urea Transporter Localization in the Bovine Gastrointestinal Tract [J]. J. Membr. Biol., 249: 77-85.

De Visser, H., Valk, H., Klop, A., et al., 1997. Nutrient fluxes in splanchnic tissue of dairy cows: Influence of grass quality [J]. J. Dairy Sci., 80: 1666-1673.

Dionissopoulos, L., AlZahal, O., Steele, M. A., et al., 2014. Transcriptomic changes in ruminal tissue induced by the periparturient transition in dairy cows [J]. American Journal of Animal and Veterinary Sciences, 9: 36.

Dyhrman, S. T., Anderson, D. M., 2003. Urease activity in cultures and field populations of the toxic dinoflagellate Alexandrium [J]. Limnol. Oceanogr., 48: 647-655.

Erb, R., Brown, C., Callahan, C., et al., 1976. Dietary urea for dairy cattle. II. Effect on functional traits [J]. J. Dairy Sci., 59: 656-667.

Farrugia, M. A., Macomber, L., Hausinger, R. P., 2013. Biosynthesis of the urease metallocenter [J]. J. Biol. Chem., 288: 13178-13185.

Fong, Y. H., Wong, H. C., Yuen, M. H., et al., 2013. Structure of UreG/UreF/UreH complex reveals how urease accessory proteins facilitate maturation of Helicobacter pylori urease [J]. PLoS Biol., 11: e1001678.

Giallongo, F., Hristov, A. N., Oh, J., et al., 2015. Effects of slow-release urea and rumen-protected methionine and histidine on performance of dairy cows [J]. J. Dairy Sci., 98: 3292-3308.

Harmeyer, J., Martens, H., 1980. Aspects of urea metabolism in ruminants with reference to the goat [J]. J. Dairy Sci., 63: 1707-1728.

Holder, V. B., Tricarico, J. M., Kim, D. H., et al., 2015. The effects of degradable nitrogen level and slow release urea on nitrogen balance and urea kinetics in Holstein steers [J]. Anim. Feed Sci. Tech., 200: 57-65.

Hu, L., Mobley, H., 1990. Purification and N-terminal analysis of urease from *Helicobacter pylori* [J]. Infect. Immun., 58: 992-998.

Huntington, G., Archibeque, S., 2000. Practical aspects of urea and ammonia metabolism in ruminants [J]. J. Anim. Sci., 77: 1-11.

Imaizumi, H., Batistel, F., de Souza, J., et al., 2015. Replacing soybean meal for wet brewer's grains or urea on the performance of lactating dairy cows [J]. Trop. Anim. Health Prod., 47: 877-882.

Jabri, E., Carr, M. B., Hausinger, R. P., et al., 1995. The crystal structure of urease from Klebsiella aerogenes [J]. Science, 268: 998.

Jin, D., Zhao, S., Wang, P., et al., 2016. Insights into Abundant Rumen Ureolytic Bacterial Community Using Rumen Simulation System [J]. Front. Microbiol., 7: 1006.

Jin, D., Zhao, S., Zheng, N., et al., 2017. Differences in Ureolytic Bacterial Composition between the Rumen Digesta and Rumen Wall Based on ureC Gene Classification [J]. Front. Microbiol., 8: 385.

Jones, G., Milligan, J., 1975. Influence on some rumen and blood parameters of feeding acetohydroxamic acid in a urea-containing ration for lambs [J]. Can. J. Anim. Sci., 55: 39-47.

Kakimoto, S., Okazaki, K., Sakane, T., et al., 1989. Isolation and taxonomie characterization of acid urease-producing bacteria [J]. Agric. Biol. Chem., 53: 1111-1117.

Kertz, A. F., 2010. Review: urea feeding to dairy cattle: a historical perspective and review [J]. Prof. Anim. Sci., 26: 257-272.

Kertz, A., Davidson, L., Cords, B., et al., 1983. Ruminal infusion of ammonium chloride in lactating cows to determine effect of pH on ammonia trapping [J]. J. Dairy Sci., 66: 2597-2601.

Kim, J. N., Henriksen, E. D., Cann, I. K., et al., 2014. Nitrogen utilization and metabolism in *Ruminococcus albus* 8 [J]. Appl. Environ. Microb., 80: 3095-3102.

Kim, M., Morrison, M., Yu, Z., 2011. Status of the phylogenetic diversity census of ruminal microbiomes [J]. FEMS Microbiol. Ecol., 76: 49-63.

Kohn, R., Dinneen, M., Russek-Cohen, E., 2005. Using blood urea nitrogen to predict nitrogen excretion and efficiency of nitrogen utilization in cattle, sheep, goats, horses, pigs, and rats [J]. J. Anim. Sci., 83: 879-889.

Lapierre, H., Lobley, G., 2001. Nitrogen recycling in the ruminant: A review [J]. J. Dairy Sci., 84: E223-E236.

Lauková, A., Koniarová, I., 1994. Survey of urease activity in ruminal bacteria isolated from domestic and wild ruminants [J]. Microbios, 84: 7-11.

Law, R. A., Young, F. J., Patterson, D. C., et al., 2009. Effect of dietary protein content on animal production and blood metabolites of dairy cows during lactation [J]. J. Dairy Sci., 92: 1001-1012.

Ligabue-Braun, R., Real-Guerra, R., Carlini, C. R., et al., 2013. Evidence-based docking of the urease activation complex [J]. J. Biomol. Struct. Dyn., 31: 854-861.

Litman, T., Søgaard, R., Zeuthen, T., 2009. Ammonia and urea permeability of mammalian aquaporins [M]. In Aquaporins. Springer, 327-358.

Liu, Q., Bender, R. A., 2007. Complex regulation of urease formation from the two promoters of the ure operon of *Klebsiella pneumoniae* [J]. J. Bacteriol., 189: 7593-7599.

Liu, Y., Hu, T., Jiang, D., et al., 2008. Regulation of urease gene of Actinomyces naeslundii in biofilms in response to environmental factors [J]. FEMS Microbiol. Lett., 278: 157-163.

Lu, Z., Gui, H., Yao, L., et al., 2015. Short-chain fatty acids and acidic pH upregulate UT-B, GPR41, and GPR4 in rumen epithelial cells of goats [J]. American Journal of Physiology-Regulatory, Integrative and Comparative Physiology, 308: R283-R293.

Lu, Z., Stumpff, F., Deiner, C., et al., 2014. Modulation of sheep ruminal urea transport by ammonia and pH [J]. American Journal of Physiology-Regulatory, Integrative and Comparative Physiology, 307: R558-R570.

Ludden, P., Harmon, D., Huntington, G., et al., 2000. Influence of the novel urease inhibitor N-(n-butyl) thiophosphoric triamide on ruminant nitrogen metabolism: Ⅱ. Ruminal nitrogen metabolism, diet digestibility, and nitrogen balance in lambs [J]. J. Anim. Sci., 78: 188-198.

Marini, J. C., Fox, D. G., Murphy, M. R., 2008. Nitrogen transactions along the gastrointestinal tract of cattle: A meta-analytical approach [J]. J. Anim. Sci., 86: 660-679.

Mehta, N., Olson, J. W., Maier, R. J., 2003. Characterization of Helicobacter pylori nickel metabolism accessory proteins needed for maturation of both urease and hydrogenase [J]. J. Bacteriol., 185: 726-734.

Milton, C., Brandt, J, R., Titgemeyer, E., 1997. Urea in dry-rolled corn diets: finishing steer performance, nutrient digestion, and microbial protein production [J]. J. Anim. Sci., 75: 1415-1424.

Mobley, H., Island, M. D., Hausinger, R. P., 1995. Molecular biology of microbial ureases [J]. Microbiol. Rev., 59: 451-480.

Mörsdorf, G., Kaltwasser, H., 1989. Ammonium assimilation in *Proteus vulgaris*, *Bacillus pasteurii*, and *Sporosarcina ureae* [J]. Arch. Microbiol., 152: 125-131.

Naeem, A., Drackley, J. K., Lanier, J. S., et al., 2014. Ruminal epithelium transcriptome dynamics in response to plane of nutrition and age in young Holstein calves [J]. Funct. Integr. Genomics, 14: 261-273.

On, S., Atabay, H., Corry, J., et al., 1998. Emended description of Campylobacter sputorum and revision of its infrasubspecific (biovar) divisions, including C. sputorum biovar paraureolyticus, a urease-producing variant from cattle and humans [J]. Int. J. Syst. Bacteriol., 48: 195-206.

Owens, F. N., Lusby, K. S., Mizwicki, K., et al., 1980. Slow ammonia release from urea: rumen and metabolism studies [J]. J. Anim. Sci., 50: 527-531.

Patra, A. K., 2015. Urea/Ammonia Metabolism in the Rumen and Toxicity in Ruminants [M]. In Rumen Microbiology: From Evolution to Revolution, eds. A. K. Puniya, S. R. & K. D. N. : Springer, 329-341.

Pisulewski, P. M., Okorie, A. U., Buttery, P. J., et al., 1981. Ammonia concentration and protein synthesis in the rumen [J]. J. Sci. Food Agr., 32: 759-766.

Polan, C., Miller, C., McGilliard, M., 1976. Variable dietary protein and urea for intake and production in Holstein cows. J. Dairy Sci., 59: 1910-1914.

Puppel, K., Kuczynska, B., 2016. Metabolic profiles of cow's blood—a review [J].J. Sci. Food Agric., 96: 4321-4328.

Reed, K. E., 2001. Restriction enzyme mapping of bacterial urease genes: using degenerate primers to expand experimental outcomes [J]. Biochem. Mol. Biol. Edu., 29: 239-244.

Reynolds, C. K., Kristensen, N. B., 2008. Nitrogen recycling through the gut and the nitrogen economy of ruminants: an asynchronous symbiosis [J]. J. Anim. Sci., 86: E293-305.

Rojek, A., Praetorius, J., Frokiaer, J., et al., 2008. A current view of the mammalian aquaglyceroporins [J]. Annu. Rev. Physiol., 70: 301-327.

Rojen, B. A., Poulsen, S. B., Theil, P. K., et al., 2011. Short communication: Effects of dietary nitrogen concentration on messenger RNA expression and protein abundance of urea transporter-B and aquaporins in ruminal papillae from lactating Holstein cows [J]. J. Dairy Sci., 94: 2587-2591.

Ryder, W., Hillman, D., Huber, J., 1972. Effect of feeding urea on reproductive efficiency in Michigan Dairy Herd Improvement Association herds [J]. J. Dairy Sci., 55: 1290-1294.

Simmons, N., Chaudhry, A., Graham, C., et al., 2009. Dietary regulation of ruminal bovine UT-B urea transporter expression and localization [J]. J. Anim. Sci., 87: 3288.

Sinclair, L. A., Blake, C. W., Griffin, P., et al., 2012. The partial replacement of soyabean meal and rapeseed meal with feed grade urea or a slow-release urea and its effect on the performance, metabolism and digestibility in dairy cows [J]. Animal, 6: 920-927.

Singh, B. K., Nunan, N., Millard, P., 2009. Response of fungal, bacterial and ureolytic communities to synthetic sheep urine deposition in a grassland soil [J]. FEMS Microbiol. Ecol., 70: 109-117.

Smith, C., Rousselet, G., 2001. Facilitative urea transporters [J]. J. Membrane Biol., 183: 1-14.

Stewart, G. S., Smith, C. P., 2005. Urea nitrogen salvage mechanisms and their relevance to ruminants, non-ruminants and man [J]. Nutr. Res. Rev., 18: 49-62.

Stewart, G., Graham, C., Cattell, S., et al., 2005. UT-B is expressed in bovine rumen: potential role in ruminal urea transport [J]. Am. J. Physiol- Reg. I., 289: R605-R612.

Su, J., Jin, L., Jiang, Q., et al., 2013. Phylogenetically diverse ure C genes and their expression suggest the urea utilization by bacterial symbionts in marine sponge Xestospongia testudinaria [J]. PLoS One, 8: e64848.

Symonds, H., Mather, D. L., Collis, K., 1981. The maximum capacity of the liver of the adult dairy cow to metabolize ammonia [J]. Brit. J. Nutr., 46: 481-486.

Upadhyay, L. S. B., 2012. Urease inhibitors: A review [J]. Indian J. Biotechnol., 11: 381-388.

Voigt, J., Krawielitzki, R., Piatkowski, B., 1980a. Studies on the effect of phosphoric phenyl ester diamide as inhibitor of rumen urease in dairy cows.3. Digestibility of the nutrients and bacterial protein synthesis [J]. Arch. Tierernahr., 30: 835-840.

Voigt, J., Piatkowski, B., Bock, J., 1980b. Studies on the effect of phosphoric phenyl ester diamide as inhibitor of the rumen urease of dairy cows. 1. Influence on urea hydrolysis, ammonia release and fermentation in the rumen [J]. Arch. Tierernahr., 30: 811–823.

Walpole, M. E., Schurmann, B. L., Gorka, P., et al., 2015. Serosal–to–mucosal urea flux across the isolated ruminal epithelium is mediated via urea transporter–B and aquaporins when Holstein calves are abruptly changed to a moderately fermentable diet [J]. J. Dairy Sci., 98: 1204–1213.

Wanapat, M., Phesatcha, K., Kang, S., 2016. Rumen adaptation of swamp buffaloes (Bubalus bubalis) by high level of urea supplementation when fed on rice straw–based diet [J]. Trop. Anim. Health Prod., 48: 1135–1140.

Weeks, D. L., Sachs, G., 2001. Sites of pH regulation of the urea channel of *Helicobacter pylori* [J]. Mol. Microbiol., 40: 1249–1259.

Whitelaw, F. G., Milne, J. S., Wright, S. A., 1991. Urease (EC 3. 5. 1. 5) inhibition in the sheep rumen and its effect on urea and nitrogen metabolism [J]. Br. J. Nutr., 66: 209–225.

Wickersham, T., Titgemeyer, E., Cochran, R., et al., 2008. Effect of rumen–degradable intake protein supplementation on urea kinetics and microbial use of recycled urea in steers consuming low–quality forage [J]. J. Anim. Sci., 86: 3079–3088.

Wilson, G., Martz, F., Campbell, J., et al., 1975. Evaluation of factors responsible for reduced voluntary intake of urea diets for ruminants [J]. J. Anim. Sci., 41: 1431–1437.

Witte, C. P., Rosso, M. G., Romeis, T., 2005. Identification of three urease accessory proteins that are required for urease activation in Arabidopsis [J]. Plant Physiol., 139: 1155–1162.

Wozny, M., Bryant, M., Holdeman, L., et al., 1977. Urease assay and urease–producing species of anaerobes in the bovine rumen and human feces [J]. Appl. Environ. Microbiol., 33: 1097–1104.

Yuan, P., Meng, K., Wang, Y., et al., 2012. Abundance and genetic diversity of microbial polygalacturonase and pectate lyase in the sheep rumen ecosystem [J]. PLoS One, 7: e40940.

Zambelli, B., Berardi, A., Martin–Diaconescu, V., et al., 2014. Nickel binding properties of Helicobacter pylori UreF, an accessory protein in the nickel–based activation of urease [J]. J. Biol. Inorg. Chem., 19: 319–334.

Zhang, Y. G., Shan, A. S., Bao, J., 2001. Effect of Hydroquinone on Ruminal Urease in the Sheep and its Inhibition Kinetics *in vitro* [J]. Asian Australas. J. Anim. Sci., 14: 1216–1220.

Zhao, S., Wang, J., Zheng, N., et al., 2015. Reducing microbial ureolytic activity in the rumen by immunization against urease therein [J]. BMC Vet. Res., 11: 94.

Zhou, J. W., Guo, X. S., Degen, A. A., et al., 2015. Urea kinetics and nitrogen balance and requirements for maintenance in Tibetan sheep when fed oat hay [J]. Small Ruminant Res., 129: 60–68.

Rumen Microbiome—Challenges and Opportunities to Manipulate and Improve its Function

Zhongtang Yu
Department of Animal Sciences, The Ohio State University, Columbus, Ohio, USA

Introduction

Ruminants depend on the symbiotic ruminal microbiome for their living and productivity. The ruminal microbiome is a complex community of microorganisms that include bacteria of different guilds, archaea of mostly methanogens, ciliate protozoa, fungi, and viruses of mostly phages of bacteria and archaea (Hobson and Stewart, 1997). It is this ruminal microbiome that drives the digestion of indigestible feed ingredients, especially fiber, to energy and nutrients that ruminant animals can assimilate. It is also because of this ruminal microbiome that ruminants became and continued to be the most populous and competitive herbivores on the Earth. The rumen microbiome has a high density of microorganisms, with the bacteria alone reaching up to 10^{11} bacteria per milliliter of rumen content while other forms of ruminal microbes being much less abundance due to low metabolic versatility. Thanks to the traditional microbiological research in the past two hundred years, a large number of the members of the ruminal microbiome have been isolated and cultured, and their basic metabolism, physiology, and ecology are understood. Understandably, most of these cultured microbes are involved in degradation of fiber, starch, and proteins. This microbiology knowledge has greatly helped the research on ruminant nutrition and improvement of rations and feed utilization efficiency.

The application of DNA sequencing technologies has convinced rumen microbiologists that the cultured members of the ruminal microbiome only represent a tip of a large ruminal microbiome iceberg. It is now well documented that the majority of the ruminal microbes remains uncultured (Creevey et al., 2014; Kim et al., 2011), and their functions remain elusive and their importance to host nutrition is yet to be determined (Firkins and Yu, 2015). The new knowledge on the diversity, composition, population and community dynamics, and function of ruminal microbiome have provided new insight into the underpinning of feed digestion, fermentation, and host productivity. It is recognized that although each cow has an individualized ruminal microbiome, a core ruminal microbiome does exist (Petri et al., 2013). However, it remains to be determined if such a core microbiome exists at species or strain levels. It is also apparent that some of the ruminal microbes are associated with each other (Henderson et al., 2015). Such association and co-occurrence suggest mutual interactions, which may be targeted to indirectly manipulate the ruminal microbes of interest. Genes involved in feed digestion appeared to be organized into gene clusters, which are regulated in response to factor other than feed (Wang et al., 2013). Hence, fine tuning of the ruminal environment may be important to enhance feed digestion.

Being responsible for feed digestion, ruminal microbes, at least some of them, can have strong positive correlation with animal productivity. Such correlation has been demonstrated between milk fat yield and the ratio of *Firmicutes* to *Bacteroidetes* in one study (Jami et al., 2014). This can be explained by acetate, the precursor of *de novo* fatty acid synthesis, being the major fermentation product of most ruminal *Firmicutes*. However, such a correlation was not found in another study (Lima et al., 2015), reflecting the variability, complexity, and redundancy of ruminal microbiome. A recent study also demonstrated that microbiome-predicted milk production and the actual milk production had a strong linear relationship (Lima et al., 2015). As we accumulate more knowledge on the ruminal microbiome and develop new models, microbiome data may help predict milk production and quality in the future.

It has been the "Holy Grail" to manipulate ruminal function to achieve enhanced feed digestion, improved nitrogen utilization efficiency, decreased methane emission and nitrogen excretion, and lowered risk of rumen acidosis. Numerous studies have demonstrated that the ruminal microbiome is responsive to dietary interventions including change in diet (Pitta et al., 2010) and supplementation of feed additives (De Nardi et al., 2016; Kim et al., 2014). The responsive nature of the ruminal microbiome provides possibilities to manipulation the ruminal microbiome and its function.

Challenges and Opportunities to Manipulate Ruminal Functions

Fiber digestion is the bottleneck of feed digestion. It is a fascinating idea to enhance fiber degradation in the rumen by repeated inoculation of cellulolytic bacteria that were isolated from the rumen back to the rumen. However, a study using *Ruminococcus* isolates did not support the original hypothesis as repeated inoculation of these isolated bacteria failed to persistently increase total ruminococci or enhance fiber degradation (Krause et al., 2001). Similarly, repeated adding strains of *Megasphaera elsdenii*, a bacterial species known to utilize lactic acid, to the rumen did not have any positive results in dairy cattle fed a high-concentrate diet (Ye, unpublished data). Given the high redundancy, resilience, and host-specificity of the ruminal microbiome (Weimer, 2015), this is not surprising. Additionally, cultivation-independent metagenomic analysis of the ruminal microbiome rarely recovered functional genes that match those of cultured ruminal bacteria (Hess et al., 2011; Wang et al., 2013). It is speculative, but the cultured ruminal bacteria may not be the predominant or competitive ones, and thus they will not have much impact even they are repeatedly inoculated back into the rumen. This call for new research efforts to isolate new bacteria and determine if they are competitive in the rumen before they are used as inoculants.

S. cerevisiae yeast has been a successful direct-fed microbial (DFM) helping improve several important aspects of dairy cow nutrition, including increasing dry matter intake and digestibility, enhancing fiber digestion, increasing volatile fatty acid production, stabilizing ruminal pH, and eventually increasing milk yield (Desnoyers et al., 2009). Because *S. cerevisiae* cannot utilize starch but glucose and other hexoses while rumen has little free sugars, the currently available live yeast products are probably not metabolic activities inside the rumen. This is corroborated by the comparable benefits between live yeast and yeast fermentation products. It is conceivable that if yeast can be metabolically active in the rumen, more benefits may be realized. If yeast can be genetically modified so that it can digest fiber and release glucose for its own use, live yeast can further improve rumen function and animal productivity. A recent study demonstrated that yeast could be armed with activities of xylanase of rumen origin (Wang et al., 2015). Although this new yeast has limited application in cattle because xylan digestion is not rate-limiting in the rumen, it showed that this approach could be used to arm yeast with other enzymes of interest to further improve feed digestion and enhancing rumen function.

Methane is produced in the rumen as a useless product. Methane emission from cattle is believed to make a significant contribution to the anthropogenic emission of greenhouse gas (GHG) and drains 2%-12% gross feed energy (Johnson and Johnson, 1995). Intensive research efforts have been made to mitigate methane emission from dairy cattle with varying success (Hristov et al., 2013). However, none of the reported studies, however, showed asignificant increase in feed efficiency after methane emission was reduced. Additionally, GHG is emitted from the production of anti-methanogenic chemicals and plant materials. Furthermore, for most countries including China and the US, methane emitted from dairy cattle probably only accounts for a minute portion of the national GHG emission. Thus, future research is warranted to comprehensively evaluate the economic and environmental cost - benefit of mitigation of methane emission from cattle.

Dietary protein is the most expensive feed ingredient, and yet dairy cattle have very low nitrogen utilization efficiency due to the extensive proteolytic and subsequent amino acid-fermenting activities in the rumen (Kohn et al., 2005).

Dietary nitrogen utilization efficiency among the dairy herds in China is also lower than that in the US. To meet the increasing demands, large quantities of soybeans are importedinto China from North and South America. Large quantities of alfalfa are also imported. Such heavy dependence on import makes the dairy products expensive in China. Therefore, research to explore new strategies to improve nitrogen utilization efficiency is well justified and urgently needed. Ruminal protozoa engulf large amounts of the ruminal microbial cells and then degrade the microbial proteins and feed proteins to oligopeptides and free amino acids, both of which are fermented to form short chain fatty acids and ammonia. A small portion of the ammonia is incorporated into microbial protein, but the majority is excreted as urea or ammonia (Firkins et al., 2007). As such, ruminal protozoa, especially the most predominant ones (*Entodinium* spp.), are considered the major culprit of low nitrogen efficiency in dairy cattle. Indeed, complete removal of ruminal protozoa through defaunation improved nitrogen efficiency (Firkins et al., 2007). However, defaunation is not feasible or practical in dairy production. Various dietary interventions have been tried, but the results are mixed. The lack of specificity of these dietary interventions may be attributable to the inconsistent results. Being the only predator in the rumen, ciliate protozoa have some unique physiology features and lifestyle. They will have to lyse the engulfed bacteria cells before they can reach the microbial protein. Through a recent transcriptomic study of *Entodinium* caudatum (unpublished data), we recently identified several genes that may be targeted to specifically control the activities and growth of this dominant rumen protozoal group so that nitrogen utilization efficiency can be improved. Orange peel was also shown to be effective in inhibiting the growth of *Entodinium caudatum in vitro* and lower ammonia nitrogen production. Being readily digestible and containing essential oils and other biologically active compounds, orange peel may be a useful feed additive helping control ruminal protozoa and improve nitrogen utilization efficiency. It should be noted that because the production of dietary nitrogen, including plant-based nitrogen, generates GHG, improved nitrogen utilization efficiency and subsequently lowered nitrogen excretion can decrease the environmental footprint of dairy production.

Future Perspectives

Previous research, especially the research conducted in the past two decades, hasgreatly advanced our knowledge of rumen microbiome andits functions and allowed some success of dietary manipulation. However, due to the vast diversity, extreme complexity, functional redundancy, and resilience of this complex system, the majority of the rumen microbes remain to be understood and their metabolism as well as functions to be elucidated. Before this system is adequately understood, rational and effective manipulation will be challenging. The rapid advancement of "~omics" technologies, including metagenomics, metatranscriptomics, metaproteomics, metabolomics, and bioinformatics will provide the unprecedented opportunities to disentangle the complex relationships between feed and rumen microbiome, rumen microbiome and its function, rumen function and host metabolism, and host metabolism and milk production. Therefore, a holistic approach incorporating nutrition, rumen microbiome, and host metabolism is needed in future research. In addition to the typical data of feed and milk, huge datasets of rumen microbiome and host metabolism will be generated from such studies. Big data analytics will be required to connect the dots among feed, rumen, host, and milk production. Team efforts, including collaborative efforts involving researchers in different related field across countries, will be needed to achieve the goal—sustainable milk production with minimal environmental impact.

References

Creevey, C. J., Kelly, W. J., Henderson, G., et al., 2014. Determining the culturability of the rumen bacterial microbiome [J]. Microb. Biotechnol., 7 (5): 467-479.

De Nardi, R., Marchesini, G., Li, S., et al., 2016. Metagenomic analysis of rumen microbial population in dairy heifers fed a high grain diet supplemented with dicarboxylic acids or polyphenols [J]. BMC Vet. Res., 12: 29.

Desnoyers, M., Giger-Reverdin, S., Bertin, G., et al., 2009. Meta-analysis of the influence of *Saccharomycescerevisiae* supplementation on ruminal parameters and milk production of ruminants [J]. J. Dairy Sci., 92 (4): 1620-1632.

Firkins, J. L., Yu., Z., 2015. Ruminant nutrition symposium: How to use data on the rumen microbiome to improve our understanding of ruminant nutrition [J]. J. Anim. Sci., 93 (4): 1450-1470.

Firkins, J. L., Yu, Z., Morrison, M., 2007. Ruminal nitrogen metabolism: Perspectives for integration of microbiology and nutrition for dairy [J]. J. Dairy Sci., 90 (13_suppl): E1-16.

Henderson, G., Cox, F., Ganesh, S., et al., 2015. Rumen microbial community composition varies with diet and host, but a core microbiome is found across a wide geographical range [J]. Sci. Rep., 5: 14567.

Hess, M., Sczyrba, A., Egan, R., et al., 2011. Metagenomic discovery of biomass-degrading genes and genomes from cow rumen [J]. Science, 331 (6016): 463-467.

Hobson, P. N., Stewart, C. S., 1997. The rumen microbial ecosystem [M]. 2nd ed. Blackie Academic and Professional, New York, NY.

Hristov, A. N., Ott, T., Tricarico, J., et al., 2013. Special topics-mitigation of methane and nitrous oxide emissions from animal operations: Iii. A review of animal management mitigation options [J]. J. Anim. Sci., 91 (11): 5095-5113.

Jami, E., White, B. A., Mizrahi, I., 2014. Potential role of the bovine rumen microbiome in modulating milk composition and feed efficiency [J]. PLoS One, 9 (1): e85423.

Johnson, K. A., Johnson, D. E., 1995. Methane emissions from cattle [J]. J. Anim Sci., 73 (8): 2483-2492.

Kim, M., Eastridge, M. L., Yu, Z., 2014. Investigation of ruminal bacterial diversity in dairy cattle fed supplementary monensin alone and in combination with fat, using pyrosequencing analysis [J]. Can. J. Microbiol., 60 (2): 65-71.

Kim, M., Morrison, M., Yu, Z., 2011. Status of the phylogenetic diversity census of ruminal microbiomes [J]. Fems. Microbiol. Ecol., 76 (1): 49-63.

Kohn, R. A., Dinneen, M. M., Russek-Cohen, E., 2005. Using blood urea nitrogen to predict nitrogen excretion and efficiency of nitrogen utilization in cattle, sheep, goats, horses, pigs, and rats [J]. J. Anim. Sci., 83 (4): 879-889.

Krause, D. O., Bunch, R. J., Conlan, L. L., et al., 2001. Repeated ruminal dosing of *Ruminococcus* spp. Does not result in persistence, but changes in other microbial populations occur that can be measured with quantitative 16s - rrna - based probes [J]. Microbiology, 147 (7): 1719-1729.

Lima, F. S., Oikonomou, G., Lima, S. F., et al., 2015. Prepartum and postpartum rumen fluid microbiomes: Characterization and correlation with production traits in dairy cows [J]. Appl. Environ. Microbiol., 81 (4): 1327-1337.

Petri, R. M., Schwaiger, T., Penner, G. B., et al., 2013. Characterization of the core rumen microbiome in cattle during transition from forage to concentrate as well as during and after an acidotic challenge [J]. PLoS One, 8 (12): e83424.

Pitta, D. W., Pinchak, E., Dowd, S. E., et al., 2010. Rumen bacterial diversity dynamics associated with changing from bermudagrass hay to grazed winter wheat diets [J]. Microb. Ecol., 59 (3): 511-522.

Wang, J. K., He, B., Du, W., et al., 2015. Yeast with surface displayed xylanase as a new dual purpose delivery vehicle of xylanase and yeast [J]. Anim. Feed Sci. Technol., 208: 44-52.

Wang, L., Hatem, A., Catalyurek, U. V., et al., 2013. Metagenomic insights into the carbohydrate-active enzymes carried by the microorganisms adhering to solid digesta in the rumen of cows [J]. PLoS One, 8 (11): e78507.

Weimer, P. J., 2015. Redundancy, resilience, and host specificity of the ruminal microbiota: Implications for engineering improved ruminal fermentations [J]. Front. Microbiol., 6: 296.

Towards Knowledge-Based Rational Strategies to Improve Nitrogen Utilization in Dairy Cattle

Zhongtang Yu

Department of Animal Sciences, The Ohio State University, Columbus, Ohio, USA

Summary

Dietary nitrogen is the most expensive feed ingredient, accounting for 40% of the total feed cost of dairy production. However, only less than 30% of the dietary nitrogen find their way to milk. Such a low nitrogen conversion ratio significantly increases feed cost and decreases the profitability of the dairy producers. The large amounts of nitrogen excreted in dairy manure also directly cause environmental pollution and enlarge the carbon footprint of dairy production. Energy efficiency and nitrogen efficiency are interdependent, and the increased use of human-edible feeds, such as cereal grains, increases milk production and improves nitrogen utilization efficiency, but the nitrogen utilization efficiency still remains low. Diet composition and animal genotype and physiological state collectively determine nitrogen utilization efficiency, but the rumen microbiome also plays a pivotal role because the majority of the nitrogen assimilable by dairy cows is the microbial proteins leaving the rumen. The amount of microbial proteins leaving the rumen and reaching the small intestines is determined by protein synthesis by various rumen microbes and proteolysis mediated by mainly rumen protozoa. The cellular protein synthesis in the rumen is affected by diet composition and digestibility, both of which affect the diversity and functionality of the rumen microbiome, but some intrinsic features of rumen microbes, such as energy spilling and synthesis of reserve carbohydrates and phosphate, also decrease cellular protein synthesis. The intraruminal recycling of microbial proteins is the main culprit decreasing microbial protein supply to the small intestines. Continued research on rumen microbiome in the past two decades has helped advance our understanding of the factors that affect the synthesis and degradation of microbial proteins in the rumen. New knowledge-based strategies are emerging to enhance the synthesis while decreasing the degradation of microbial proteins in the rumen to achieve improved nitrogen utilization efficiency in dairy cows.

Low Nitrogen Utilization Efficiency in Dairy Cattle and it Has Consequences

Dietary nitrogen is the most expensive feed ingredient, and yet, dairy cattle only use no more than 30% (even lower among the dairy cattle in China) of the dietary nitrogen in producing milk. Such low nitrogen efficiency significantly increases feed cost and decreases the profitability and sustainability of the dairy industries. Additionally, because of the low nitrogen utilization efficiency, large amounts of dietary nitrogen are excreted in dairy manure, directly creating environmental pollution and increasing the carbon footprint of dairy production, the later of which is attributed to the fact that all dietary nitrogen is fixed from atmospheric N_2 by either farming crops or chemical processes, both of which release greenhouse gasses (CO_2 and nitrous oxide). As estimated by the US EPA, 86% of the national ammonia emissions in the US in 2006 were from livestock and fertilizer (http://epa.gov/climatechange). Moreover, more dairy cattle are also needed to meet the national needs for milk, contributing further to the greenhouse gas (GHG) emission share from dairy production.

In the rumen, dietary nitrogen is primarily (about 70%) converted to microbial proteins, which is the major direct protein source available to ruminants. However, a large portion of the microbial proteins in the form of bacterial and other microbial cells are engulfed and subsequently degraded by rumen protozoa, predominantly species of Entodinium. Rumen protozoa are predators and collectively account for up to 50% of the total microbial biomass (Newbold et al., 2015). They break down the engulfed microbial proteins into oligopeptides and free amino acids, which are rapidly fermented to short chain fatty acids and ammonia by certain ruminal microbes, including hyper-ammonia producing bacteria (Rychlik and Russell, 2000). Ruminants themselves cannot use ammonia nitrogen (NH_3-N) to synthesize milk or muscle proteins. Although a small proportion of the rumen ammonia can be reused by some ruminal microbes to resynthesize microbial proteins, most of the ruminal NH_3-N is excreted as urea or fecal ammonia. Collectively, rumen protozoa could engulf as much as 24% of the total ruminal bacteria per day (Hespell et al., 1997), and up to 50% of the engulfed bacterial proteins is excreted by protozoa as oligopeptides and free amino acid (Jouany, 1996). Because of such bacterivory, protozoa recycle up to 50% of the microbial nitrogen in the rumen, decreasing the outflow of microbial proteins to the small intestines. The guilt of rumen protozoa in lowering dietary nitrogen utilization efficiency was clearly demonstrated by a significant increase in nitrogen utilization efficiency (by 34% compared to the control) in a sheep model (Belanche et al., 2011).

Rumen protozoa also "waste" metabolic energy through the synthesis of storage carbohydrates (Hackmann and Firkins, 2015). Rumen protozoa synthesize large amounts of storage carbohydrates as glycogen (Figure 1) (Park et al., 2017) and store them as carbon and energy reserves after feeding and then utilize them when carbohydrates become limited (toward before feeding). The synthesis of glycogen (polysaccharides with the same chemical composition as starch) requires ATPs, two ATPs per glucose (almost the same number of ATPs produced from glucose fermentation). Indeed, protozoa account for the most accumulation of reserve carbohydrates found in the rumen, and a competition experiment showed that protozoa accumulated nearly 35-fold more reserve carbohydrates than rumen bacteria (Denton et al., 2015). Because nitrogen efficiency and energy efficiency are correlated (Phuong et al., 2013), the waste of metabolic energy during glycogen synthesis directly lowers nitrogen efficiency.

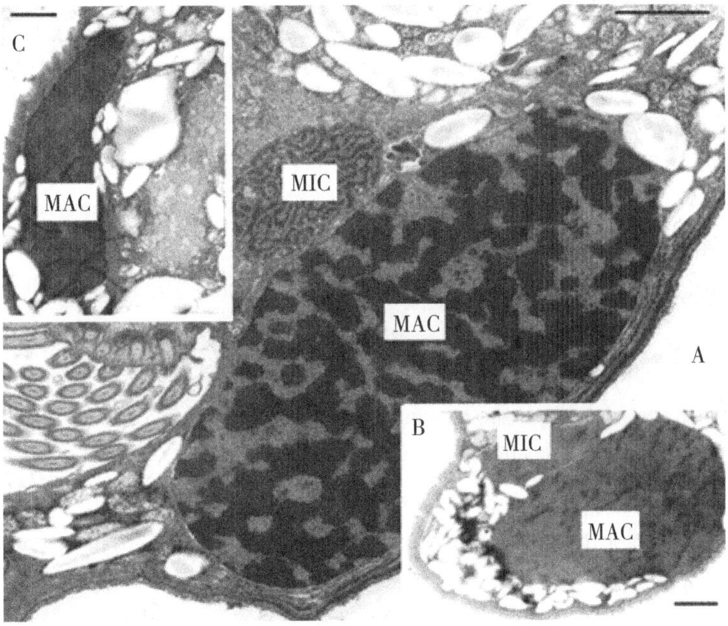

Figure 1　Glycogen accumulated as storage carbohydrates inside of *entodinium caudatum* cells. adapted from part et al. (2017)

Proteolytic and deaminating bacteria also contribute to the low nitrogen utilization efficiency in dairy cattle (Anantasook et al., 2013). A small group of rumen bacteria that have particularly high deaminating activities (Eschenlauer et al., 2002) are shown to ferment free amino acids to ammonia and short - chain fatty acids, turning microbial nitrogen into ammonia nitrogen that cow cannot utilize. Although these so-called hyper-ammonia producing bacteria have received more research attention, other rumen bacteria including species of *Prevotella* (Wallace, 1996) and *Megasphaera* (Rychlik et al., 2002) also deaminate amino acids to ammonia. Both protozoa and proteolytic/deaminating bacteria are targets that have been aimed at to improve nitrogen utilization efficiency in dairy cattle.

The Past, Present, and Future in Research to Improve Dietary Nitrogen Utilization Efficiency in Dairy Cattle

Most of the metabolic energy in dairy cattle, as in other ruminants, is provided as volatile fatty acids (VFA) that are produced from feed fermentation by rumen microbiome. The synthesis of microbial proteins in the rumen also require energy (ATP). Thus, energy efficiency and nitrogen efficiency are correlated ($r = 0.62$) (Phuong et al., 2013). It is well established that synchronized nitrogen and energy supplies can improve nitrogen efficiency in dairy cows (Godden et al., 2001; Kolver et al., 1998; Piao et al., 2012). Supplementation of certain essential amino acids such as lysine and methionine, which are rumen-protected, also improves nitrogen retention (Dinn et al., 1998; Pereira et al., 2017).

Other feed additives have been examined for their ability to improve nitrogen efficiency. Monensin, which is an ionophore, can inhibit Gram-positive bacteria including those hyper-ammonia producing bacteria, has been used in North America and some other countries to improve nitrogen utilization (Akins et al., 2014; Lee et al., 2019). Plant bioactive compounds and substances, such as palm oil (Anantasook et al., 2013), can also improve nitrogen efficiency. In a very recent study, lysophospholipids, a family of simple phospholipids that signal through G-protein-coupled receptors (GPCRs) and are involved in a broad range of biological processes of mammals, were found effective in increasing milk production and improve nitrogen efficiency (Lee et al., 2019). However, the improvement achieved by the aforementioned means are limited in magnitude, and research continues to gain a better understanding of the cows and their rumen microbiome to further improve nitrogen efficiency in dairy production.

Hypothetically, effective improvement in nitrogen utilization can be better achieved by specifically controlling the proteolytic/deaminating bacteria and protozoa in the rumen. Several studies have shown that hops could inhibit hyper-ammonia producing bacteria and lower amino acid deamination *in vitro* (Flythe, 2009). Future research may explore the enzymes that mediate amino acid-deamination and develop specific inhibitors to decrease deamination of amino acids in the rumen.

Protozoa are among the major barriers to efficiency nitrogen utilization in dairy cows. Being the sole predators in the rumen, they rely on several enzymes to lyse the engulfed microbial cells and then degrade their cellular proteins. These enzymes included lysozyme, chitinase, and lysosomal peptidases (Morgavi et al., 1994, 1996). In a recent study, we found that these enzymes were expressed at high level in *Entodinium caudatum*, one of the most predominant and bacterivorous rumen protozoan. We showed that specific inhibition of these enzymes almost completely inhibited the growth of *Entodinium caudatum* and significantly decreased ammonia concentration but exerted no adverse effect on feed digestion or fermentation *in vitro*. Specific inhibition of rumen protozoa may be an effective targeted control mechanism by which dietary nitrogen utilization efficiency can be improved without adversely affecting the overall rumen microbiome and its function.

References

Akins, M. S., Perfield, K. L., Green, H. B., et al., 2014. Effect of monensin in lactating dairy cow diets at 2 starch concentrations [J]. J. Dairy Sci., 97 (2): 917-929.

Anantasook, N., Wanapat, M., Cherdthong, A., et al., 2013. Effect of plants containing secondary compounds with palm oil on feed intake, digestibility, microbial protein synthesis and microbial population in dairy cows [J]. Asian-Australas J. Anim Sci., 26 (6): 820-826.

Belanche, A., Abecia, L., Holtrop, G., et al., 2011. Study of the effect of presence or absence of protozoa on rumen fermentation and microbial protein contribution to the chyme [J]. J. Anim. Sci., 89 (12): 4163-4174.

Denton, B. L., Diese, L. E., Firkins, J. L., et al., 2015. Accumulation of reserve carbohydrate by rumen protozoa and bacteria in competition for glucose [J]. Appl. Environ. Microbiol., 81 (5): 1832-1838.

Dinn, N. E., Shelford, J. A., Fisher, L. J., 1998. Use of the Cornell net carbohydrate and protein system and rumen-protected lysine and methionine to reduce nitrogen excretion from lactating dairy cows [J]. J. Dairy Sci., 81 (1): 229-237.

Eschenlauer, S. C., McKain, N., Walker, N. D., et al., 2002. Ammonia production by ruminal microorganisms and enumeration, isolation, and characterization of bacteria capable of growth on peptides and amino acids from the sheep rumen [J]. Appl. Environ. Microbiol., 68 (10): 4925-4931.

Flythe, M. D., 2009. The antimicrobial effects of hops (*Humulus lupulus* L.) on ruminal hyper ammonia-producing bacteria [J]. Lett. Appl. Microbiol., 48 (6): 712-717.

Godden, S. M., Lissemore, K. D., Kelton, D. F., et al., 2001. Relationships between milk urea concentrations and nutritional management, production, and economic variables in Ontario dairy herds [J]. J. Dairy Sci., 84 (5): 1128-1139.

Hackmann, T. J., Firkins, J. L., 2015. Maximizing efficiency of rumen microbial protein production [J]. Front. microbiol., 6: 465.

Kolver, E., Muller, L. D., Varga, G. A., et al., 1998. Synchronization of ruminal degradation of supplemental carbohydrate with pasture nitrogen in lactating dairy cows [J]. J. Dairy Sci., 81 (7): 2017-2028.

Lee, C., Morris, D. L., Copelin, J. E., et al., 2019. Effects of lysophospholipids on short-term production, nitrogen utilization, and rumen fermentation and bacterial population in lactating dairy cows [J]. J. Dairy Sci., 102 (4): 3110-3120.

Morgavi, D. P., Sakurada, M., Tomita, Y., et al., 1994. Presence in rumen bacterial and protozoal populations of enzymes capable of degrading fungal cell walls [J]. Microbiol., 140 (3): 631-636.

Morgavi, D. P., Sakurada, M., Tomita, Y., et al., 1996. Electrophoretic forms of chitinolytic and lysozyme activities in ruminal protozoa [J]. Curr. Microbiol., 32 (3): 115-118.

Newbold, C. J., de la Fuente, G., Belanche, A., et al., 2015. The role of ciliate protozoa in the rumen [J]. Front. Microbiol., 6: 1313.

Park, T., Meulia, T., Firkins, J. L., et al., 2017. Inhibition of the rumen ciliate *Entodinium caudatum* by antibiotics [J]. Front. Microbiol., 8 (1189): 1189.

Pereira, A. B. D., Whitehouse, N. L., Aragona, K. M., et al., 2017. Production and nitrogen utilization in lactating dairy cows fed ground field peas with or without ruminally protected lysine and methionine

[J]. J. Dairy Sci., 100 (8): 6239–6255.

Phuong, H. N., Friggens, N. C., de Boer, I. J., et al., 2013. Factors affecting energy and nitrogen efficiency of dairy cows: A meta–analysis [J]. J. Dairy Sci., 96 (11): 7245–7259.

Piao, M. Y., Kim, H. J., Seo, J. K., et al., 2012. Effects of synchronization of carbohydrate and protein supply in total mixed ration with korean rice wine residue on ruminal fermentation, nitrogen metabolism and microbial protein synthesis in Holstein steers [J]. Asian–Australas J. Anim Sci., 25 (11): 1568–1574.

Rychlik, J. L., LaVera, R., Russell, J. B., 2002. Amino acid deamination by ruminal megasphaera elsdenii strains [J]. Curr. Microbiol., 45 (5): 340–345.

Rychlik, J. L., Russell, J. B., 2000. Mathematical estimations of hyper–ammonia producing ruminal bacteria and evidence for bacterial antagonism that decreases ruminal ammonia production[J]. Fems. Microbiol. Ecol., 32 (2): 121–128.

Wallace, R. J., 1996. Ruminal microbial metabolism of peptides and amino acids [J]. J. Nutr., 126 (4 Suppl): 1326S–1334S.

Nutritional Effects on Immune Function and Mastitis in Dairy Cows

Bill Weiss

Department of Animal Sciences, The Ohio State University, Wooster, OH

Abstract

Mastitis is arguably the health problem with the greatest economic impact to dairy farmers. In addition to treatment costs and reduced milk yield, milk quality is negatively affected, and the resulting pain from the infection is an animal welfare concern. Proper nutrition can improve immune function making cows better able to defend themselves against mastitis pathogens. Nutritional intervention can be via reducing immune-suppresors or by enhancing immunity. Ketosis (specifically B-hydroxyl butyrate) inhibits function of different immune cells putting cows at increased risk for ketosis. Feeding dry cows adequate, but not excessive energy and feeding fresh cows adequate glucose precursors (starch and protein) should reduce ketosis and risk of mastitis. Hypocalcemia can also inhibit immune function and increase mastitis. Proper mineral and vitamin nutrition (Ca, P, Mg, K, vitamin D) and in many situations the use of supplemental anions (negative DCAD diets) can reduce hypocalcemia and may reduce mastitis. Proper supplementation of antioxidant nutrients should enhance immunity and reduce mastitis; however excess supplementation may have the opposite effect. Inadequate selenium intake increases mastitis and somatic cell count (SCC). Adequate supplemental vitamin E, especially during the dry period and transition period can reduce mastitis. Both Cu and Zn supplementation has also been shown to either reduce clinical mastitis or reduce SCC. Vitamin A, β-carotene, vitamin C, and vitamin D affect immunity but data showing reduced mastitis when those nutrients are supplemented is limited. In most cases, supplementation rates that approximate current NRC standard are adequate to maintain good immune function and reduce the risk of mastitis.

Introduction

The risk that a cow will develop mastitis is a function of pathogen load at the teat end and the cow's ability to prevent a bacterial infection from becoming established in the mammary gland. Nutrition indirectly affects teat end exposure via changes in the amount of manure produced and by altering characteristics of manure (e.g., moisture concentration, pH), but effects on mastitis would probably be small. Conversely, nutrition can have significant effects on the immune system thereby affecting infection rate and severity of mastitis. The highest rates of mastitis generally occur at or shortly after parturition (Smith et al., 1985). Early lactation is also the time when most cows experience short-term malnutrition, i.e., intake of nutrients does not meet nutrient requirements. The immune system, as any physiological system, does not function optimally during periods of malnutrition. In addition, the immune system has high requirements for specific nutrients and when these nutrients are not provided in adequate amounts, immune function may suffer. This review will concentrate on nutritional influences on immune function and mastitis during the periparturient period.

Energy and protein

During late gestation and early lactation, dry matter intake (DMI) by dairy cows is quite low whereas nutrient demand, especially post-partum is extremely high. This leads to cows being in negative protein and energy balance. Body fat and protein are mobilized by the cow to provide the energy and amino acids needed for maintenance functions and to produce milk. The protein deficient is short-lived because: (1) protein intake by cows can be increased easily by increasing the concentration of protein in the diet, and (2) labile body protein reserves are depleted quickly and once they are exhausted, milk production will decrease to match protein supply. An immune response can include antibody production and cellular proliferation both of which require amino acids. However, compared to the kilogram quantity of milk protein produced daily by early lactation cows, the amino acid needs of the immune system are small. No direct data are available showing that mitigating the moderate protein deficiency that occurs in early lactation improves immune function and increases resistance to mastitis. However, one study reported very modest beneficial effects on immune function when peripartum cows were infused with 300 g of glutamine per day (Doepel et al., 2006) but this likely has little practical significance. If protein nutrition is adequate for milk production in early lactation, it likely is adequate for proper immune function.

The energy deficient experienced by most cows lasts much longer than the protein deficient and usually starts a few days before calving and continues for several weeks after parturition. Body energy reserves in a cow are usually much greater than body protein reserves, and it is very difficult to increase energy intake in early lactation via diet changes. Normal, healthy cows lose 0.25 to 0.5 body condition score (BCS) units in early lactation and reach their BCS nadir by 4 to 7 week of lactation. Some cows start losing body condition several day or even a few weeks before calving, continue losing condition after calving and lose more than 1 BCS unit in early lactation. This severe negative energy balance is either a consequent of health disorders (e.g., milk fever, retained fetal membranes, or metritis) or will lead to health problems (e.g., ketosis and displaced abomasum). Negative energy balance has also been identified as a risk factor for mastitis.

The degree of negative energy balance experienced by cows is correlated with immune function. Various measures of energy balance such as calculated energy balance, plasma concentrations of non-esterified fatty acids (NEFA) and B-hydroxy-butyrate (BHBA) were negatively correlated with concentrations of antibodies in plasma and milk SCC in early lactation cows (van Knegsel et al., 2007). In that study, all treatment average energy balances were reasonable and based on BHBA and NEFA cows were not suffering from clinical ketosis. Experimentally - induced negative energy balance in steers (DMI was severely restricted) did not negatively affect neutrophil function (Perkins et al., 2001) but neutrophils from cows naturally afflicted with subclinical or clinical ketosis had reduced functionality (Zerbe et al., 2000). An epidemiological study found that high concentrations of plasma ketones or a loss of more than 0.5 BCS units were significant risk factors for the development of udder edema which then was a risk factor for the development of clinical mastitis (Compton et al., 2007); however, they also found that low concentrations of NEFA was associated with increased risk of mastitis. In support of that finding, (Berry et al., 2007) reported that increased BCS loss was associated with lower SCC. During the peripartum period, negative energy balance and elevated concentrations of NEFA and BHBA coincides with numerous other events including hormonal changes, hypocalcemia, and changes in vitamin status, therefore it is not possible to determine unequivocally that energy balance direct affect on immune function. However, enough data are available to strongly suggest that excessive mobilization of body fat and the associated increase in NEFA and BHBA during the peripartum period contributes to immunosuppression. Management and dietary practices that should help reduce excessive body condition loss include:

1. Prevent cows from becoming too fat in late lactation and the dry period. This may require a pen dedicated to fat lactating cows so that they can be fed a low energy diet. Excess energy consumption is a common problem during the dry period because dry cows only require about 14 Mcal of NEL/d. To meet, but not exceed, the energy requirement a diet based on less digestible feeds is needed so that the rumen gets full before overconsumption of NEL occurs.

2. Avoid a large decrease in DMI during the prepartum period. DMI can decrease by more than 20% during the last 1 - 2 week of gestation. This large drop in intake causes cows to mobilize fat which can infiltrate the liver cause fatty liver and ketosis. The drop in intake can be mitigated by feeding a less digestible diet to far-off dry cows so that average DMI for a Holstein cow during the dry period is around 25-26 lbs/day (~12 kg). Cows with high DMI during the early dry period tend to have a greater decrease in DMI during late gesta-

tion than do cows that have more moderate DMI during the early dry period (Douglas et al., 2006). The peripartum decrease in DMI can also be moderated by feeding a well-balanced prefresh diet (e.g., 30% to 35% NDF, 30% to 40% concentrate with good forage). Intake by specific animals can be reduced when pens are overcrowded. Make sure pens containing prefresh animals have adequate bunk space and stalls.

3. Promote a rapid increase in energy intake post calving which usually requires a rapid increase in DMI. Feeding excessive grain (i.e., starch) or fat to increase the energy density of diets (i.e., Mcal/kg) usually is counterproductive because it often reduces DMI. Feeding a well-balanced diet based on high quality forage, that contains moderate concentrations of fiber (approximately 30% NDF) and starch (22% to 25%) and < 5% total fat improves DMI. Overcrowding fresh cows also restricts their intake.

Energy source (Specific fatty acids)

Neutrophils and other types of immune cell have high concentrations of polyunsaturated fatty acids (PUFA) in their membranes and higher concentrations of specific PUFA are related to improved neutrophil function. In nonruminants, fatty acid profiles of cells reflect the diet composition but in ruminants, dietary unsaturated fatty acids are often biohydrogenated to saturated fatty acids making it difficult to substantially change fatty acid profiles of cells. In two separate studies with transition cows from the same group (Lessard et al., 2003; Lessard et al., 2004) the exact opposite response to fat supplements was observed. In one study lymphocyte proliferation was enhanced when flax seed was fed (a source of n-3 PUFA) compared with cows fed soybeans (a source of n-6 PUFA) but in the other study, cows fed soybeans had enhanced lymphocyte proliferation. At this time, no compelling data are available to support feeding specific types of fat to improve mammary gland health and reduce mastitis.

Calcium and Other Minerals Related to Hypocalcemia

Cows with milk fever are much more likely to get clinical mastitis than cows without milk fever (Curtis et al., 1985) because:

1. Calcium is required for muscle contractions and the teat sphincter of cows with hypocalcemia may not contract as quickly or as completely as for cows with normal blood Ca increasing the risk of bacterial invasion.

2. Cows with hypocalcemia spend more time lying down which increase teat end exposure.

3. Cows with milk fever have higher concentrations of plasma cortisol than normal cows (Horst and Jorgensen, 1982) and cortisol suppresses immune function.

4. Ca status of monocytes is impared in cows with milk fever (Kimura et al., 2006). When monocytes are activated intracellular Ca is released but the amount of Ca released is less

in cows with milk fever. This reduces the ability of the monocyte to function properly.

Available data clearly shows that preventing subclinical and clinical milk will reduce the prevalence of mastitis in early lactation. Dietary concentrations of Ca, phosphorus, magnesium, potassium, chloride, sulfur, and vitamin D are related to milk fever. One approach is to feed slightly less Ca to dry cow than their requirement. The marginal Ca deficiency increases mobilization of Ca from bone. Another approach is to feed an anionic diet (elevated concentrations of chloride and sulfur without elevated concentrations of sodium and potassium. This induces metabolic acidosis which is then compensated by mobilizing phosphate from the bone bringing Ca with it. If possible, avoid feeding diets with excessive concentrations of K and make sure dietary Mg is adequate (>0.25% of diet DM).

Antioxidant Nutrients

Reviews are available discussing relationships between immune function and minerals and vitamin of ruminants in greater detail than is presented here (Sordillo, 2005; Weiss and Spears, 2005). Substantial amounts of free radical are produced during an inflammatory response such as that which occurs when the mammary gland becomes infected. When adequate antioxidants are present, free radicals are kept in check which increases the life span of certain immune cells. When antioxidant capacity is limited the lifespan of those immune cells is reduced and the infection can become established or severity of the infection can increase. Cells and animals have developed sophisticated systems to control oxidative stress. Components of the antioxidant system include enzymes (many of which contain metal cofactors), vitamins, and numerous other compounds. A simplified version of the antioxidant system is shown in Table 1 and Figure 1.

Table 1 Some of the antioxidant systems found in mammalian cells

Component (location in cell)	Nutrients Involved	Function
Superoxide dismutase (cytosol)	Copper and zinc	An enzyme that converts superoxide to hydrogen peroxide
Superoxide dismutase (mitochondria)	Manganese and zinc	An enzyme that converts superoxide to hydrogen peroxide
Ceruloplasmin (water phase)	Copper	An antioxidant protein, may prevent copper and iron from participating in oxidation reactions
Glutathione peroxidase (cytosol)	Selenium	An enzyme that converts hydrogen peroxide to water
Catalase (cytosol)	Iron	An enzyme (primarily in liver) that converts hydrogen peroxide to water
Ascorbic acid (cytosol)	Vitamin C	Reacts with several types of ROM

(Continued)

Component (location in cell)	Nutrients Involved	Function
α-tocopherol (membranes)	Vitamin E	Breaks fatty acid peroxidation chain reactions
β-carotene (membranes)	β-carotene	Prevents initiation of fatty acid peroxidation chain reactions

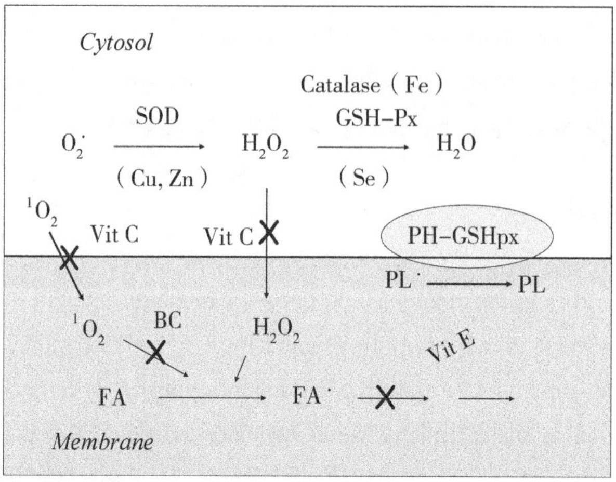

Figure 1 Simplified depiction of the cellular antioxidant system showing relationships with antioxidant nutrients. enzymes: SOD = superoxide dismutase, GSH-px = glutathione peroxidase, PH-GSHpx = phospholipid hydroperoxide glutathione peroxidase. nutrients: BC = B-carotene, virt C = vitamin C (ascorbic acid), vit E = vitamin E (tocopherol). other: FA = fatty acid. FA• = FA radical, $O_2^{\bullet}$ = superoxide, $^1O_2$ = singlet oxygen; PL = phospholipid; PL• = PL radical

Vitamin A and B-carotene

The effects of vitamin A and B-carotene on mastitis measures have been inconsistent. Some studies have found positive effects on neutrophil and lymphocyte function when cows are supplemented with approximately 70,000 IU/d of vitamin A or 300 mg to 600 mg of β-carotene (Michal et al., 1994) but in a clinical study similar treatments had no effect on mammary gland health (Oldham et al., 1991). A likely reason for different responses among studies is differences in vitamin A and β-carotene status of the control cows. Jukola et al (1996) suggested that plasma concentrations of β-carotene in dairy cows should be > 3 mg/L to optimize udder health. Currently available data does not support feeding vitamin A in excess of the current NRC requirement (approximately 70,000 IU/d) to improve mammary gland health. Supplemental β-carotene may have some benefit if cows are in low β-carotene status (i.e., fed a diet based largely on weathered, low quality hay).

Copper and Zinc

Cows and heifers fed diets with 20 mg/kg supplemental copper had less severe mastitis following a mammary gland challenge (*E. coli*) and fewer natural infection than animals fed diets with about 8 mg/kg (Harmon and Torre, 1994; Scaletti et al., 2003). Tomlinson et al. (2002) summarized results of 12 experiments and reported an overall significant reduction (196,000 vs. 294,000) in SCC when Zn-met was supplemented (between 200 and 380 mg of Zn/d). In that summary, 4 of the experiments used a control diet that did not meet NRC (2001) requirements for Zn. Whitaker et al. (1997) compared supplemental Zn from a mixture of Zn proteinate and inorganic Zn or from all inorganic sources. Source of Zn had no effect on infection rate, new infections, clinical mastitis and SCC. Currently available data suggests that diets should contain about 15 to 20 mg/kg of copper (assuming no antagonists) and 50 to 60 mg/kg of Zn. Obtaining at least a portion of the supplemental zinc from zinc methionine may be beneficial.

Selenium and Vitamin E

Supplemental vitamin E and/or Se has been shown to reduced prevalence and severity of mastitis (Smith et al., 1997). Based on mammary challenge experiments, the positive effects of Se were greater when clinical responses are more severe (i. e., *E. coli* vs. *S. aureus* challenge) (Erskine et al., 1989; Erskine et al., 1990). The positive effects of supplemental Se on mammary gland health are well-established; a more recent question concerns source of supplemental Se. In the U.S. supplemental Se can be provided by sodium selenate or selenite (inorganic) or by Se-yeast (organic). Cows fed Se-yeast usually have higher concentrations of Se in plasma, whole blood, and milk compared with cows fed an equal amount of inorganic Se but neutrophil function has not been affected by Se source (Weiss and Hogan, 2005). When Se antagonists are present (e.g., sulfate) obtaining a portion of Se from Se-yeast, especially during the dry period and early lactation should be beneficial. The exact quantity of vitamin E needed by peripartum cows is not known; however feeding more than 1,000 IU/d during this period probably is beneficial.

The NRC requirement for vitamin E is about 500 IU/d for lactating cows and 1000 IU/d for dry cows. Essentially no new research has been conducted evaluating vitamin E requirements for lactating cows but several experiments have been conducted with dry cows. Increasing vitamin E supplementation during the transition period (2 or 3 week prepartum until 1 or 2 week post partum) has improved measures of immune function or improved mammary gland health in several, but not all, studies (Weiss et al., 1997; Baldi et al., 2000; Politis et al., 2004; Persson Waller et al., 2007) (Figure 2). Supplementation rates during the transition period ranged from 2,000 to 4,000 IU/d. No study has shown any negative effects of high supplementation rates in the prefresh period. Therefore the only known cost is the cost of the additional vitamin E but because the supplementation period is short (a few weeks) and the potential payoff is high (reduced mastitis and reduced retained placenta) increasing vitamin E supplementation during the prefresh period is justified. Increasing vitamin E supplementation during the entire dry period, however is not justified. A recent study (Bouwstra et al., 2010) evaluated the effects of feeding 3,000 IU of vitamin E/d (controls were fed approximately 130 IU/d) during the dry period. On 3 of 5 farms, more cases of mastitis occurred when cows were fed high vitamin E and on 2 farms little difference in mastitis was observed between treatments (Figure 3). Overall, they reported that feeding high vitamin E increased the risk of mastitis by 1.7×compared with the control. Although I have some technical concerns regarding the paper (e.g., methods used to diagnose mastitis), the paper clearly shows no benefit of increasing amounts of vitamin E and potentially it might have negative effects. Vitamin E supplementation during the dry period should be limited to 1,000 IU/d.

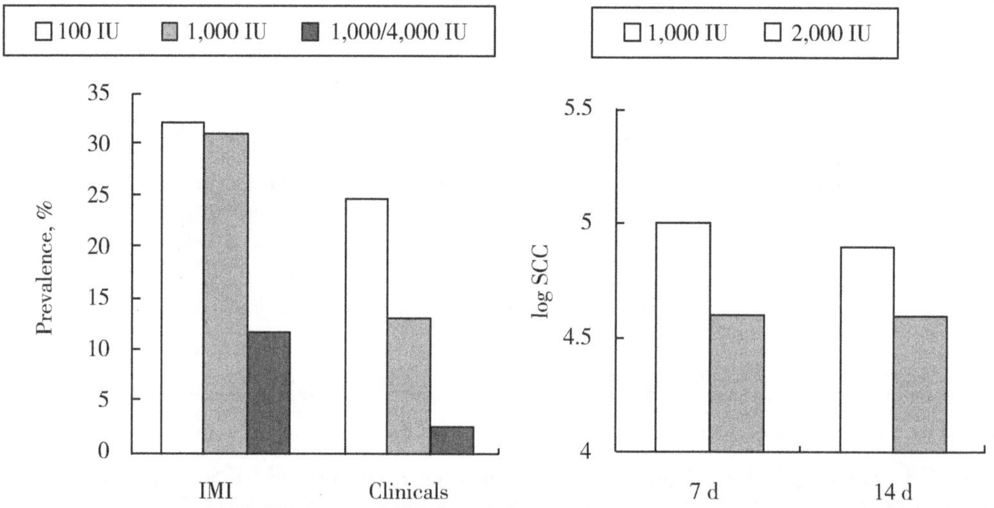

Figure 2 Effects of extra vitamin E during the prefresh (approximately last 2 wk of gestation) on prevalence of intramamary gland infection (IMI), clinical mastitis, and somatic cell count (SCC). The panel on the left (Weiss et a., 1976) has data from an experiment where cows were fed either 100 IU/d or 1,000 IU of supplemental vitamin E during the entire dry period or 1,000 IU/d until 14 d prepartum and then fed 4,000 IU/d. The extra vitamin E greatly reduced IMI and clinical mastitis in early lactation (first 21 days). the panel on the right (Baldi et al., 2000) is from an experiment in which cows were fed either 1000 or 2000 IU of supplemental vitamin E/d during the last 14 d of gestation. The extra vitamin E reduced SCC during the first 14 d of the following lactation

No new data are available refuting current NRC requirements for vitamin E except during the prefresh period. Increasing vitamin E intake to between 2,000 and 4,000 IU/d can help re-

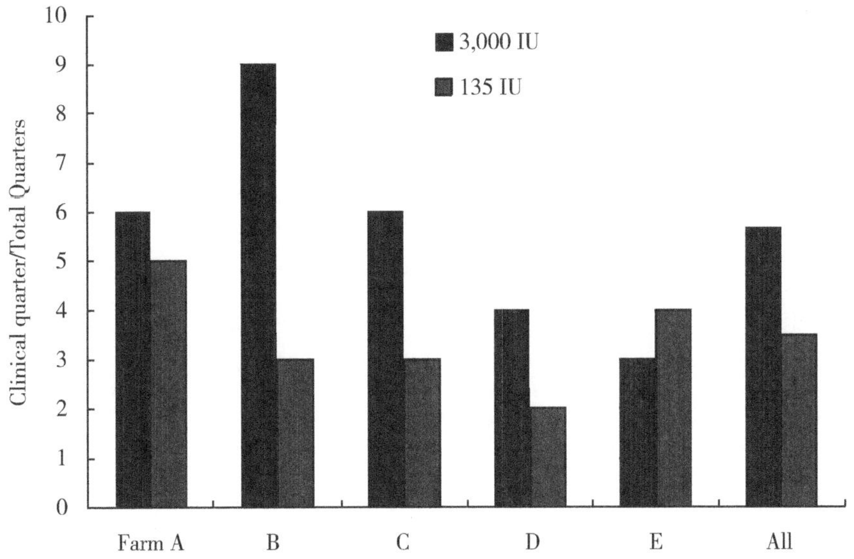

Figure 3 Rate of clinical mastitis in cows fed 135 or 3,000 IU of supplemental vitamin E during the dry period on 5 different farms. Figure derived from bouwstra et al. (2010)

duce mastitis, retained fetal membranes and perhaps metritis. New data suggest over supplementation for the entire dry period (3,000 IU/d) may be a risk factor for mastitis and should be avoided.

Vitamin C

Vitamin C (ascorbic acid) is probably the most important water soluble antioxidant in mammals. Most forms of vitamin C are extensively degraded in the rumen, but cows can synthesize vitamin C and it is not considered an essential nutrient for cattle. The concentration of ascorbic acid is high in neutrophils and increases as much as 30-fold when the neutrophil is stimulated. Within a limited range (67,000 to 158,000 cells/mL), SCC was not correlated with plasma vitamin C concentrations in cows (Santos et al., 2001). Injecting ascorbic acid following intramammary challenge with endotoxin had very limited effects on inflammation and other clinical signs in cows (Chaiyotwittayakun et al., 2002). We found significant correlations between measures of ascorbic acid status and clinical signs of mastitis caused by *E. coli* challenge (Figure 4) (Weiss et al., 2004). That does not mean that increasing vitamin C status of cows will reduce the prevalence or severity of mastitis. A follow up experiment was conducted to determine whether feeding supplemental vitamin C to periparturient cows would enhance neutrophil function and reduce the inflammatory response following an endotoxin challenge (Weiss and Hogan, 2007). We were successful in enhancing ascorbic acid status of cows, but supplemental vitamin C had no effect on neutrophil function or inflammation. Based on current data, vitamin C is not recommended for either prophylactic or therapeutic treatment of mastitis.

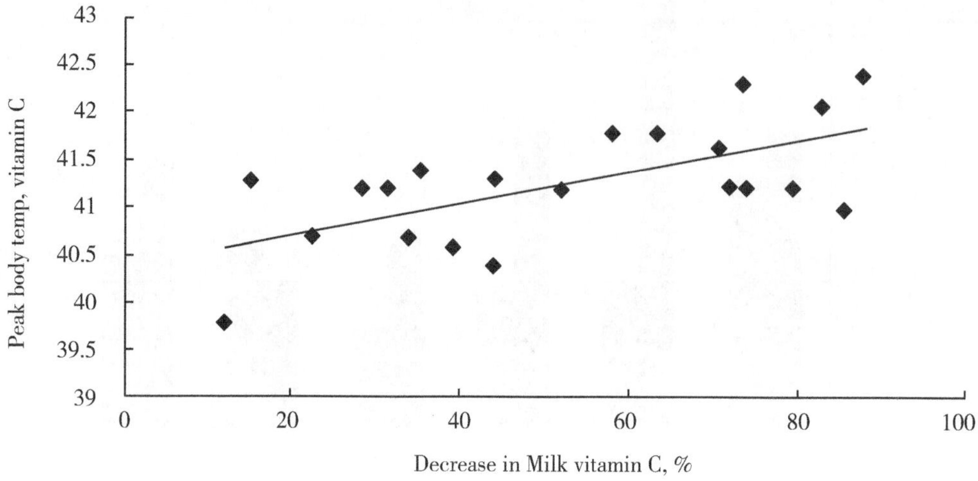

Figure 4 Relationship between concentration of ascorbic acid in milk and body temperature of cows following an infusion of *E. coli* into the mammary gland. as concentration of vitamin C in milk decreased more, febrile response was greater (Weiss et al., 2004)

Vitamin D

Because of the interest in vitamin D for human health, vitamin D for dairy cows is also being re-evaluated. Although adequate data are not available to quantitatively adjust the current NRC vitamin D requirement (approximately 20,000 IU/d), data are available suggesting potential benefits from increasing supplementation rates. Studies with humans and limited research with bovine cells have shown that vitamin D has important roles in immune function and that blood concentrations of 25-OH vitamin D (humans) required for maximal immune response was greater than concentrations required for optimal Ca metabolism (Lippolis, 2011). Whether this is true for dairy cows will require new studies. Dairy cows housed inside without exposure to sun and fed vitamin D at NRC recommendations had significantly lower plasma concentrations of 25-OH vitamin D than cows fed no supplemental vitamin D but housed outside in the summer with extensive sun exposure (Hymøller et al., 2009). We do not know the optimal concentration of plasma 25-OH vitamin D but current supplementation rates may not provide for maximal concentrations (Nelson et al., 2016).

Conclusions

To improve mammary gland health:

1. Feed and manage late lactation and dry cows to maintain proper body condition. Avoid a large decrease in feed intake around parturition and a large loss in BCS in early lactation.

2. Prevent hypocalcemia via proper mineral nutrition for dry cows.

3. Feed adequate, but not excessive amounts of trace minerals and vitamins. Selenium and vitamin E are especially critical. Consider increasing vitamin E supplementation during the pre-fresh period.

References

Berry, D. P., Lee, J. M., Macdonald, K. A., et al., 2007. Associations Among Body Condition Score, Body Weight, Somatic Cell Count, and Clinical Mastitis in Seasonally Calving Dairy Cattle [J]. J. Dairy Sci., 90: 637-648.

Compton, C. W. R., McDougall, S., Parker, K., et al., 2007. Risk Factors for Peripartum Mastitis in Pasture-Grazed Dairy Heifers [J]. J. Dairy Sci., 90: 4171-4180.

Curtis, C. R., Erb, H. N., Sniffen, C. J., et al., 1985. Path Analysis of Dry Period Nutrition, Postpartum Metabolic and Reproductive Disorders, and Mastitis in Holstein Cows [J]. J. Dairy Sci., 68: 2347-2360.

Doepel, L., Lessard, M., Gagnon, N., et al., 2006. Effect of Postruminal Glutamine Supplementation on Immune Response and Milk Production in Dairy Cows [J]. J. Dairy Sci., 89: 3107-3121.

Douglas, G. N., Overton, T. R., Bateman, H. G., et al., 2006. Prepartal plane of nutrition, regardless of dietary energy source, affects periparturient metabolism and dry matter intake in Holstein cows [J]. J. Dairy Sci., 89: 2141-2157.

Erskine, R. J., Eberhart, R. J., Grasso, P. J., et al., 1989. Induction of Escherichia coli mastitis in cows fed selenium - deficient or selenium - supplemented diets [J]. Amer J. Vet. Res., 50: 2093-2100.

Erskine, R. J., Eberhart, R. J., Scholz, R. W., 1990. Experimentally induced Staphylococcus aureus mastitis in selenium-deficient and selenium-supplemented dairy cows [J]. J. Amer. Vet. Med. Assoc., 51: 1107-1111.

Harmon, R. J., Torre, P. M., 1994. Copper and zinc: do they influence mastitis [J]. Proc. Natl. Mast. Council., 54-65.

Horst, R. L., Jorgensen, N. A., 1982. Elevated Plasma Cortisol during Induced and Spontaneous Hypocalcemia in Ruminants [J]. J. Dairy Sci., 65: 2332-2337.

Jukola, E., Hakkarainen, J., Saloniemi, H., et al., 1996. Blood selenium, vitamin E, vitamin A, and B-carotene concentrations and udder health, fertility treatments and fertility [J]. J. Dairy Sci., 79: 838-845.

Kimura, K., Reinhardt, T. A., Goff, J. P., 2006. Parturition and hypocalcemia blunts calcium signals in immune cells of dairy cattle [J]. J. Dairy Sci., 89: 2588-2595.

Lessard, M., Gagnon, N., Godson, D. L., et al., 2004. Influence of parturition and diets enriched in n-3 or n-6 polyunsaturated fatty acids on immune response of dairy cows during the transition period [J]. J. Dairy Sci., 87: 2197-2210.

Lessard, M., Gagnon, N., Petit. H. V., 2003. Immune response of postpartum dairy cows fed flaxseed [J]. J. Dairy Sci., 86: 2647-2657.

Michal, J. J., Heirman, L. R., Wong, T. S., et al., 1994. Modulatory effects of dietary B-carotene on blood and mammary leukocyte function in periparturient dairy cows [J]. J. Dairy Sci., 77: 1408-1421.

Nelson, C. D., Lippolis, J. D., Reinhardt, T. A., et al., 2016. Vitamin D status of dairy cattle: Outcomes of current practices in the dairy industry [J]. J. Dairy Sci., 99: 10150-10160.

Oldham, E. R., Eberhart, R. J., Muller, L. D., 1991. Effects of supplemental vitamin A and B-carotene during the dry period and early lactation on udder health [J]. J. Dairy Sci., 74: 3775-3781.

Perkins, K. H., Vandehaar, M. J., Tempelman, R. J., et al., 2001. Negative energy balance does not decrease expression of leukocyte adhesion or antigen-presenting molecules in cattle [J]. J. Dairy Sci., 84: 421-428.

Santos, M. V., Lima, F. R., Rodrigues, P. H. M., et al., 2001. Plasma ascorbate concentrations are not correlated with milk somatic cell count and metabolic profile in lactating and dry cows [J]. J. Dairy Sci., 84: 134-139.

Scaletti, R. W., Trammell, D. S., Smith, B. A., et al., 2003. Role of dietary copper in enhancing resistance to Escherichia coli mastitis [J]. J. Dairy Sci., 86: 1240-1249.

Smith, K. L., Hogan, J. S., Weiss, W. P., 1997. Dietary vitamin E and selenium affect mastitis and milk quality [J]. J. Anim. Sci., 75: 1659-1665.

Smith, K. L., Todhunter, D. A., Schoenberger, P. S., 1985. Environmental Mastitis: Cause, Prevalence, Prevention [J]. J. Dairy Sci., 68: 1531-1553.

Sordillo, L. M., 2005. Factors affecting mammary gland immunity and mastitis susceptibility [J]. Livest Prod Sci., 98: 89-99.

van Knegsel, A. T. M., de Vries Reilingh, G., Meulenberg, S., et al., 2007. Natural Antibodies Related to Energy Balance in Early Lactation Dairy Cows [J]. J. Dairy Sci., 90: 5490-5498.

Weiss, W. P., Hogan, J. S., 2007. Effects of vitamin C on neutrophil function and responses to intramammary infusion of lipopolysaccharide in periparturient dairy cows [J]. J. Dairy Sci., 90: 731-739.

Weiss, W. P., Hogan, J. S., Smith, K. L., 2004. Changes in vitamin C concentrations in plasma and milk from dairy cows after an intramammary infusion of *Escherichia coli* [J]. J. Dairy Sci., 87: 32-37.

Weiss, W. P., Spears, J. W., 2005. Vitamin and trace mineral effects on immune function of ruminants [R]. Pages 473-496 in 10th International Symp. on Ruminant Physiology. Wageningen, Denmark, Copenhagen, Denmark.

Zerbe, H., Schneider, N. R., Leibold, W., et al., 2000. Altered functional and immunophenotypical properties of neutrophilic graulocytes in postpartum cows associated with fatty liver [J]. Theriogenology, 54: 771-786.

Nutrition, Metabolism, and Immune Dysfunction Affecting Reproduction in Postpartum Dairy Cows

Matthew C. Lucy

Department of Animal Sciences University of Missouri, Columbia 65211

Abstract

Milk and milk solids production per cow is increasing annually in dairy systems. Peak milk production is in early lactation when the uterus and ovary are recovering from the previous pregnancy. The competing processes of milk production, uterine involution, the restoration of ovarian activity and pregnancy can be at odds, particularly if the unique homeorhetic processes that typify early lactation become imbalanced and the cow experiences negative energy balance and (or) metabolic disease. Although glucose is a major product of carbohydrate digestion in the rumen it is rapidly fermented to volatile fatty acids (VFA). Glucose, therefore, must be resynthesized in the liver of the postpartum cow via gluconeogenesis. An early lactation cow will produce approximately 50 kg of milk per day. This equates to a glucose requirement for milk production alone of 3.6 kg per day. Homeorhesis leads to an increase in the synthesis of glucose via gluconeogenesis that is irreversibly lost to milk lactose. Irreversible loss of glucose during lactation can invoke an endocrine and metabolic state that impinges upon postpartum uterine health, estrous cyclicity and subsequent establishment of pregnancy. The first 30 days postpartum may be most critical in terms of the impact that metabolites and metabolic hormones have on reproduction. Immune dysfunction caused in part by the postpartum metabolic profile leads to a failure in uterine involution and uterine disease. Understanding homeorhetic mechanisms that involve glucose and collectively affect postpartum uterine health, cyclicity and pregnancy should lead to methods to improve postpartum fertility in dairy cows.

Introduction

The cow undergoes a series of homeorhetic mechanisms postpartum that increase glucose supply through hepatic gluconeogenesis (Bauman and Currie, 1980). The cow also assumes a state of insulin resistance that redirects glucose to the mammary gland. In spite of these mechanisms, the postpartum cow has chronically low blood glucose concentrations because she fails to meet the glucose requirement. Negative energy balance and inadequate blood glucose during early lactation compromises the function of tissues that control reproduction. The first 30 days postpartum may be the most-critical in terms of the impact that metabolites and metabolic hormones have on reproduction. Two essential processes occur during the first 30 days postpartum—the restoration of ovarian cyclicity and uterine involution. These two essential processes may be directly affected by glucose supply.

Restoration of Ovarian Cyclicity Postpartum

LeRoy et al. (2008) concluded that glucose and insulin were the most-likely molecules to exert an effect on hypothalamic GnRH and pituitary LH secretion in the postpartum dairy cow. Velazquez et al. (2008) found that there is a positive association between insulin, IGF1, and the day postpartum that the cow begins to cycle. At the level of the ovary, both insulin and IGF1 promote the proliferation, differentiation, and survival of follicular cells and act synergistically with LH and FSH (Lucy, 2008, 2011). Glucose controls insulin secretion in the whole animal and ultimately controls hepatic IGF1 secretion viainsulin release. Circulating glucose and the insulin/IGF1 systems, therefore, are functionally linked to the reproduction of the cow through their effects on cyclicity (Lucy, 2011).

The associations between postpartum hormone and metabolites and subsequent reproduction are found early postpartum when the most-extreme homeorhetic states are known to occur. The early postpartum metabolic profile, therefore, may have the capacity to imprint ovarian tissue either through permanent effects on the genome (epigenetic mechanisms) or by changing the chemical composition of the cells themselves. Perhaps the best-studied example of the metabolic imprint is the relationship between early postpartum NEFA and its effect on the composition of the oocyte and function of follicular cells (Leroy et al., 2011).

Uterine Health and Immune Function

The re-initiation of ovarian activity postpartum is a traditional focus of studies of postpartum metabolism. Recently, however, greater emphasis has been placed on uterine health and the central place that uterine immune cell function occupies in determining the reproductive success of the postpartum cow (LeBlanc, 2012; Wathes, 2012). Under normal circumstances, uterine involution is completed during the first month postpartum. During involution, the uterus shrinks in size, re-establishes the luminal epithelium, and immune cells (primarily polymorphonuclear neutrophils or PMN) infiltrate the uterus to clear residual placental tissue as well as infectious microorganisms (LeBlanc et al., 2011).

Nearly all cows have pathogenic bacteria in the uterus postpartum. Some cows develop puerperal (acute) metritis which is clinically diagnosed by fever, loss of appetite, fetid discharge, and inflammation of the uterus. Other cows develop clinical metritis (fetid discharge without fever or loss of appetite) or subclinical endometritis (superficial inflammation of the endometrium). One-half to two-thirds of cows remain healthy (no uterine disease). Cows that develop uterine disease suffer from long-term infertility (Pinedo et al., 2015).

The postpartum cow has a dysfunctional immune system particularly during the first month after calving. The current theory is that the metabolic environment in postpartum cows suppresses the innate immune system through effects on PMN function (Graugnard et al., 2012; LeBlanc, 2012). In most cases, changes in circulating concentrations of nutrients and metabolites that occur in the postpartum cow are exactly opposite to those that would benefit the function of PMN. Epidemiological evidence that indicates that an abnormal metabolic profile during the periparturient period predisposes the cow to uterine disease during the early postpartum period and infertility later postpartum (Chapinal et al., 2012).

Glucose is the primary metabolic fuel that PMN use to generate the oxidative burst that leads to killing activity. The glucose is stored as glycogen within the PMN. Galvão et al. (2010) observed that cows developing uterine disease had lesser glycogen concentration in their PMN. Their conclusion was that the lesser glycogen reserve led to a reduced capacity for oxidative burst in PMN that predisposed the cow to uterine disease. A cow's homeorhetic capacity (i.e., capacity for gluconeogenesis, lipid mobilization, etc.) and her inherent resistance to disease are largely manifested after calving but the underlying biology is theoretically in place before she calves.

Uterine Involution As an Inflammatory Process

Inflammation is an important process in both healthy and diseased individuals that must be held in check (Buckley et al., 2011). Too little inflammation in response to infection, for example, leads to a failed response to the pathogen, a heightened disease state and possible death. Too much inflammation leads to pain, swelling, and tissue damage (scarring and fibrosis) that may have long-term consequences in terms of tissue function. Detachment of the placenta and the expulsion of both placenta and calf cause hemorrhage and physical trauma in the uterus. At the same time, the lumen of the uterus is exposed to environmental pathogens that can rapidly cause disease. The tissue damage and infection that occurs postpartum lead to a massive inflammatory response within the uterus; this is particularly true for cows that mount an unsuccessful disease defense and develop metritis (clinical disease of the postpartum uterus).

We propose the following model (Figure 1). Injury and disease lead to inflammation through the release of pathogen-associated molecular patterns (PAMPs), damage-associated molecular patterns (DAMPs) and reactive oxygen species (ROS). Cytokine release from immune cells then stimulates stem cell proliferation and adult cell differentiation. The extent of the inflammatory response to the microbiome determines the success of the regenerative process. For example, too little inflammation and the tissue may be lost; too much inflammation and the tissue may become either fibrotic (scarring; Stramer et al., 2007) or cancerous (West et al., 2015). We believe that different populations of clonal cell lines develop within a healthy compared with diseased postpartum uterus and give rise to uniquely imprinted cells within the endometrium. This mechanism gives rise to long-term changes in the transcriptome of the uterus and affects the capacity of the uterus to establish and also maintain a pregnancy.

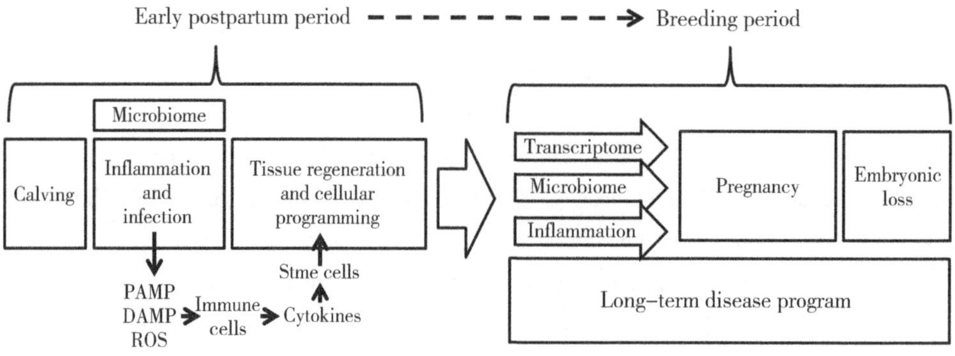

Figure 1 Model for disease programming in the postpartum cow uterus

Conclusions

The endocrine and metabolic environment of the lactating cow affects the capacity of the cow to become pregnant postpartum. The hormones responsible for the homeorhetic mechanisms that support lactation can also act on the ovary, uterus, and immune system to affect their function prior to and during the breeding period. The specific mechanism through which the metabolic environment of early lactation deposits a lasting imprint on ovarian and uterine function is less clear. Also less clear are the mechanisms that link lactation to a predisposition for pregnancy loss in the lactating cow. Uterine inflammation in lactating cows may be an important mechanism explaining pregnancy loss.

References

Bauman D. E., Currie W. B., 1980. Partitioning of nutrients during pregnancy and lactation: a review of mechanisms involving homeostasis and homeorhesis [J]. J. Dairy Sci., 63: 1514-1529.

Buckley C. D., Gilroy D. W., Serhan C. N., et al., 2013. The resolution of inflammation [J]. Nat. Rev. Immunol., 13: 59-66.

Chapinal N., Leblanc S. J., Carson M. E., et al., 2012. Herd-level association of serum metabolites in the transition period with disease, milk production, and early lactation reproductive performance [J]. J. Dairy Sci., 95: 5676-5682.

Galvão K. N., Flaminio M. J., Brittin S. B., et al., 2010. Association between uterine disease and indicators of neutrophil and systemic energy status in lactating Holstein cows [J]. J. Dairy Sci., 93: 2926-2937.

Graugnard D. E., Bionaz M., Trevisi E., et al., 2012. Blood immunometabolic indices and polymorphonuclear neutrophil function in peripartum dairy cows are altered by level of dietary energy prepartum [J]. J. Dairy Sci., 95: 1749-1758.

LeBlanc S. J., 2012. Interactions of metabolism, inflammation, and reproductive tract health in the postpartum period in dairy cattle. Reprod [J]. Domest. Anim., 47 (5): 18-30.

LeBlanc S. J., Osawa T., Dubuc J., 2011. Reproductive tract defense and disease in postpartum dairy cows [J]. Theriogenology, 76: 1610-1618.

Leroy J. L., Rizos D., Sturmey R., et al., 2011. Intrafollicular conditions as a major link between maternal metabolism and oocyte quality: a focus on dairy cow fertility [J]. Reprod. Fertil. Devel., 24: 1-12.

Leroy J. L., Vanholder T., Van Knegsel A. T., et al., 2008. Nutrient prioritization in dairy cows early postpartum: mismatch between metabolism and fertility [J]. Reprod. Domest. Anim., 43 (2): 96-103.

Lucy M. C., 2008. Functional differences in the growth hormone and insulin-like growth factor axis in cattle and pigs: implications for post-partum nutrition and reproduction [J]. Reprod. Domest. Anim., 43: 31-39.

Lucy M. C., 2011. Growth hormone regulation of follicular growth [J]. Reprod. Fertil. Devel., 24: 19-28.

Pinedo P. J., Velez J. S., Bothe H., et al., 2015. Effect of intrauterine infusion of an organic-certified product on uterine health, survival, and fertility of dairy cows with toxic puerperal metritis [J]. J. Dairy Sci., 98: 3120-3132.

Stramer B. M., Mori R., Martin P., 2007. The inflammation-fibrosis link? A Jekyll and Hyde role for blood cells during wound repair [J]. J.Invest.Dermatol, 127: 1009-1017.

Velazquez M. A., Spicer L. J., Wathes D. C., 2008. The role of endocrine insulin-like growth factor-I (IGF-I) in female bovine reproduction [J]. Domest. Anim. Endocrinol., 35: 325-342.

Wathes D. C., 2012. Mechanisms linking metabolic status and disease with reproductive outcome in the dairy cow [J]. Reprod. Domest. Anim., 47 (4): 304-312.

West N. R., McCuaig S., Franchini F., et al., 2015. Emerging cytokine networks in colorectal cancer [J]. Nat. Rev. Immunol., 15: 615-629.

Role of Nutrition During Pregnancy on Cow and Calf Health and Performance

Juan J. Loor[1] and Danielle N. Coleman[2]

[1]Professor of Animal Sciences and Nutritional Sciences, [2]PhD Candidate,
Department of Animal Sciences, University of Illinois, Urbana, IL 61801

Abstract

The periparturient and neonatal periods are the most challenging in the life cycle of dairy cattle. As a result, understanding the mechanisms coordinating metabolic and physiologic adaptations in keys organs of the cow and calf during those life stages remains an active area of research. Nutritional management approaches and the environment can impact the pregnant cow and the developing calf. Ample evidence from research with non-ruminant species underscores the potential for nutrition during pregnancy to induce "developmental/nutritional programming" of the offspring, in part through "epigenetic" mechanisms that can alter gene transcription and organ function. Epigenetic changes can occur irrespective of the genetic code of the animal. Level of dietary crude protein, energy supply, heat stress, or the supply of nutrients such as methionine, choline, folic acid (i.e. "methyl donors") during pregnancy can trigger epigenetic alterations in the offspring. Some of these factors have been studied extensively in the context of health and productivity of the peripartum cow; however, there is a paucity of information as it pertains to the calf. With the growing emphasis on nutritional management of the calf during the pre-weaning period in the context of productivity in later life, it is important to ascertain what (if any) role maternal nutrition has on the neonatal response. In the context of nutritional management to help cow and calf, recent work with rumen-protected methyl donors underscores the dual benefit of enhancing the supply of essential nutrients for the cow and calf.

Introduction

It is well-accepted among dairy nutritionists and physiologists that the period around parturition ("periparturient" or "transition" period) is one of the most-challenging in the life cycle of dairy cows (Loor et al., 2013; Bradford et al., 2015). Nutrient requirements of dairy cows increase as gestation progresses, largely due to the exponential growth of the gravid uterus and fetus (NRC, 2001). Conditions such as inflammation and oxidative stress and fat depot mobilization in the latter stages of pregnancy contribute to reducing voluntary dry matter intake (DMI) with a consequent shortfall in nutrient availability for the cow and fetus (Loor et al., 2013ab; Bradford et al., 2015). From an efficiency standpoint, improving nutrient utilization around the time of parturition by keeping cows healthy has been a major area of research for over 40 years (for example, Coppock et al., 1972). In particular, the reduction of DMI during transition is central to the increased risk of metabolic disorders (ketosis, fatty liver, milk fever), but also immune-related disorders that cows face. Recognition of the complexity of biological interactions occurring in the transition cow has resulted in a substantial amount of research focused on identifying mechanisms underlying metabolic, physiologic, and immune adaptations in key organs (see reviews by Loor et al., 2013ab, Roche et al., 2013; Bradford et al., 2015). Therefore, the main objectives of the present paper are to briefly summarize adaptations in metabolic events in key organs of the cow during transition, critical physiological events in the neonatal period, linkages between certain nutrients and some of the metabolic events, introduce the concept of "nutritional programming" in the context of transition cow nutrition, molecular mechanisms involved, and the potential "dual benefit" of nutritional management for cow and calf.

Biological Adaptations in the Transition Cow

The focus of the dairy industry in most countries over the past several decades has been maximising milk yield/cow, thereby creating a "nutrient highway" from the daily ration and mobilization of body reserves [~0.6 kg fat/d, ~0.04 kg protein/d, and ~0.15 kg water/d during the first 8-weeks of lactation (Tamminga et al., 1997)] directly to the udder to sustain milk production. In the transition dairy cow a series of biological mechanisms bring about the prioritization of milk production at the cost of body reserves (Bauman and Currie, 1980); for example, insulin concentrations are drastically reduced and the response of hormone-sensitive lipase in adipose tissue to lipolytic stimuli in high-yielding dairy cows (e.g. low insulin, high growth hormone and catecholamines, or high glucocorticoid concentrations) is greater, thus, facilitating lipid mobilisation and transport to tissues. The transition period is also characterized by a state of inflammation that can be measured by an increase in the production of positive acute-phase proteins (APP) such as haptoglobin and serum amyloid A (SAA), and a concomitant decrease in the production of negative APP such as albumin (Bertoni et al., 2008). At the level of the liver, the well-established triggers of these responses are the pro-inflammatory cytokines IL-6, IL-1β, and TNF-α (Bradford et al., 2015). On the other hand, oxidative stress is driven by the imbalance between the production of reactive oxygen metabolites (ROM), reactive nitrogen species (RNS), and the neutralizing capacity of antioxidant mechanisms in tissues and blood. Some of the well-established cellular antioxidants include glutathione, taurine, superoxide dismutase (SOD), and vitamins A and E (Bertoni et al., 2008). When oxidative stress overwhelms cellular antioxidant capacity, the ROM can induce an inflammatory response which is controlled via changes in mRNA abundance of transcription regulators (e.g. STAT3, NFKB). A summary of the changes in concentrations of a number of blood biomarkers associated with metabolism and health status of the cow are presented in Table 1 (modified from Loor et al., 2013).

Table 1 Trends in concentrations of some metabolic and inflammatory biomarkers during the periparturient period. when available the relative changes over time post-partum (days, d) are indicated. arrows denote decrease (⇓) or increase (⇑) in concentration. Modified from Loor et al. (2013)

Biomarker	Relative change between late-pregnancy and early lactation (days, d)
Metabolism	
NEFA	⇑ to ~14 d, then ⇓ ~14–40 d
Glucose	⇑ at 0 d, then ⇓ by 7 d, then ⇑ by 63 d
BHBA	⇑ to 7 d, then ⇓ by ~14–35 d
Urea	⇑ at 0 d, then ⇓ 7 d through ~14 d, then ⇑ 63 d
Lactate	⇑ at 0 d, then ⇓ by 7 d and changes little
Inflammation and liver function	
Ceruloplasmin	⇑ to ~4 d, then ⇓ by 14 d
Haptoglobin	⇑ to 4–7 d, then ⇓ by ~14–18 d through 28 d
IL-6	⇓ to 0 d, then ⇑ by 4 d and ⇓ by 9 d
TNF-α	⇓ to 0 d, then ⇑ to 8 d, then ⇓ by 30 d
Albumin	⇓ to 7 d, then ⇑ by 28 d
Bilirubin	⇑ to 4 d ⇓ 21 d
Globulin	⇓ to 0 d ⇑ 63 d
Reactive oxygen metabolites	⇑ to 7 d ⇓ 28 d
Paraoxonase	⇓ to 0 d, then ⇓ to 0–7 d ⇑ 63 d
Cholesterol	⇓ to 0 d, then ⇓ to 0–7 d ⇑ 28 d
Globulin	⇓ to 0 d, then ⇑ to ~30 d, and changes little
Vitamin A	⇓ to 0 d, then ⇑ to ~30 d, and changes little
Alpha-tocopherol[1]	⇓ to 0 d, then ⇑ to ~51 d

(Continued)

Biomarker	Relative change between late-pregnancy and early lactation (days, d)
Alkaline phosphatase	⇓ to 0–28 d
Lactate dehydrogenase	⇑ to 0–28 d
Gamma-glutamyl transferase	⇑ to 0–28 d
Glutamic oxaloacetic acid transferase	⇑ to 0–28 d
Glutathione peroxidase[2]	⇓ to 0–51 d
Selenium	If no supplementation ⇓ to 0 d, then unchanged; If supplemented from dry-off stable to 0 d, then ⇑ to 21 d and remains elevated to 51 d

[1] Concentration of alpha - tocopherol decreases whether diet is supplemented with selenium or not. However, concentration increases to a greater extent with selenium supplementation from dry off to peak lactation. [2] Activity of glutathione peroxidase decreases whether diet is supplemented with selenium or not. However, supplementing selenium from dry off to peak lactation reduces the rate of the decrease in activity.

Most research evidence, although derived mainly from rodent work, indicates that changes in abundance of mRNA exert a major influence on physiological function. Many of the homeorhetic changes (i.e. chronic changes in multiple body tissues to support a dominant physiological state) correspond with changes in transcript abundance (i. e. abundance of mRNA) for key genes, indicating that molecular changes underpin the physiological perturbations associated with the transition from pregnancy to lactation (Loor, 2010). In the case of homeorhesis, tissue responses are coordinated by a complex network of proteins that 'share' information arising from cues (e. g. hormones and metabolites) from within the organ or from the external milieu (e.g. the blood). These networks have evolved so that tissues can accurately respond to external signals and either maintain homeostasis or provide priority to certain physiological functions (e. g. milk production); hence, regulation of mRNA transcription plays a pivotal role in coordinating physiological function during the transition period.

With advancing technology enabling more targeted molecular assays simultaneously across multiple tissues (Bionaz and Loor, 2012), knowledge of the mechanisms underpinning periparturient homeorhetic changes and, arguably, more importantly, the effect of environmental factors (e. g. nutrition, thermal stress) on metabolism and other physiologic functions (e. g. immune system) continues to increase (Table 2). In general, it is noteworthy that changes in mRNA abundance for a number of key genes encoding important enzymes or proteins correspond with published biochemical studies of enzyme activity. With the quick development and lower costs for utilizing "high-throughput" molecular technologies, it is expected that additional mechanistic information pertaining to other levels of regulation of tissue function [e.g. non-protein coding RNA (microRNA)] will be generated in the near future.

Table 2 Trends in key metabolic pathways in liver, adipose, and mammary tissue of dairy cattle from late-pregnancy to early post-partum. Modified from Loor (2010) and roche et al. (2013)

Organ and pathway	Biological process	Change between late-pregnancy and	
		First week postpartum	Second to fifth week postpartum
Liver			
Ureagenesis	Arginine biosynthesis	No change to decrease	Modest increase
Glucose metabolism	Glycolysis and TCA cycle	Modest increase	No change to increase
Growth hormone signalling	IGF-1 binding/transport	Decrease	No change to decrease
Gluconeogenesis	Glucose synthesis	No change to modest increase	No change to increase
Lipoprotein metabolism	Synthesis of lipoprotein	Decrease	Decrease
Cholesterol metabolism	Synthesis and transport	No change to decrease	Modest increase
Fatty acid transport	Cellular uptake	Modest increase	Modest increase
Fatty acid oxidation	Mitochondrial and peroxisomal degradation of long-chain fatty acids	Modest increase	Modest increase
Fatty acid esterification	Long-chain fatty acid transfer into triacylglycerol	Increase	Increase
Ketogenesis	Synthesis of ketone bodies	Decrease	Increase
Lipid droplet formation	Desaturation and cytosolic lipid storage	Modest increase	No change to modest increase
Adipose			
Lipid and carbohydrate metabolism	Lipogenesis and adipogenesis; transcriptional regulation; glucose uptake	Marked decrease	Sustained decrease
Lipolysis	Hormone-stimulated and basal lipolysis	Modest increase	Sustained modest increase
Insulin signalling	IRS-1 phosphorylation	Decrease	Not assessed
Insulin signaling pathway	Gene expression	Decrease	Modest to moderate increase

(Continued)

Organ and pathway	Biological process	Change between late-pregnancy and	
		First week postpartum	Second to fifth week postpartum
Fatty acid transport and nutrient use	Long - chain fatty acid uptake, transport, and lactate utilisation	Modest decrease	Modest to moderate increase
Mammary			
Lipid metabolism	Fatty acid synthesis, triacylglycerol synthesis, cholesterol and sphingolipid synthesis, desaturation	Marked increase	Marked increase
Carbohydrate metabolism	Lactose synthesis	Marked increase	Marked increase
Energy metabolism	Oxidative phosphorylation and Krebs cycle	Increase	Increase
Amino acid metabolism	His, Val, Leu, and Ile metabolism	Increase	Increase

"Functional Roles" of Nutrients: Relevance to the Transition Period

The fact that certain nutrients serve other functions besides being building blocks of macromolecules is well-established in monogastrics. Among the essential nutrients, poly-unsaturated fatty acids, essential amino acids (AA), B vitamins, choline, and trace minerals are recognized as precursors for molecules that have inflammatory activity (eicosanoids), antioxidant activity (glutathione, taurine) or serve an essential role in the catalytic activity of important enzymes (vitamin B_{12}, Zn, Mn). The 1-carbon metabolism pathway is particularly important in the context of functional nutrition because it links the metabolism of methionine (Met), choline, B vitamins, and folic acid through various routes that not only generate antioxidants but also carnitine (via Tri-CH_3-Lysine metabolism) and phosphatidylcholine (PC), which are important in hepatic lipid metabolism (Figure 1). All these biological processes clearly are important in the context of dairy cattle efficiency, as exemplified, for example, by the fact that an infectious challenge in sheep (McNeil et al., 2016) or beef cattle (Burciaga - Robles, 2009) caused marked changes in tissue AA and protein metabolism. Under such conditions, AA utilization by the liver increases dramatically and diverts them from anabolic purposes such as milk protein synthesis in lactating cows or muscle deposition in growing steers or heifers. In addition, recent research has suggested that cells of the immune system and the reproductive tract have unique requirements for AA, particularly during early lactation (Jafari et al., 2006; Garcia et al., 2016; Noleto et al., 2017). The role of choline in the function of immune cells has also been recently highlighted, both in lactating cows (Garcia et al., 2018) and neonatal calves (Abdelmegeid et al., 2017). Application of the functional concept for these nutrients requires rumen-protection technology, and there already are several commercial products that can deliver AA, choline, and some B vitamins post-ruminally.

The liver is the main organ where the 1-carbon metabolism operates (Speckmann et al., 2017), with PC synthesis (at least in non-ruminants) being the largest 1-carbon sink and necessary for the packaging and export of triacylglycerol into very low-density lipoproteins (VLDL; Corbin and Zeisel, 2012). At least in rodents, a deficiency in both Met and choline can induce hepatic steatosis, oxidative stress, and inflammation (Jha et al., 2014). McCarthy et al. (1968) first hypothesized that Met deficiency in ruminants may limit hepatic VLDL synthesis and be a causative factor of ketosis. Rate of hepatic VLDL synthesis was subsequently demonstrated to be lower in ruminants than monogastrics (Pullen et al., 1990). This inherent feature of ruminants is particularly important at parturition when homeorhetic adaptations lead to marked increases in blood NEFA which are taken up by liver, hence, increasing the susceptibility for hepatic lipidosis (Grummer, 1993). Several studies since then have assessed the role of Met as a potentially limiting AA in the regulation of hepatic fatty acid metabolism. Because of extensive ruminal degradation, early work evaluating Met utilized intravenous infusions or a "hydroxyl analog" of Met (Bertics and Grummer, 1999. McCarthy et al., 1968. Piepenbrink et al., 2004). The analog offers some protection against ruminal metabolism but there are currently other protection technologies that ensure a greater level of ruminal "by-pass" as well as high intestinal Met bioavailability (Berthiaume et al., 2006). A review of the literature also indicated that rumen-protected choline around parturition can reduce liver triacylglycerol concentration in some (Shahsavari et al., 2016), but not all instances (Zenobi et al., 2018).

Grummer (1993) proposed that utilization of TAG for VLDL synthesis after parturition is impaired when the level of hepatic Met is insufficient. More recent work has established an association between low levels of serum Met during the first 14 days postpartum and severe hepatic lipidosis (Shibano and Kawamura, 2006). The work of Dalbach et al. (2011) demonstrated that it is feasible to increase the serum concentration of Met during the first 2 weeks postpartum by feeding rumen-protected Met (RPM). The rate of hepatic metabolism in high-producing cows nearly doubles after parturition (Reynolds et al., 2003), which could be one reason explaining the increase in net liver uptake of Met (Larsen and Kristensen, 2013). In fact, other than histidine, Met was the only AA for which net uptake by the liver increased between pre and postpartum (Larsen and Kristensen, 2013). Aspects of Met metabolism in liver via the 1-carbon metabolism (Figure 1) are well-described in monogastrics, and to some extent in classical studies with sheep (Snoswell and Xue, 1987). In a recent study from our group, Zhou et al. (2017) provided the first demonstration that activity of betaine - homocysteine methyltransferase (BHMT), methionine synthase (MTR), and cystathionine synthase (CBS) (Figure 1) increases around parturition and might be responsive to the supply of Met.

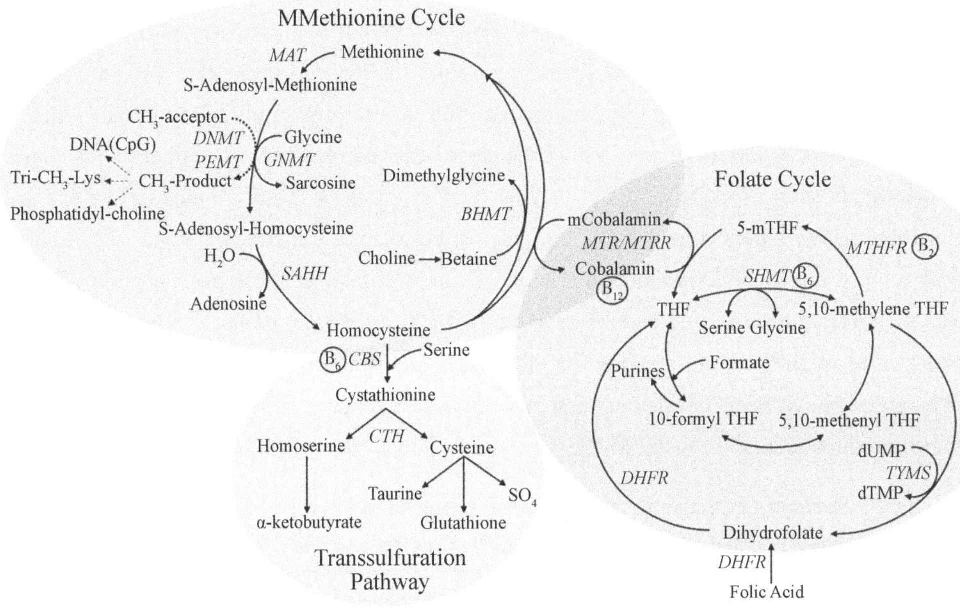

Figure 1 1 - carbon metabolic pathways. enzymes (in bold): BHMT = betaine homocysteine methyltransferase; CBS = cystathionine beta synthase; CTH = cystathionine gama-lyase; DHFR = dihydrofolate reductase; DNMT = DNA methyltransferase; GNMT = glycine N-methyltransferase; MAT = methionine adenosyltransferase; MTHFR = methylenetetrahydrofolate reductase; MTR = 5-methyltetrahydrofolate-homocysteine methyltransferase; MTRR = 5-methyltetrahydrofolate-homocysteine methyltransferase reductase; PEMT = phosphatidylethanolamine N-methyltransferase; SAHH = s-adeonsoyl-homocysteine hydrolase; SHMT = serine hydroxymethyltransferase; TYMS = thymidylate synthetase. Substrates: 5-mTHF = 5-methyl-tetrahydrofolate; dUMP = deoxyuridine monophosphate; dTMP = thymidine monophosphate; THF = tetrahydrofolate. B vitamins: B_2 = riboflavin; B_6 = Pyridoxal 5′-phosphate; B_{12} = cobalamin (Atmaca, 2004)

Rumen-protected Methionine in Transition Cows

Increasing the delivery of Met to the liver via supplementation of rumen-protected sources is particularly important for the animal, not only because of the key role of Met in milk protein synthesis but also for production of glutathione and taurine (intracellular antioxidants, Figure 1), and provision of methyl groups (Finkelstein, 1990). Thus, although not the only essential nutrient with such a role, Met is one example of an AA with a clear functional role (Figure 1). A series of recently published studies with dairy cows has demonstrated the unique role of Met, beyond serving as a source of AA for protein synthesis, in the context of allowing cows to maintain more consistent rates of dry matter intake around parturition, reduce inflammation and oxidative stress, have a better innate immune function, remain healthier, and optimize production of milk (Osorio et al., 2013a; Zhou et al., 2016ac; Batistel et al., 2017b, 2018).

Supplementation of Met during the peripartal period concomitantly increases milk yield, milk protein, and milk fat soon after calving (Ordway et al., 2009; Osorio et al., 2013a). These responses are in large part driven by enhancing Met availability and by additional flux of Met through the Met cycle in liver, which consequently increases the production of downstream compounds such as cysteine (Cys). Just as Met, Cys is a sulfur-containing AA and both contribute sulfur bonds during milk protein synthesis in the mammary gland (Pocius et al., 1981). In terms of milk production, work in our lab feeding RPM has detected positive responses in terms of maintaining consistent rates of DMI prepartum (last 21 days) and faster and greater rates of DMI during the first 30 to 60 days after calving (Table 3). The milk production responses have been consistent with research from other groups demonstrating benefits of postpartum supplementation of RPM (St-Pierre and Sylvester, 2005).

Table 3 Summary of major production responses in cows fed a rumen-protected Met supplemented diet during the transition period and early lactation

Item	Experiment		
	Osorio et al. (2013a)	Zhou et al. (2016a)	Batistel et al. (2017b)
Diet crude protein, %	15 (pre), 17.5 (post)	14.6 (pre), 17.3 (post)	15.7 (pre), 17.4 (post)
Lys, % of metabolizable protein (MP)	6.1-6.6	6.2-6.6	6.3-6.5
Lys, grams in MP	81 (pre), 112 (post)	85 (pre), 148 (post)	89 (pre), 156 (post)
Met, % of MP	2.15-2.35 (1.8-1.9)[2]	2.3-2.4 (1.8-1.9)[2]	2.2-2.3 (1.7-1.8)[2]
Met, grams in MP	29 (pre), 40 (post)	33 (pre), 55 (post)	32 (pre), 56 (post)
	Average response relative to the control diet[1]		
DMI, kg/d			
Prepartum	+0.5	+1.1	+1.2
Postpartum	+2.1	+2.0	+1.6
Milk yield, kg/d	+3.4	+3.8	+4.3
Energy-corrected milk, kg/d	+3.9	+4.1	+4.6

[1]Corn silage-based diet. Rumen-protected Met was supplemented from-21 through 30 days in milk in Osorio et al. (2013a) and Zhou et al. (2016a) or until 60 days in milk in Batistel et al. (2017b).[2] Range in the control unsupplemented diet. [3]Amounts in the prepartum (pre) or postpartum (post) diet.

The transient inflammatory-like status around parturition appears to be a "normal" aspect of the adaptations to lactation (Bradford et al., 2015), with its positive or negative impact depending on its degree. Cows that approach parturition with a greater (but still subclinical) level of circulating cytokines have greater inflammation and oxidative stress, and lower liver function, often through 30 days in milk, along with lower milk yield and lower postpartum DMI (Bertoni et al., 2008). In addition to their fundamental function in immunity, cytokines (ILs), interferons (IFNs) and TNF-α also elicit pathophysiological effects. This leads to what is commonly known as "sickness behavior", whose primary manifestation is satiety. Similar to how cows react during an inflammatory state around parturition, the reduction in DMI around calving is an example of this behavior. In mice, these cytokines have been shown to reduce meal size and duration, as well as decrease meal frequency and prolong inter-meal intervals (Plata-Salaman, 1995). Furthermore, cytokines directly affect the hypothalamus. IL-1β and IFN act directly and specifically on the glucose-sensitive neurons in the brain "satiety" and "hunger" sites (Plata-Salaman, 1995). Thus, the increased DMI observed when feeding RPM can be partly explained by a reduction in inflammation, as it directly (at the hepatic level and by dampening the immune cell overresponse) and indirectly (reducing oxidative stress) decreases circulating pro-inflammatory cytokines. In our studies with supplementation of RPM, we have detected consistent responses in a number of biomarkers in plasma and liver tissue, indicating that Met helps reduce inflammatory and oxidative stress status of the cows.

In research from our group and others with RPM supplementation during the peripartal period we have consistently detected improvements in immunometabolic status (Osorio et al., 2014ab; Sun et al., 2016; Zhou et al., 2016a; Batistel et al., 2018). Furthermore, compared with rumen-protected choline, greater supply of Met resulted in increased antioxidant concentration in liver tissue despite a lower concentration of PC (Zhou et al., 2017). Those responses were due to the greater abundance of phosphatidylethanolamine methyltransferase and CBS (Zhou et al., 2017). A greater supply of choline did not change the mRNA abundance of BHMT and MTR in cows with a greater supply of choline. A summary of major effects of RPM on immunometabolic biomarkers is in Table 4.

Table 4 Summary of additional beneficial effects of feeding rumen-protected methionine during the transition period and early lactation. ↑ = beneficial increase; ↓ = beneficial decrease; ↔ = no change in concentration

Biomarker	Response[1]	Biological function
Metabolism		
Carnitine	↑ (liver)	β-oxidation of Fatty Acids
Cholesterol	↑ ↑ (plasma)	Lipoprotein metabolism
Inflammation		

(Continued)

Biomarker	Response[1]	Biological function
IL-1beta	↓ (plasma)	Pro-inflammatory cytokine
Haptoglobin	↓↓ (plasma)	Inflammation signal
Albumin	↑↑ (plasma)	Acute-phase response
Oxidative stress		
Reactive oxygen metabolites (ROM)	↔/↓ (plasma)	Peroxides, superoxide, OH-radicals
Glutathione	↑↑ (liver, blood)	Antioxidant
Taurine	↔/↑ (plasma)	Antioxidant
Antioxidant capacity	↔/↑ (plasma)	Total antioxidants in blood
Paraoxonase	↑↑ (plasma)	Antioxidant enzyme

[1]Relative to a control or rumen-protected choline supplemented diet (Osorio et al., 2013a; Zhou et al., 2017; Sun et al., 2016; Batistel et al., 2018).

Maternal Nutrition and the Developing Calf

The concept of nutritional programming

The concept that differences in nutritional experiences at critical periods in early life, both pre-and postnatally, can program an individual's development, metabolism and health for the future is generally referred to as "nutritional programming" (for example, Wu et al., 2004). There is ample evidence in non-ruminant species that maternal dietary methyl donors, such as Met, choline, folic acid, and betaine, play a role in nutritional programming (for example, Ji et al., 2015). These nutrients can elicit a programming effect, partly through "epigenetics", i.e. phenotypic changes that do not involve alterations in the DNA sequence. Epigenetic changes can occur through methylation of DNA, RNA, and histones. Dietary methyl donors serve as precursors of S-adenosyl-methionine (SAM, Figure 1) that could be used via methyltransferases (for example, GNMT) to methylate DNA, RNA and histones (Hollenbeck, 2012; Lin et al., 2014). In newborn piglets, it was demonstrated that maternal folic acid supplementation altered the expression of genes associated with immunity, oxidative stress response and hepatic energy metabolism (Liu et al., 2013). In addition, supplementing betaine to sows during pregnancy resulted in alterations in the expression of gluconeogenic genes in the liver of newborn piglets, partly through changes in DNA methylation (Cai et al., 2014).

Other epigenetic changes unrelated to methylation involve acetylation of histones and/

or changes in abundance of microRNA (non-protein coding RNA). Alterations in microRNA abundance (at least in non-ruminants) are particularly important for fine-tuning regulation of several cellular process (Aguilera et al., 2010) that modulate innate immune function, including regulation of senescence, differentiation, adherence capacity and cytokine production (Gantier, 2013). These observations could be of interest if linked to the fact that epigenetic marks are candidates for bearing the memory of specific intrauterine nutritional exposures causing alterations in long-term programming of mRNA abundance, and consequently inducing developmental adaptations in physiology and metabolism. Although the exact mechanisms whereby mature microRNA can repress or activate translational activity (Morales et al., 2017), promote destabilization of target mRNA, and regulate the abundance level of target genes remains under debate (Eulalio et al., 2008), their biological role seems unequivocal. To add more complexity, it is now understood that epigenetic regulation of histones (i.e. methylation or acetylation) also can impact microRNA abundance (Morales et al., 2017). We have recently demonstrated alterations in a number of microRNA that are expressed in adipose tissue during the periparturient period (Table 5). Both, BCS at calving and prepartum overfeeding can alter the abundance of certain microRNA involved in inflammation and fat deposition (e.g. miR-99a, miR-145, miR-155).

Table 5 Details and functions of microRNA expressed in adipose tissue of periparturient cows (Vailati-Riboni et al., 2016)

microRNA	Function or expression pattern
Infiltration of immune cells	
miR-26b	Abundance is associated with the number of macrophages infiltrating the fat depot Affected by levels of circulating TNF-α, leptin and resistin
miR-126	Directly inhibits *CCL2* mRNA abundance
miR-132	Abundance is associated with the number of macrophages infiltrating fat depots Activates inflammation via NF - κB signalling and the transcription of *IL8* and *CCL2* Lower abundance is associated with increased secretion of IL-6
miR-155	Abundance is associated with the number of macrophages infiltrating fat depots
miR-193	Indirectly inhibits *CCL2* abundance through a network of transcription factors
Inflammation and lipolysis	
miR-99a	Negative correlation with secretion of IL-6 and level of free fatty acids
miR-145	Affects secretion of TNF-α, regulating lipolysis
miR-221	Lower abundance is associated with high levels of TNF-α
Proadipogenic	

(Continued)

microRNA	Function or expression pattern
miR-103	Regulates mRNA abundance of *PPARG*, *PANK*1, *CAV*1, *FASN*, *ADIPOQ* and *FABP*4
miR-143	Regulates mRNA abundance of *ERK*5, *SLC2A*4, *TFAP2A*, *LIPE*, *PPARG*, *CEBPA*, and *FABP*4
miR-378	Targets *PPARG* mRNA abundance through the MAPK1 pathway

Nutritional programming in ruminants

The available literature on nutritional programming of various physiological aspects specifically in ruminants (for example, embryo development, placental function, muscle and fat deposition) has been reviewed in at least 3 recent comprehensive papers (Sinclair et al., 2013, 2016; Chavatte-Palmer et al., 2018). For the purpose of the present short-review we list in Table 6 nutritional factors relevant to ruminants for which there are known epigenetics roles. We also list available information on programming effects of body composition and appetite. Most available data come from sheep and beef cattle, and there is some evidence that mammary development, fertility, welfare and behavior, and immune function might be susceptible to nutritional programming (Sinclair et al., 2016).

Table 6 Dietary/nutritional factors and known epigenetic effects

Nutrient category	Nutritional factor	Epigenetic mechanism
Fatty acids	Butyrate (from gastrointestinal fermentation)	Histone modifications
One-carbon metabolism	Methionine, choline, betaine, folic acid, arginine, Vitamins B_2, B_6, and B_{12}	DNA methylation
Polyphenols	Genistein, quercetin, resveratrol	microRNA, histone modifications
Minerals	Cu, Zn, Co, Ni	microRNA, histone modifications
Dietary changes	Level of dietary energy	microRNA
	Methyl donor deficiency	DNA methylation, histone modifications
	Protein restriction	DNA methylation, histone modifications
Programming event	Biological response	
Skeletal muscle	Muscle fiber formation, mass or size	DNA methylation, microRNA
Fat	Increased deposition	DNA methylation

(Continued)

Nutrient category	Nutritional factor	Epigenetic mechanism
Appetite regulation	Alteration of neural circuits (hypothalamus)	DNA methylation
	Alterations in endocrine factors	Orexigenic gene upregulation
		Anorexigenic gene downregulation

Utero-placental nutrient transport and metabolism

In dairy cows, the final trimester of gestation is characterized by marked fetal growth, and proper placental transfer of nutrients is required to ensure adequate fetal development (NRC, 2001). Besides maternal nutrient availability, it is well-established in non-ruminants that expression and activity of specific transporters in the placenta influence the transport of nutrients from maternal to fetal circulation (Jones et al., 2007). The anatomical structure of the placenta precludes direct contact of maternal and fetal blood, emphasizing the importance of protein transporters, concentration gradients and diffusion channels for membrane nutrient exchange (Brett et al., 2014).

In bovine, the areas of maternal-fetal interface are limited to discrete round structures named placentomes, which consist of maternal caruncles interdigitating with fetal cotyledons (Brett et al., 2014; Bridger et al., 2007). In order to reach the fetal blood, nutrients from maternal circulation must surpass the syncytiotrophoblasts (Firth, 1966), which contain two polarized membranes: the microvillus membrane facing maternal circulation and the basal plasma membrane facing the fetal vascular structure (Brett et al., 2014). Both membranes have low permeability, hence, constituting rate-limiting steps for the transport of medium and large substrates into fetal circulation (Brett et al., 2014). Consequently, nutrients are predominantly absorbed and translocated into fetal circulation by nutrient-specific transport proteins situated within the microvillous membrane and basal plasma membrane (Lager and Powell, 2012).

Besides performing an indispensable function delivering essential nutrients for fetal growth, placental tissue (as other mammalian cells) has an array of additional nutrient-sensing signaling pathways, such as the mammalian target of rapamycin (mTOR) complex (Jansson et al., 2013). For instance, in humans, mTOR alters the activity of placental AA transporters in response to the level of nutrient supply (Roos et al., 2009). These nutrient-signaling mechanisms are also sensitive to hormones such as insulin, thus, shifts in the endocrine environment resulting from alterations in dietary energy density could alter nutrient delivery to the placenta (and fetus), as well as the abundance of various nutrient transporters. A recent study from our laboratory provided some of the first data demonstrating the presence of various nutrient transporters in term placenta from dairy cows (Table 7). These data are important in the context of understanding potential linkages between dietary nutrient supply and availability to the fetus.

Table 7 Transporters of amino acids, fatty acids, glucose and vitamins detected in term placentome from Holstein cows (Batistel et al., 2017a)

Transporter and gene symbol	Transporter and gene symbol
Glutamate (SLC1A1)	Glucose, galactose, mannose (SLC2A1)
Neutral AA (SLC1A5)	Glucose, galactose, mannose, maltose (SLC2A3)
Heavy chain AA (SLC3A2)	Glucose/fructose (SLC2A4)
Taurine transporter (SLC6A6)	Glucose (SLC2A5)
Branched-chain and aromatic AA (SLC7A5)	Glucose (SLC2A6)
Branched-chain and aromatic AA (SLC7A8)	Glucose and fructose (SLC2A8)
Neutral AA (SLC38A1)	Glucose and galactose (SLC2A9)
Neutral AA (SLC38A2)	Glucose (SLC2A10)
Sodium-dependent AA (SLC38A6)	Glucose (SLC2A11)
Glu, Gln, His, Ser, Ala, Asn (SLC38A7)	Glucose (SLC2A12)
Sodium - dependent AA, Ala, GLn, Glu, Asp (SLC38A10)	Glucose cotransporter (SLC2A13)
Sodium-dependent AA (SLC38A11)	Glucose, galactose, mannose (SLC5A11)
Sodium - independent, Leu, Phe, Val, Met (SLC43A2)	Multivitamin (SLC5A6)
Long chain fatty acid (SLC27A1)	Betaine (SLC6A12)
Long chain fatty acid (SLC27A2)	Thiamine (SLC19A2)
Long chain fatty acid (SLC27A3)	Thiamine (SLC19A3)
Choline (SLC44A1)	
Choline (SLC44A3)	
Folate (SLC46A1)	

Maternal Prepartum Nutrition and Dairy Calf Development

General

Modern feeding systems for dairy cattle recommend sufficient feed in late-pregnancy (i.e. last trimester of gestation) to provide ~1.5 times maintenance energy requirements to maintain maternal body condition and fetal growth (Quigley and Drewry, 1998). There are comprehensive reviews published on the potential for maternal plane of nutrition (for example, Quigley and Drewry, 1998; Funston and Summers, 2013; Khanal and Nielsen, 2017) or micronutrient supply (Reynolds et al., 2010; Bach, 2012) to alter fetal and neonatal growth of cattle and sheep. The consensus seems to be that, unless nutrition is markedly inadequate (e.g. feed deprivation, nutrient imbalances, poor quality forage in the early dry period), prepartum nutrition should not markedly affect fetal development or chemical composition. Although the current approach for feeding high-producing dairy cows during the last 3-4 weeks prepartum is designed to provide additional energy to support fetal growth (and adapt the rumen to a lactation diet), the specific requirements for essential nutrients (AA, trace minerals, B vitamins, choline) by the cow, conceptus and fetus remain unknown. However, there is evidence from bovine studies that a limitation in the supply of methyl donors can impact not only physiological responses of the cows, but also important developmental aspects in the calf that could have long-term consequences (discussed below).

Effects of maternal dietary energy level on calf development

Although there are few published studies in dairy cows designed to address the relationship between prepartal energy intake and calf development, there is some evidence that over-and under-feeding energy can have aneffect. For example, cows fed a lower-energy diet (1.25 vs. 1.55 Mcal/kg dry matter; ~13% crude protein) for the last 21 days prior to parturition delivered calves that were lighter (39.2 vs. 43.9 kg), shorter (74.7 vs. 78.0 cm), and had lower body length (72.6 vs. 74.2 cm) (Gao et al., 2012). In contrast, feeding a higher-energy diet (1.47 vs. 1.24 Mcal/kg; ~15% crude protein) for the last 3 weeks prepartum resulted in lighter calves (44.0 vs. 48.6 kg) at birth (Osorio et al., 2013b). In cattle, fetal skeletal muscle matures during late-gestation, hence, prenatal plane of nutrition of the cow at this time would impact muscle growth of the calf (Sinclair et al., 2016). Such response explains in part the lower body weight at birth and muscle mass when dams are nutrient-restricted during gestation. Despite this evidence, it should be noted that a number of studies in which the energy density (or source of energy) of the prepartum diet has been modified to address metabolic responses of the cow did not report differences in calf birth weight (for example, Rabelo et al., 2003; Dann et al., 2005, 2006; Guo et al., 2007; Janovick and Drackley, 2010). Although it is challenging to draw major conclusions across the available published experiments with dairy cows (for example, confounding effects of environment, diet composition), data from pigs (for example, folic acid and betaine) underscores the potential for specific nutrients in the pregnant cow diet to potentially alter calf development. Hence, variations in micronutrient availability to the cow and calf in the published studies dealing with the role of dietary "energy density" might account for lack of consistency in terms of measurable effects on the calf, particularly birth body weight. Without longitudinal performance data for the calves it would be impossible to determine programming effects that might been induced by maternal energy nutrition.

Programming of Neonatal Calf Immune Function

Maternal plane of nutrition, body condition score, and metabolic stress

Plane of dietary energy prepartum altered the profiles of T lymphocytes in neonatal calves (Gao et al., 2012). As discussed in a previous section, the lower energy diet prepartum led to lower calf birth weight, but also a lower ratio of $CD4^+$ to $CD8^+$ T cells, suggesting those calves were less "immunocompetent". Additional support for a negative effect on immune function was the lower concentration of interleukin-4 (IL-4), a key cytokine involved in the synthesis of immunoglobulins and also the profile of $CD4^+$ cells, which would be important in conferring the young calf immune protection against pathogens. Other aspects of immune function such as the oxidant status of the calves were altered, including lower total antioxidant capacity and superoxide dismutase activity, while activity of glutathione peroxidase and concentrations of lipid peroxide products were increased. Although this study did not address epigenetic mechanisms per se, data generated by our group indicated that maternal dietary energy could alter neonatal calf mRNA abundance of subsets of genes with important roles in the innate immune response (by neutrophils) including the control of cytokine production (Osorio et al., 2013b). These molecular effects were not associated with differences in colostrum quality.

The use of body condition score (BCS) to assess body fat and muscle reserves is well-established (for example, Roche et al., 2013), with low values reflecting emaciation and high values obesity. The BCS at which a cow calves, her nadir BCS, and the amount of BCS lost after calving are associated with milk production, reproduction, and health. Furthermore, data from our laboratory has demonstrated associations between BCS at calving and altered profiles of systemic (plasma) and molecular (liver, adipose) biomarkers of inflammation, oxidative stress, and metabolism (for example, Akbar et al., 2015). For instance, compared with a calving BCS of 3.0-3.25, calving at BCS greater or lower than that not only increases risk of metabolic disorders (for example, fatty liver and ketosis), but could also trigger a longer period of inflammation and oxidative stress. Hence, besides body reserves, BCS at calving represents an evaluation of "stress status" in dairy cows. The use of an "oxidant status index" (OSi) has recently been proposed as way to determine stress status of pregnant cows in late-pregnancy (Ling et al., 2018). A priori, the OSi (based on plasma NEFA, haptoglobin, ROS, RNS) seems to capture metabolic stress regardless of BCS; however, higher plasma concentrations of NEFA and haptoglobin during the last 28 days prior to calving also were associated with degree of metabolic stress regardless of BCS (Ling et al., 2018).

The fact that exposure to stresses (for example, heat stress) during late-pregnancy causes impaired immunefunction (Tao et al., 2012), milk and reproductive performance (Monteiro et al., 2016), and mammary development (Skibiel et al., 2018) in offspring has highlighted the importance of evaluating calf development in the context of suitable biomarkers of stress. In a recent study classifying cows according to OSi in the last 28 days prior to calving, it was reported that calves born to cows classified as experiencing higher metabolic stress had lower body weight at birth and throughout the first 4 weeks of age (Ling et al., 2018). Those calves also had greater ROS and RNS concentrations, along with greater inflammation biomarkers (plasma haptoglobin, TNF-α), indicating greater basal inflammatory status. In contrast, lipopolysaccharide (LPS)-induced inflammatory responses were less robust in calves exposed to higher maternal biomarkers of metabolic stress, suggesting a compromised immune response. Whether these negative maternal effects of metabolic stress were caused via epigenetics mechanisms is unknown.

Maternal dietary micronutrients

Nutritional supplementation with vitamins, minerals, and other micronutrients during pregnancy have been intensively studied as effectors for immune system activation, not only for the cow, to face the transition period, but also for the offspring to adapt to the extra-uterine life (Girard et al., 1995; Thornton, 2010). In the context of B vitamins and folic acid, and although requirements in dairy cows and calves are unknown, it is commonly thought that synthesis of those compounds by ruminal microorganisms is sufficient to meet demands of cow and fetus (for example, Girard et al., 1995; Ragaller et al., 2009). The absence of differences in birth body weight when folic acid supply increased during late-pregnancy is opposite to data from non-ruminants (Girard et al., 1995), underscoring the fact that dairy cows fed typical diets during late-pregannoy are in "adequate" folic acid status. Trace mineral elements such as Cu, Cr and Zn have important roles in the health and immunity of peripartal dairy cows (Spears and Weiss, 2008). The implications of trace mineral deficiency or impaired placental transfer of these minerals to fetal and neonatal ruminant metabolism have been studied for more than 30 years (Hidiroglou, 1980). For instance, dairy calves supplemented with an injectable trace mineral complex containing Se, Cu, Zn, and Mn increased PMN and glutathione peroxidase activity, while reducing incidence of diarrhea, pneumonia, and otitis (Teixeira et al., 2014). These constitute an example of the innate immune response of the animal, one in which cells such as PMN are partly regulated via signaling pathways and changes in mRNA expression.

Although changes in mRNA abundance are known to partly control adaptations in PMN due to inflammation, more recent studies have concluded that epigenetic modifications through the activity of microRNA are an important part of the regulation of several cellular process (Aguilera et al., 2010) that modulate PMN function, including regulation of senescence, differentiation, adherence capacity, and cytokine production (Gantier, 2013).

These observations and the fact that epigenetic markers are candidates for bearing the memory of specific intrauterine nutritional exposures, causing alterations in long-term mRNA abundance and cell function, led us to study the potential role of maternal organic trace mineral supplementation during late-pregnancy on microRNA profiles in the PMN from neonatal calves (Jacometo et al., 2015). Despite the lack of effect of organic (ORG) vs. inorganic (INO) trace mineral supplementation on calf birth weight, the abundance of toll-like receptor pathway genes indicated a pro-inflammatory state in INO calves, with greater abundance of various inflammatory mediators. The lower abundance of miR-155 and miR-125b in ORG calves indicated the potential for maternal ORG trace minerals in regulating the PMN inflammatory response, at least via alterations in mRNA and microRNA abundance. Because these data were from "basal" non-stimulated cells, further studies would be helpful in evaluating the benefit of these responses under challenged conditions, e.g. after an inflammatory or pathogen challenge.

Maternal Dietary Methyl Donors

Effects on the placenta

Fetal growth is greatly increased during the final third trimester of gestation, and proper placental transfer of nutrients is required to ensure adequate development (Borowicz et al., 2007). Besides blood flow, the expression and activity of specific transporters in the placenta, e.g. glucose, AA, choline can limit nutrientdelivery to the fetus (Jones et al., 2007; Regnault et al., 2005). Although not all micronutrients have been studied in the context of maternal transport mechanisms, it is well-established in non-ruminants that placental AA transport is dependent on maternal circulating AA profiles and transport capacity. In fact, both of these are affected by the composition and amount of AA in the diet (Brown et al., 2011). Among the essential AA, the gradual decrease in plasma Met concentration between-21 and 10 d relative to parturition in dairy cows suggested that it could be limiting not only for the cow but also the calf during the last stages prior to calving (Zhou et al., 2016b). A classical study demonstrated the essentiality of Met for normal embryo development (Coelho et al., 1989), i.e. embryos cultured in medium containing serum from cows supplemented with RPM, which increased serum Met from 4.8 μg/mL to ~75 μg/mL after 14 days of supplementation, developed normally *in vitro* compared with embryos cultured in serum from cows not receiving supplemental RPM (Coelho et al., 1989). Thus, in the context of maternal nutrition and fetal development, our lab has placed special interest on the link between supply of Met and choline. As such, we have begun to study not only responses to changes in the post-ruminal supply of these nutrients at the cow and calf level, but also in terms of placental metabolism.

In two recent studies, we have reported associations between supplemental RPM during the last 3 weeks prior to calving and molecular mechanisms in placenta that may account for greater calf birth body weight (Batistel et al., 2017a, 2019). In Figure 2, we depict the alterations in mRNA abundance of nutrient transporters in response to supplemental RPM. Clearly, enhancing the supply of Met in late-pregnancy can upregulate abundance of AA, glucose, and vitamin transporters, potentially enhancing the availability of those nutrients to the developing fetus. We have speculated that these responses at the placental level are partly due to the greater DMI in cows fed RPM (Table 2).

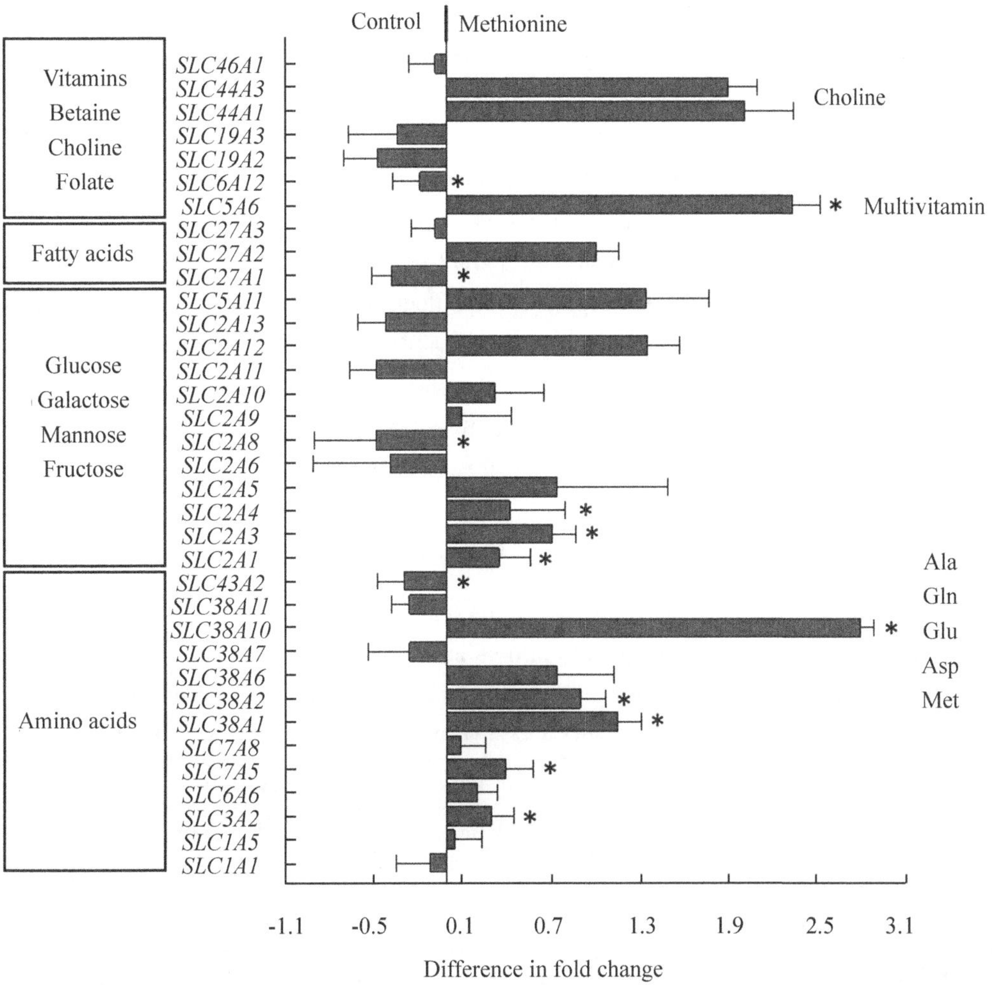

Figure 2 Fold-change difference (Methionine vs. control; negative values denote downregulation in mRNA abundance and positive values upregulation) of nutrient transporters reported in Table 4. Cows were fed a basal control diet or the basal diet plus ethyl-cellulose rumen-protected methionine (0.09% of diet dry matter during the last 28 days of pregnancy (Batistel et al., 2016)

There is evidence in non-ruminants that nutrients or dietary changes can trigger changes in epigenetic marks in placenta, with some of those changes being associated with the sex of the fetus. For example, high-fat diets triggered sex-specific gene expression in mouse placenta (Gallou-Kaban et al., 2010; Gabory et al., 2012). Those data led us to explore whether the enhanced supply of methyl donors in late-pregnancy could also alter epigenetic (for example, DNA methylation) marks in the placenta in relation to calf sex (Batistel et al., 2019). We not only evaluated total DNA methylation but also used metabolomics and gene expression analyses to assess metabolism through the 1-carbon metabolism pathway. Selected results are summarized in Table 8. A major finding from that analysis was that 1-carbon metabolism was affected differently by Met supply in placenta from cows delivering female compared with male calves. For instance, compared with placenta from male Controls, male Met placenta had greater concentrations of glutathione (derived from transsulfuration, Figure 1) and also greater MTR activity. In contrast, compared with Control, concentrations of Met and SAM, DNA methylation, and mRNA abundance of DNA methyltransferases was greater in placenta from cows fed Met and carrying female calves. The lower global DNA methylation and the greater abundance of *DNMT3A* and *DNMT3B* in placenta from cows carrying female calves and receiving a greater supply of Met underscored the complex interaction between metabolism and epigenetic modifications. A possible explanation for these seemingly-opposite results is that the extra energy available to the placenta due to greater DMI in cows carrying female calves and fed Met stimulated active DNA methylation (Bochtler et al., 2017) which was followed by *de novo* methylation (*DNMT3A* and *DNMT3B*). The DNA methylation changes detected may be one of the mechanisms behind the differences in calf growth through 9 weeks of age (discussed below).

Table 8 Selected metabolite concentrations, enzyme activity, DNA methylation, and mRNA abundance of DNA methyl transferases in term placenta from Holstein cows fed a basal control diet or the basal diet plus ethyl-cellulose rumen-protected methionine (0.09% of diet dry matter during the last 28 days of pregnancy (Batistel et al., 2016)

Items measured in placenta	Maternal dietary groups			
	Male calves		Female calves	
	Control	Methionine	Control	Methionine
Metabolites (relative units)				
Methionine	646×10^{3}	663×10^{3}	627×10^{3b}	711×10^{3a}
S-Adenosyl-methionine (SAM)	123×10^{3}	159×10^{3}	123×10^{3b}	190×10^{3a}
Folic acid	97	89	76	109
Betaine	32×10^{3}	33×10^{3}	31×10^{3}	35×10^{3}
Choline	355×10^{3}	391×10^{3}	364×10^{3}	382×10^{3}

(Continued)

Items measured in placenta	Maternal dietary groups			
	Male calves		Female calves	
	Control	Methionine	Control	Methionine
Cysteine	13× 10^3	15× 10^3	13× 10^3	14× 10^3
Taurine	31× 10^3	31× 10^3	31× 10^3	30× 10^3
Glutathione	190× 10^{3b}	383× 10^{3a}	206× 10^3	267× 10^3
Vitamin B_{12}	35^b	66^a	26	24
Enzyme activity (nmol/h/mg protein)				
BHMT	4.0	8.2	4.0	7.8
MTR	8.1^b	13.2^a	8.6	9.8
CBS	6.6	8.7	7.1	8.9
Global DNA methylation (%)	3.2	3.5	4.6^a	3.2^b
DNMT3A mRNA	0.63	0.59	0.48^b	0.61^a
DNMT3B mRNA	0.41	0.40	0.59^b	0.72^a

[ab]Means differ (P<0.10) between control and methionine groups within calf sex.

Effects on calf development in utero and the first weeks of life

The recent studies of Batistel et al. (2017b, 2018) not only confirmed the benefits of enhancing the supply of Met to dairy cows during the periparturient period on DMI, production, and health but also underscored the responsiveness of the placenta. More importantly, they demonstrated that calf development in utero and growth during the first 9 weeks of life responded to an increase in maternal supply of Met (Alharthi et al., 2018). That study allowed us to also look at a potential interaction between colostrum and maternal supply of Met (Table 9). A total of 39 calves were in Control (n = 22 bulls, 17 heifers) and 42 in Met (n = 20 bulls, 22 heifers). At birth, calves were randomly allocated considering dam treatment and colostrum as follows: (1) calves from Control cows and colostrum from Control cows (n = 21); (2) calves from Control cows and colostrum from Met cows (n = 18); (3) calves from Met cows and colostrum from Met cows (n = 22); and (4) calves from Met cows and colostrum from Control cows (n = 20). All calves were housed, managed, and fed individually during the first 9 weeks of life.

Despite greater daily DMI pre-partum in cows fed Met [15.7 kg/d vs. (14.4± 0.12) kg/d], colostrum quality and quantity were not affected by maternal diet. At birth, Met calves had greater body weight [44.1 kg vs. (42.1±0.70) kg], hip height [81.3 cm vs. (79.6±0.53) cm] and wither height [77.8 cm vs. (75.9±0.47) cm]. Regardless of colostrum source, the greater body weight, hip height, and wither height in Met calves at birth persisted through 9 weeks of age, resulting in average responses of +3.1 kg body weight, +1.9 cm hip height, and +1.8 cm wither height compared with Controls. Average daily gain during the 9 weeks was (0.72±0.02) kg/d in Met calves compared with (0.67±0.02) kg/d in Control calves. Respiratory scores were normal and did not differ due to maternal Met supply or colostrum source. However, fecal scores tended to be lower in Met calves regardless of colostrum source.

Table 9 Weekly growth parameters (1–9 weeks of age), daily starter intake and average daily gain (1–56 d of age) in calves born to Holstein cows offered a control diet (CON) or CON supplemented with ethyl-cellulose rumen-protected Met (Mepron ® at 0.09% of diet DM; Evonik Nutrition & Care GmbH, Germany) during the last 28 d of pregnancy

Item	Maternal diet		Colostrum type		SEM	*P* value[1]	
	CON	MET	CON	MET		M	C
Body weight, kg	59.3[b]	62.4[a]	61.1	60.5	1.9	0.02	0.66
Hip height, cm	86.9[b]	88.8[a]	87.5	88.1	0.68	<0.01	0.34
Hip width, cm	20.3	20.6	20.3	20.6	0.32	0.26	0.17
Wither height, cm	82.7[b]	84.5[a]	83.3	83.9	0.67	<0.01	0.31
Body length, cm	126	128	127	126	1.01	0.17	0.68
Daily starter intake, kg	0.79	0.85	0.80	0.84	0.09	0.19	0.39
Average daily gain, kg	0.67[b]	0.72[a]	0.68	0.71	0.02	0.03	0.21
Rectal Temperature, ℃	38.3	38.7	38.4	38.7	0.26	0.32	0.39
Fecal score[2]	1.83[a]	1.71[b]	1.80	1.74	0.09	0.07	0.34
Respiratory score[3]	1.08	1.06	1.07	1.06	0.03	0.61	0.79

[1]Effect of maternal diet (M) and colostrum type (C). There was a time effect ($P < 0.01$) for all these measurements except respiratory score. None of the potential two-way interactions were statistically significant ($P > 0.10$).

[2]Fecal score based on appearance: 1 = Firm well formed; 2 = Soft, pudding like; 3 = Runny, package batter; 4 = Liquid, splatters.

[3]Respiratory score based on appearance: 1 = Normal; 2 = Runny rose; 3 = Heavy breathing; 4 = Cough moist; 5 = Cough dry.

[ab]Means differ ($P \leq 0.05$)

In addition to the differences in calf development and postnatal growth, earlier studies focusing on the effects of enhanced maternal Met supply in late-pregnancy or in the first 70 days post-partum revealed a number of molecular alterations in the embryo or neonatal calf (Table

10). Whether those molecular changes involve epigenetic events remains to be determined. However, the fact that they encompass biological processes beyond metabolism (for example, immune function) seems to underscore novel biological roles for Met, and potentially other methyl donors. Further epigenetic analysis, e.g. DNA methylation, of these targets could provide more concrete evidence for a link with met supply. Of particular relevance is the long-term functional ramifications of these molecular changes in terms of health, fertility, and milk production ability.

Table 10 Summary of selected studies evaluating the link between enhanced methionine supply and potential programming effects on whole embryo, liver, or innate immune cells

Objective	Key biological responses	Reference
Evaluate the effect of maternal rumen-protected methionine supplementation during the first 70 days of lactation on the transcriptome of bovine pre-implantation embryos	Increasing methionine supply altered expression of genes related to embryonic development (e.g., VIM, IFI6, BCL2A1, and TBX15) and immune response (e. g., NKG7, TYROBP, SLAMF7, LCP1, and BLA-DQB) in pre-implantation embryos	Peñagaricano et al., 2013. PLoS One., 8 (8): e72302.
Determine the effect of maternal rumen-protected methionine during late-pregnancy on blood biomarkers and the liver metabolic transcriptome in neonatal calves	Transcriptome results indicated that calves from methionine-supplemented cows underwent a faster maturation of gluconeogenesis and fatty acid oxidation in the liver, which would be advantageous for adapting to the metabolic demands of extrauterine life	Jacometo et al., 2016. J. Dairy Sci., 99 (8): 6753-6763.
Determine the effect of maternal rumen-protected methionine during late-pregnancy on blood biomarkers and hepatic 1-carbon metabolism	Transcriptome data indicates that calves from methionine-supplemented cows underwent alterations in Met, choline, and homocysteine metabolism partly to synthesize taurine and glutathione, which would be advantageous for controlling metabolic-related stress	Jacometo et al., 2017. J. Dairy Sci., 100 (4): 3209-3219.
Determine the effect of maternal rumen-protected methionine during late-pregnancy on immune function and epigenetic markers in blood neutrophils	microRNA and mRNA data from isolated blood neutrophils indicated that maternal methionine could alter innate immune function via changes in abundance of targets associated with cell adhesion and chemotaxis, oxidative stress, Toll-like receptor signaling, and Met metabolism	Jacometo et al., 2018. J. Dairy Sci., 101 (9): 8146-8158.

Current Limitations, Outstanding Issues, and Future Research

It is evident that nutrition and the metabolic, or stress state of the cow during late-pregnancy can impact the final stages of development of the calf, which in turn can induce chronic changes in growth, metabolism, and immune function. Clearly, these three factors are interrelated, and we often rely on nutrition to tailor the metabolism of the cow in a way that minimizes stress. Given the impact of micronutrients and functional nutrients such as methyl donors, on key pathways related to cow health (e.g. oxidant, inflammatory, and immune status) and the programming of the calf, there is urgent need to define "adequacy thresholds" for those compounds. An example of that is the work our lab has performed with methyl donors, particularly Met supply during the periparturient period. At the onset of those efforts we sought to enhance Met supply to evaluate changes in the antioxidant and inflammatory state of the cow (Osorio et al., 2013a). Our guide was to use the well-established lysine to Met ratio of 3 : 1 in the MP to formulate experimental diets in terms of the amounts (grams per day) that cows were fed (Osorio et al., 2013a). Although using that relationship yielded novel physiological responses (recall this "ideal" ratio was determined in lactating cows) and confirmed the original hypothesis (Met supply enhances antioxidant synthesis), the "optimal" levels of Met, other essential AA, and micronutrients for cows in the periparturient period is still unknown. It is expected that the actual grams of these nutrients that must be supplied beyond what the basal diet (or microbes) provides will differ according to the characteristic of the basal diet (e.g. forage type). However, the lack of knowledge in terms of functionalnutrient adequacy is a major current limitation.

One of the outstanding issues pertaining to the concept of "feeding for dual purpose with dual benefit" is to what extent molecular or epigenetic effects that have been reported in relation to maternal nutrition in dairy cows translate to long-term changes in production phenotypes. It is also unknown the degree to which changes in DNA methylation might be associated with micronutrients or functional nutrients. Clearly, fertility and milk production are two of the most important outcomes for the dairy industry and both are related to health. Application of "high-throughput" or "next-generation" DNA sequencing technologies to map the entire "epigenome" would be helpful in potentially identifying CpG islands in specific genomic regions. Although it remains to be determined, the potential exists to identify loci (or genetic markers) in the fetus (or embryo) that are altered by maternal diet or the supply of specific functional nutrients at an epigenome level.

There also needs to be greater emphasis in the future on assessing the role of nutritional programming of the young calf during the neonatal or pre-weaning period. In the context of nutrition and developmental programming, there is evidence that improving the pre-weaning plane of nutrition of the dairy calf can serve as a major environmental factor influencing the expression of the genetic capacity of the animal for milk yield once it enters lactation (Soberon et al., 2012). Doubling the birth weight by 60 days of age through changes in rate and length of feeding of a higher protein (28% crude protein) milk replacer was significantly correlated with first-lactation milk yield, such that for every 1 kg of pre-weaning average daily gain, heifers produced on average 850 kg more milk during their first lactation. Some of those responses could be related with enhanced mammary development as demonstrated in studies comparing control (0.45 kg/d; 20% crude protein, 20% fat milk replacer) or "enhanced" milk replacers (1.13 kg/d; 28% crude protein; 25% fat) during the pre-weaning phase (Geiger et al., 2016). Clearly, there are unique opportunities for enhancing the productive efficiency of dairy cattle through nutrition in utero or prior to weaning. Additional research should try to establish mechanisms and also explore the unique roles of micronutrients and functional nutrients.

The role for "vertical transfer" of microbes from mother into the fetal hindgut is an unexplored area that merits greater emphasis. The hindgut microbiome in the neonate is crucial for proper regulation of host metabolism, immune response and other key physiological processes via the production of numerous bioactive metabolites, such as volatile fatty acids, essential AA, vitamins and neurotransmitters that can impact signaling pathways and metabolism (Thursby and Juge, 2017). There is growing recognition that these coordinated processes could promote growth and development in dairy calves (Malmuthuge and Guan, 2017). Whether it can be programmed during pregnancy or early life in ruminants remains largely unknown.

Conclusions

This review highlights the need for additional work establishing the adequacy of functional nutrients during pregnancy and to perform epigenome analyses to provide the mechanistic insights required in the field of nutritional programming of dairy cattle production and health. This is best exemplified by the extent of knowledge in epigenetic programming through micronutrients and functional nutrients in non-ruminant species and small ruminants (mainly sheep). Because of the obvious importance of dairy cattle to worldwide agriculture and as a source of nutrition for humans, more emphasis should be placed on studying traits of economic importance where animals are offered more thoughtfully formulated diets that facilitate the study of specific micro and functional nutrients. In that context, it would also be important to develop robust technologies to "protect" functional nutrients from ruminal metabolism while allowing a high bioavailability at the level of the small intestine.

References

Abdelmegeid, M. K., Vailati-Riboni, M., Alharthi, A., et al., 2017. Supplemental methionine, choline, or taurine alter *in vitro* gene network expression of polymorphonuclear leukocytes from neonatal Holstein calves [J]. J. Dairy Sci., 100 (4): 3155-3165.

Aguilera, O., Fernandez, A. F., Munoz, A., et al., 2010. Epigenetics and environment: a complex relationship [J]. J. Appl. Physiol., 109 (1): 243-251.

Alharthi, A. S., Batistel, F., Abdelmegeid, M. K., et al., 2018. Maternal supply of methionine during late-pregnancy enhances rate of Holstein calf development in utero and postnatal growth to a greater extent than colostrum source [J]. J. Anim. Sci. Biotechnol., 9: 83.

Akbar, H., Grala, T. M., Vailati Riboni, M., et al., 2015. Body condition score at calving affects systemic and hepatic transcriptome indicators of inflammation and nutrient metabolism in grazing dairy cows [J]. J. Dairy Sci., 98 (2): 1019-1032.

Atmaca, G., 2004. Antioxidant effects of sulfur-containing amino acids [J]. Yonsei Med. J., 45: 776-788.

Bach, A., 2012. Ruminant Nutrition Symposium: Optimizing Performance of the Offspring: nourishing and managing the dam and postnatal calf for optimal lactation, reproduction, and immunity [J]. J. Anim Sci., 90 (6): 1835-1845.

Bauman, D. E., Currie, W. B., 1980. Partitioning of nutrients during pregnancy and lactation: a review of mechanisms involving homeostasis and homeorhesis [J]. J. Dairy Sci., 63: 1514-1529.

Batistel, F., Alharthi, A. S., Wang, L., et al., 2017a. Placentome nutrient transporters and mammalian target of rapamycin signaling proteins are altered by the methionine supply during late gestation in dairy cows and are associated with newborn birth weight [J]. J. Nutr., 147 (9): 1640-1647.

Batistel, F., Arroyo, J. M., Bellingeri, A., et al., 2017b. Ethyl-cellulose rumen-protected methionine enhances performance during the periparturient period and early lactation in Holstein dairy cows [J]. J. Dairy Sci., 100 (9): 7455-7467.

Batistel, F., Arroyo, J. M., Garces, C. I. M., et al., 2018. Ethyl-cellulose rumen-protected methionine alleviates inflammation and oxidative stress and improves neutrophil function during the periparturient period and early lactation in Holstein dairy cows [J]. J. Dairy Sci., 101 (1): 480-490.

Batistel, F., Alharthi, A. S., Yambao, R. R. C., et al., 2019. Methionine Supply During Late-Gestation Triggers Offspring Sex-Specific Divergent Changes in Metabolic and Epigenetic Signatures in Bovine Placenta [J]. J. Nutr., 149 (1): 6-17.

Berthiaume, R., Thivierge, M. C., Patton, R. A., et al., 2006. Effect of ruminally protected methionine on splanchnic metabolism of amino acids in lactating dairy cows [J]. J. Dairy Sci., 89: 1621-1634.

Bertics, S. J., Grummer, R. R., 1999. Effects of fat and methionine hydroxy analog on prevention or alleviation of fatty liver induced by feed restriction [J]. J. Dairy Sci., 82: 2731-2736.

Bertoni, G., Trevisi, E., Han, X., et al., 2008. Effects of inflammatory conditions on liver activity in puerperium period and consequences for performance in dairy cows [J]. J. Dairy Sci., 91: 3300-3310.

Bionaz, M., Loor, J. J., 2012. Ruminant Metabolic Systems Biology: Reconstruction and Integration of Transcriptome Dynamics Underlying Functional Responses of Tissues to Nutrition and Physiological State [J]. Gene Regul. Syst. Biol., 6: 109-125.

Bochtler, M., Kolano, A., Xu, G. L., 2017. DNA demethylation pathways: Additional players and regulators [J]. Bioessays, 39 (1): 1-13.

Bradford, B. J., Yuan, K., Farney, J. K., et al., 2015. Invited review: Inflammation during the transition to lactation: New adventures with an old flame [J]. J. Dairy Sci., 98 (10): 6631-50.

Brett, K. E., Ferraro, Z. M., Yockell-Lelievre, J., et al., 2014. Maternal-fetal nutrient transport in pregnancy pathologies: the role of the placenta [J]. Int. J. Mol. Sci., 15: 16153-16185.

Bridger, P. S., Menge, C., Leiser, R., et al., 2007. Bovine caruncular epithelial cell line (BCEC-1) isolated from the placenta forms a functional epithelial barrier in a polarised cell culture model [J]. Placenta., 28: 1110-1117.

Borowicz, P. P., Arnold, D. R., Johnson, M. L., et al., 2007. Placental growth throughout the last two thirds of pregnancy in sheep: vascular development and angiogenic factor expression [J]. Biol.Reprod., 76: 259-267.

Brown, L. D., Green, A. S., Limesand, S. W., et al., 2011. Maternal amino acid supplementation for intrauterine growth restriction [J]. Front.Biosci. (Schol Ed) 3: 428-444.

Burciaga-Robles, L. O., 2009. Effects of bovine respiratory disease on immune response, animal performance, nitrogen balance, and blood and nutrient flux across total splanchnic in beef steers [D]. Department of Animal Science, Oklahoma State Univ., Stillwater.

Cai, D., Jia, Y., Song, H., et al., 2014. Betaine supplementation in maternal diet modulates the epigenetic regulation of hepatic gluconeogenic genes in neonatal piglets [J]. PLoS One, 9: e105504.

Chavatte-Palmer, P., Velazquez, M. A., Jammes, H., et al., 2019. Review: Epigenetics, developmental programming and nutrition in herbivores [J]. Animal, 12 (s2): s363-s371.

Coelho, C. N., Weber, J. A., Klein, N. W., et al., 1989. Whole rat embryos require methionine for neural tube closure when cultured on cow serum [J]. J. Nutr., 119: 1716-1725.

Coppock, C. E., Noller, C. H., Wolfe, S. A., et al., 1972. Effect of forage - concentrate ratio in complete feeds fed ad libitum on feed intake prepartum and the occurrence of abomasal displacement in dairy cows [J]. J. Dairy Sci., 55 (6): 783-9.

Corbin, K. D., Zeisel, S. H., 2012. Choline metabolism provides novel insights into nonalcoholic fatty liver disease and its progression [J]. Curr. Opin. Gastroenterol., 28: 159-165.

Dalbach, K. F., Larsen, M., Raun, B. M., et al., 2011. Effects of supplementation with 2-hydroxy-4-(methylthio) -butanoic acid isopropyl ester on splanchnic amino acid metabolism and essential amino acid mobilization in postpartum transition Holstein cows [J]. J. Dairy Sci., 94: 3913-3927.

Dann, H. M., Morin, D. E., Bollero, G. A., et al., 2005. Prepartum intake, postpartum induction of ketosis, and periparturient disorders affect the metabolic status of dairy cows [J]. J. Dairy Sci., 88 (9): 3249-64.

Dann, H. M., Litherland, N. B., Underwood, J. P., et al., 2006. Diets during far-off and close-up dry periods affect periparturient metabolism and lactation in multiparous cows [J]. J. Dairy Sci., 89 (9): 3563-77.

Eulalio, A., Huntzinger, E., Izaurralde, E., 2008. Getting to the root of miRNA - mediated gene silencing [J]. Cell, 132 (1): 9-14.

Finkelstein, J. D., 1990. Methionine metabolism in mammals [J]. J. Nutr. Biochem., 1: 228-237.

Firth, D. R. Interspike interval fluctuations in the crayfish stretch receptor [J]. Biophys J., 6: 201-15.

Funston, R. N., Summers, A. F., 2013. Epigenetics: setting up lifetime production of beef cows by managing nutrition [J]. Annu. Rev. Anim. BioSci., 1: 339-363.

Gallou-Kabani, C., Gabory, A., Tost, J., et al., Sex-and diet-specific changes of imprinted gene expression and DNA methylation in mouse placenta under a high - fat diet [J]. PLoS One, 5 (12): e14398.

Gabory, A., Ferry, L., Fajardy, I., et al. Maternal diets trigger sex - specific divergent trajectories of gene expression and epigenetic systems in mouse placenta [J]. PLoS One, 7 (11): e47986.

Gantier, M. P., 2013. The not-so-neutral role of microRNAs in neutrophil biology [J]. J. Leuk. Biol., 94 (4): 575-583.

Garcia, M., Elsasser, T. H., Juengst, L., et al., 2016. Short communication: Amino acid supplementation and stage of lactation alter apparent utilization of nutrients by blood neutrophils from lactating dairy cows *in vitro* [J]. J. Dairy Sci., 99: 3777-3783.

Garcia, M., Mamedova, L. K., Barton, B., et al., 2018. Choline regulates the function of bovine immune cells and alters the mRNA abundance of enzymes and receptors involved in its metabolism *in vitro* [J]. Front. Immunol., 9: 2448.

Geiger, A. J., Parsons, C. L., Akers, R. M., 2016. Feeding a higher plane of nutrition and providing exogenous estrogen increases mammary gland development in Holstein heifer calves [J]. J. Dairy Sci., 99 (9): 7642-7653

Girard, C. L., Matte, J. J., Tremblay, G. F., 1995. Gestation and lactation of dairy cows: a role for folic acid? [J]. J. Dairy Sci., 78 (2): 404-411.

Grummer, R. R., 1993. Etiology of lipid-related metabolic disorders in periparturient dairy cows [J]. J. Dairy Sci., 76: 3882-3896.

Hollenbeck, C. B., 2012. An introduction to the nutrition and metabolism of choline [J]. Cent. Nerv. Syst. Agents. Med. Chem., 12: 100-113.

Gao, F., Liu, Y. C., Zhang, Z. H., et al., 2012. Effect of prepartum maternal energy density on the growth performance, immunity, and antioxidation capability of neonatal calves [J]. J. Dairy Sci., 95: 4510-4518.

Guo, J., Peters, R. R., Kohn, R. A., 2007. Effect of a transition diet on production performance and metabolism in periparturient dairy cows [J]. J. Dairy Sci., 90 (11): 5247-5258.

Hidiroglou, M., 1980. Trace elements in the fetal and neonate ruminant: a review [J]. Can. Vet. J., 21 (12): 328-335.

Jacometo, C. B., Osorio, J. S., Socha, M., et al., 2015. Maternal consumption of organic trace minerals alters calf systemic and neutrophil mRNA and microRNA indicators of inflammation and oxidative stress [J]. J. Dairy Sci., 98: 7717-7729.

Jafari, A., Emmanuel, D. G., Christopherson, R. J., et al., 2006. Parenteral administration of glutamine modulates acute phase response in postparturient dairy cows [J]. J. Dairy Sci., 89: 4660-4668.

Janovick, N. A., Drackley, J. K., 2010. Prepartum dietary management of energy intake affects postpartum intake and lactation performance by primiparous and multiparous Holstein cows [J]. J. Dairy Sci., 93 (7): 3086-3102.

Jansson, N., Rosario, F. J., Gaccioli, F., et al., Activation of placental mTOR signaling and amino acid transporters in obese women giving birth to large babies [J]. J Clin Endocrinol Metab., 98: 105-113.

Jha, P., Knopf, A., Koefeler, H., et al., 2014. Role of adipose tissue in methionine-choline-deficient model of non-alcoholic steatohepatitis (NASH) [J]. Biochim. Biophys. Acta., 1842: 959-970.

Ji, Y., Wu, Z., Dai, Z., et al., 2015. Nutritional epigenetics with a focus on amino acids: implications for the development and treatment of metabolic syndrome [J]. J. Nutr. Biochem., 27: 1-8.

Jones, H. N., Powell, T. L., Jansson, T., 2007. Regulation of placental nutrient transport—a review [J]. Placenta., 28: 763-774.

Khanal, P., Nielsen, M. O., 2017. Impacts of prenatal nutrition on animal production and performance: a focus on growth and metabolic and endocrine function in sheep [J]. J. Anim. Sci., Biotechnol., 8: 75.

Lager, S., Powell, T. L., 2012 Regulation of nutrient transport across the placenta [J]. J. Pregnancy, 2012: 179827.

Larsen, M., Kristensen, N. B., 2013. Precursors for liver gluconeogenesis in periparturient dairy cows [J]. Animal, 7: 1640-1650.

Lin, G., Wang, X., Wu, G., et al., 2014. Improving amino acid nutrition to prevent intrauterine growth

restriction in mammals [J]. Amino Acids, 46: 1605-1623.
Ling T, Hernandez-Jover, M., Sordillo, L. M., et al., 2018. Maternal late-gestation metabolic stress is associated with changes in immune and metabolic responses of dairy calves [J]. J. Dairy Sci., 101 (7): 6568-6580.
Liu, J., Yao, Y., Yu, B., et al., 2013. Effect of maternal folic acid supplementation on hepatic proteome in newborn piglets [J]. Nutrition, 29: 230-234.
Loor, J. J., 2010. Genomics of metabolic adaptations in the peripartal cow [J]. Animal, 4: 1110-1139.
Loor, J. J., Bertoni, G., Hosseini, A., et al., 2013. Functional welfare-using biochemical and molecular technologies to understand better the welfare state of peripartal dairy cattle [J]. Anim.Prod.Sci., 53: 931-953.
Loor, J. J., Bionaz, M., Drackley, J. K., 2013. Systems physiology in dairy cattle: nutritional genomics and beyond [J]. Annu. Rev. Anim. BioSci., 1: 365-392.
Malmuthuge, N., Guan, L. L., 2017. Understanding the gut microbiome of dairy calves: Opportunities to improve early-life gut health [J]. J. Dairy Sci., 100 (7): 5996-6005.
McCarthy, R. D., Porter, G. A., Griel, L. C., 1968. Bovine ketosis and depressed fat test in milk: a problem of methionine metabolism and serum lipoprotein aberration [J]. J. Dairy Sci., 51: 459-462.
McNeil, C. J., Hoskin, S. O., Bremner, D. M., et al., 2016. Whole-body and splanchnic amino acid metabolism in sheep during an acute endotoxin challenge [J]. Br. J. Nutr., 116 (2): 211-222.
Morales, S., Monzo, M., Navarro, A., 2017. Epigenetic regulation mechanisms of microRNA expression [J]. Biomol.Concepts, 8 (5-6): 203-212.
Noleto, P. G., Saut, J. P., Sheldon, I. M., 2017. Short communication: Glutamine modulates inflammatory responses to lipopolysaccharide in ex vivo bovine endometrium [J]. J. Dairy Sci., 100: 2207-2212.
NRC, 2001. Nutrient Requirements of Dairy Cattle [M]. 7th ed. Natl. Acad. Press, Washington, DC.
Ordway, R. S., Boucher, S. E., Whitehouse, N. L., et al., 2009. Effects of providing two forms of supplemental methionine to periparturient Holstein dairy cows on feed intake and lactational performance [J]. J. Dairy Sci., 92: 5154-5166.
Osorio, J. S., Ji, P., Drackley, J. K., et al., 2013a. Supplemental Smartamine M or MetaSmart during the transition period benefits postpartal cow performance and blood neutrophil function [J]. J. Dairy Sci., 96: 6248-6263.
Osorio, J. S., Trevisi, E., Ballou, M. A., et al., 2013b. Effect of the level of maternal energy intake prepartum on immunometabolic markers, polymorphonuclear leukocyte function, and neutrophil gene network expression in neonatal Holstein heifer calves [J]. J. Dairy Sci., 96: 3573-3587.
Osorio, J. S., Ji, P., Drackley, J. K., et al., 2014a. Smartamine M and MetaSmart supplementation during the peripartal period alter hepatic expression of gene networks in 1-carbon metabolism, inflammation, oxidative stress, and the GH/IGF-1 axis pathways [J]. J. Dairy Sci., 97 (12): 7451-7464.
Osorio, J. S., Trevisi, E., Ji, P., et al., 2014b. Biomarkers of inflammation, metabolism, and oxidative stress in blood, liver, and milk reveal a better immunometabolic status in peripartal cows supplemented with Smartamine M or MetaSmart [J]. J. Dairy Sci., 97 (12): 7451-7464.
Piepenbrink, M. S., Marr, A. L., Waldron, M. R., et al., 2004. Feeding 2-hydroxy-4-(methylthio)-butanoic acid to periparturient dairy cows improves milk production but not hepatic metabolism [J]. J. Dairy Sci., 87: 1071-1084.
Plata-Salaman, C. R., 1995. Cytokines and feeding suppression: an integrative view from neurologic to molecular levels [J]. Nutrition, 11: 674-677.
Pocius, P. A., Clark, J. H., Baumrucker, C. R., 1981. Glutathione in bovine blood: possible source of amino acids for milk protein synthesis [J]. J. Dairy Sci., 64: 1551-1554.
Pullen, D. L., Liesman, J. S., Emery, R. S., 1990. A species comparison of liver slice synthesis

and secretion of triacylglycerol from nonesterified fatty acids in media [J]. J. Anim. Sci., 68: 1395-1399.

Quigley J. D., Drewry, J. J., 1998. Nutrient and immunity transfer from cow to calf pre-and postcalving [J]. J. Dairy Sci., 81 (10): 2779-2790.

Rabelo, E., Rezende, R. L., Bertics, S. J., et al., 2003. Effects of transition diets varying in dietary energy density on lactation performance and ruminal parameters of dairy cows [J]. J. Dairy Sci., 86 (3): 916-925.

Ragaller, V., Hüther, L., Lebzien, P., 2009. Folic acid in ruminant nutrition: a review [J]. Br.J. Nutr., 101 (2): 153-164.

Regnault, T. R., Marconi, A. M., Smith, C. H., et al., 2005. Placental amino acid transport systems and fetal growth restriction—a workshop report [J]. Placenta, 26 (Suppl A): S76-80.

Reynolds, C. K., Aikman, P. C., Lupoli, B., et al., 2003. Splanchnic metabolism of dairy cows during the transition from late gestation through early lactation [J]. J. Dairy Sci., 86: 1201-1217.

Reynolds, L. P., Borowicz, P. P., Caton, J. S., et al., 2010. Developmental programming: the concept, large animal models, and the key role of uteroplacental vascular development [J]. J. Anim Sci., 88 (13 Suppl): E61-72.

Roche, J. R., Kay, J. K., Friggens, N. C., et al., 2013. Assessing and managing body condition score for the prevention of metabolic disease in dairy cows [J]. Vet. Clin. North Am. Food Anim.Prac., 29 (2): 323-336.

Roos, S., Powell, T. L., Jansson, T., 2009. Placental mTOR links maternal nutrient availability to fetal growth [J]. Biochem.Soc.Trans., 37: 295-298.

Soberon, F., Raffrenato, E., Everett, R. W., et al., 2012. Preweaning milk replacer intake and effects on long-term productivity of dairy calves [J]. J. Dairy Sci., 95 (2): 783-793.

Spears, J. W., Weiss, W. P., 2008. Role of antioxidants and trace elements in health and immunity of transition dairy cows [J]. Vet. J, 176 (1): 70-76.

St-Pierre, N. R., Sylvester, J. T., 2005. Effects of 2-hydroxy-4-(methylthio) butanoic acid (HMB) and its isopropyl ester on milk production and composition by Holstein cows [J]. J. Dairy Sci., 88: 2487-2497.

Shahsavari, A., D'Occhio, M. J., Al Jassim, R., 2016. The role of rumen-protected choline in hepatic function and performance of transition dairy cows [J]. Br. J. Nutr., 116 (1): 35-44.

Shibano, K., Kawamura, S., 2006. Serum free amino acid concentration in hepatic lipidosis of dairy cows in the periparturient period [J]. J. Vet. Med. Sci., 68: 393-396.

Sinclair, K. D., Watkins, A. J., 2013. Parental diet, pregnancy outcomes and offspring health: metabolic determinants in developing oocytes and embryos [J]. Reprod. Fertil. Dev., 26 (1): 99-114.

Sinclair, K. D., Rutherford, K. M., Wallace, J. M., et al., 2016. Epigenetics and developmental programming of welfare and production traits in farm animals [J]. Reprod. Fertil. Dev., 28 (10) 1443-1478.

Skibiel, A. L., Dado-Senn, B., Fabris, T. F., et al., 2018. In utero exposure to thermal stress has long-term effects on mammary gland microstructure and function in dairy cattle [J]. PLoS One., 13 (10): e0206046.

Snoswell, A. M. Xue, G. P., 1987. Methyl group metabolism in sheep [J]. Comp. Biochem. Physiol. B, 88: 383-394.

Speckmann, B., Schulz, S., Hiller, F., et al., 2017. Selenium increases hepatic DNA methylation and modulates one-carbon metabolism in the liver of mice [J]. J. Nutr. Biochem., 48: 112-119.

Sun, F., Cao, Y., Cai, C., et al., 2016. Regulation of Nutritional Metabolism in Transition Dairy Cows: Energy Homeostasis and Health in Response to Post-Ruminal Choline and Methionine [J]. PLoS

One., 11 (8): e0160659.

Tamminga, S., Luteijn, P.A., Meijer, R. G. M., 1997. Changes in composition and energy content of liveweight loss in dairy cows with time after parturition [J]. Liv. Prod. Sci., 52: 31-38.

Tao, S., Monteiro, A. P., Thompson, I. M., et al., 2012. Effect of late-gestation maternal heat stress on growth and immune function of dairy calves [J]. J. Dairy Sci., 95 (12): 7128-7136.

Thursby, E., Juge, N., 2017. Introduction to the human gut microbiota [J]. Biochem. J., 474 (11): 1823-1836.

Teixeira, A. G., Lima, F. S., Bicalho, M. L., et al., 2014. Effect of an injectable trace mineral supplement containing selenium, copper, zinc, and manganese on immunity, health, and growth of dairy calves [J]. J. Dairy Sci., 97 (7): 4216-4226.

Thornton, C. A., 2010. Immunology of pregnancy [J]. Proc. Nutr. Soc., 69 (3): 357-365.

Vailati-Riboni, M., Kanwal, M., Bulgari, O., et al., 2016. Body condition score and plane of nutrition prepartum affect adipose tissue transcriptome regulators of metabolism and inflammation in grazing dairy cows during the transition period [J]. J. Dairy Sci., 99 (1): 758-770.

Wu, G., Bazer, F. W., Cudd, T. A., et al., 2004. Maternal nutrition and fetal development [J]. J. Nutr., 134: 2169-2172.

Zenobi, M. G., Gardinal, R., Zuniga, J. E., et al., 2018. Effects of supplementation with ruminally protected choline on performance of multiparous Holstein cows did not depend upon prepartum caloric intake [J]. J. Dairy Sci., 101 (2): 1088-1110.

Zhou, Z., Bulgari, O., Vailati-Riboni, M., et al., 2016a. Rumen-protected methionine compared with rumen-protected choline improves immunometabolic status in dairy cows during the peripartal period [J]. J. Dairy Sci., 99 (11): 8956-8969.

Zhou, Z., Loor, J. J., Piccioli-Cappelli, F., et al., 2016b. Circulating amino acids in blood plasma during the peripartal period in dairy cows with different liver functionality index [J]. J. Dairy Sci., 99 (3): 2257-2267.

Zhou, Z., Vailati-Riboni, M., Trevisi, E., et al., 2016c. Better postpartal performance in dairy cows supplemented with rumen-protected methionine compared with choline during the peripartal period [J]. J. Dairy Sci., 99 (11): 8716-8732.

Zhou, Z., Garrow, T. A., Dong, X., et al., 2017. Hepatic activity and transcription of betaine-homocysteine methyltransferase, methionine synthase, and cystathionine synthase in periparturient dairy cows are altered to different extents by supply of methionine and choline [J]. J. Nutr., 147 (1): 11-19.

Amino Acid Metabolism in Support of Lactation

M. D. Hanigan, X. Huang, J. M. Prestegaard,
A. C. Hruby, L. Marra Campos, I. A. Teixeira, V. L. Daley
Dept. of Dairy Science Virginia Tech

Abstract

The world population will continue to expand for the remainder of the century putting pressure on the food supply. Land area is finite and thus food produced for the growing population must increase through improved efficiency. When measured as human edible protein or energy produced per unit of human edible protein or energy consumed (Wilkinson, 2011), dairy cattle are relatively efficient when byproduct use is maximized. Still, further progress in protein efficiency is possible if diets are balanced to provide only those essential amino acids (AA) required to support production. The challenge to achieve the latter is a lack of knowledge required to predict AA supply from the diet and the milk protein and body function responses to varying supplies of absorbed AA. We have developed an isotope dilution-based *in vivo* technique to assess the absorbed supply of each of the essential AA contributed from individual ingredients in mixed diets. This technique has errors of determination of less than 15%, which is approximately half of the error of determination of total ruminal outflow. We have used the method to assess AA bioavailabilties for alfalfa hay, grass hay, maize silage, maize grain, maize distillers grain, brewers grain, soybean hulls, blood meal, and feather meal, and are currently using it to screen various ruminally protected AAs. We have also used the method to assess dietary AA supply in a production study so that we can calculate the efficiency of transfer of absorbed AA to milk protein. Regarding the milk protein response problem, we have now demonstrated at the cellular, tissue, and animal levels both with experimental work and by modeling that there is not one single-limiting AA. The mammary supply of at least 8 of the essential AA acts both independently and additively with energy supply to set the overall level of milk protein production, likely via activation of the mTOR pathway (although other pathways and mechanisms cannot be ruled out). In the case of valine, the response is negative, suggesting it is antagonistic towards one or more of the other AA. The response to individual AA is mitigated by feedback inhibition on AA transporters in the mammary tissue where an oversupply of an AA results in a reduction in the transport activity for that AA. Thus, the tissue is able to modify transport activity to maintain production in the case of low AA supplies and mitigate production increases when provided with an excess AA supply. The critical point from this work is that high energy diets are required to maximize AA efficiency, and that currently available models are inadequate to determine true AA requirements, limiting the maximum AA efficiency potential of the lactating dairy cow.

Introduction

Providing sufficient, high-quality protein to the world is a universal challenge. The global population is projected to approach 10 billion people by 2050 and plateau at 11.2 billion by 2100 (UN, 2017). Thus, the inevitable food energy and protein demand warrants a substantial increase in food production. In addition to increased world population, greater prosperity in regions that were previously less affluent is resulting in greater animal product consumption (Delgado, 2003). This adds to the challenge of maintaining food sufficiency, as some loss in efficiency is generally realized with conversion of feed ingredients to animal products. Given finite land availability, the needed increases in food production will largely have to be delivered via improved efficiency, allowing more human edible food production from a fixed land mass. These gains likely must occur across the food production system including gains in food animal efficiency.

Food production must also be sustainable. Nitrogen (N), phosphorus, ammonia, and greenhouse gas production per unit of food produced must decline as food production increases over the next 100 years if we are to maintain a similar or improved environmental footprint. Thus, animal production systems must become more efficient at converting dietary carbon, N, and phosphorus to animal products than they are today.

Milk protein production can be quite efficient when measured as human edible dietary protein consumed per unit of human edible protein produced. In fact, this ratio can be greater than one, resulting in a netcontribution to the food supply if dietary inputs to the cows are restricted to forage and byproducts of human food and energy production systems (Ertl et al., 2015ab). However, gross protein or N efficiency of dairy cattle is poor relative to swine and poultry (Bouwman et al., 2013), which contributes to environmental pollution (Uuml et al., 2001; Agle et al., 2008).

In a survey carried out on 103 large-scale dairies across the US [(613±46) cows; (34.5±0.3) kg of milk per cow per day], nutritionists reported feeding diets with (17.8±0.1)% crude protein (CP; Caraviello et al., 2006). A meta-analysis of 846 experimental diets found a similar mean diet CP content and identified that conversion efficiencies for dietary and metabolizable N (based on NRC, 2001) to milk protein averaged 24.6% and 42.6%, respectively (Hristov et al., 2004). Assuming the same intake and diet composition (22.1 kg/d DMI and 17.8% CP), over a 10-month lactation, the 9 million dairy cattle in the US consume 1.73 million metric tons (mmt) of N and excrete 1.30 mmt in manure (Livestock, Dairy, and Poultry Outlook: August 2012, LDPM-218, Dairy Economic Research Service, USDA). The difference of 0.425 mmt is captured in milk and meat. Dividing by milk production yields 13.8 g of N excreted per liter of milk produced.

If the US dairy cattle population were fed diets with 13% CP, without a loss in production as demonstrated by Haque et al. (2012), they would consume 1.26 mmt and excrete 0.837 mmt in manure. This represents a 27% reduction in N intake and a 35% reduction in waste N. Dividing by milk produced yields 8.8 g of N excreted in manure per liter of milk produced. Thus at 13% dietary CP, 27% more milk could be produced with no change in N input and a reduction in environmental impact with respect to N loading.

Such an improvement in efficiency would also reduce the cost of producing milk as protein is an expensive dietary nutrient, representing approximately 42% of the cost of a lactating cow ration in the United States (St-Pierre, 2012). The reduction of dietary protein could result in decreased demand for high-protein ingredients, reduced price of those ingredients, and diversion of existing cropland use to higher-yielding crops, such as maize instead of oilseeds (Lambin and Meyfroidt, 2011).

Combining improved N efficiency with greater use of byproduct feeding would yield greater human edible energy and protein per hectare of land, reduced environmental impact, and reduced overall cost. Thus, achieving improved N efficiency is an important goal for dairy production scientists.

Achieving Improved N Efficiency

Dairy cattle diets are currently balanced to meet metabolizable protein (MP) requirements, which were established using dose-response methods. However, the cow does not require MP; it requires the essential AA delivered by the MP. This is perhaps most apparent when feeding diets constructed largely from maize products which are inherently low in lysine (Polan et al., 1991). Such a diet could be created to meet MP requirements, but animals may still respond to the addition of a protected lysine source or more protein that also provides lysine to the ration (Vyas and Erdman, 2009).

Given the heterogeneity in the AA composition of dietary protein and the variation in the flow of both undegraded dietary protein and microbial protein from the rumen, it is self-evident that the MP requirements must be greater than truly required by the animal if the mix of AA in the MP were perfectly matched to true requirements in all cases. For example, trials which contained a "poor" mix of AA would result in poor production at a given level of MP intake thus pulling the average response per unit of MP down. Regression of milk production on

MP intake for a collection of such trials will result in a prediction that represents the average (Figure 1). Diets containing better AA mixes will lie above the line. Had the collection of studies been restricted to only those lying above the line in Figure 1, the requirement for MP at a given level of milk protein would have been less. If restricted to only those studies at the very top end of observed production per unit of MP, then the MP requirement would be even lower. Additionally, those diets generated much greater MP efficiency as compared to those lying below the line. Thus, the data demonstrate that significant improvements in efficiency are possible, but they cannot be achieved using models based solely on MP.

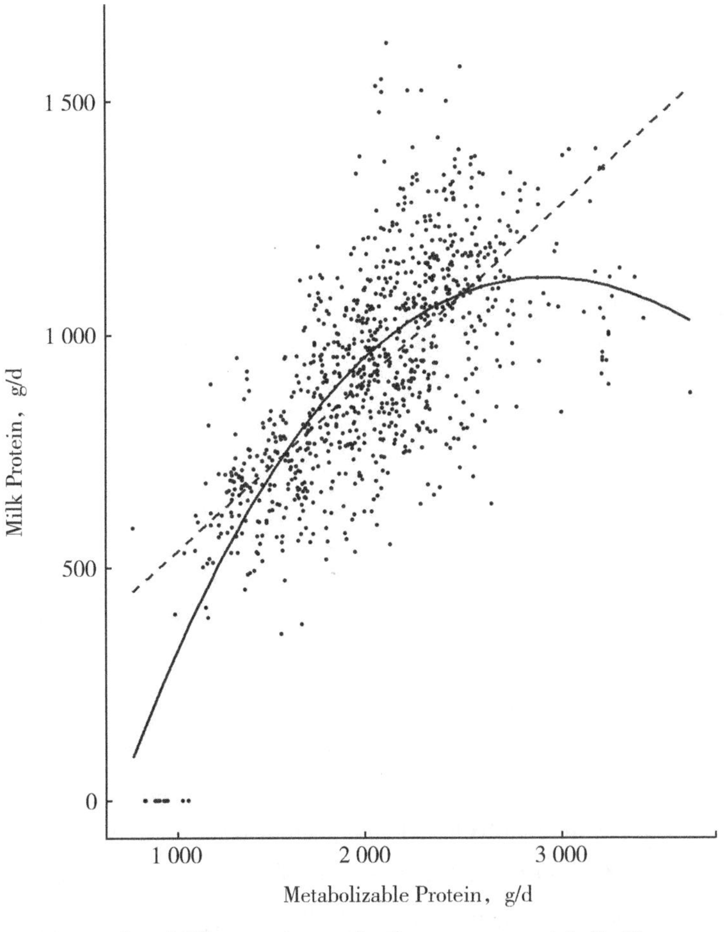

Figure 1 Milk protein production versus metabolizable protein supply. Data represent treatment means collected from the literature. The dashed line represents a linear regression and the solid line a quadratic model

The existing NRC (2001) model provided recommendations for the dietary content of lysine (Lys) and methionine (Met), however, those recommendations were relative to MP intake. Thus, diets must be balanced to meet MP requirements and the Lys and Met recommendations used to avoid AA imbalance problems at that particular level of MP intake. The (NRC, 2001) does not provide guidance for the design of diets based solelyon AA, so balancing for low MP diets is not possible with this system.

To move from an MP-based model to one based on individual AA, one must define the relationships among the supply of AA and milk protein output. This requires knowledge of the supply of individual AA and the integrative response to those AA and the relationship with energy supply.

Predicting AA Supply

Rumially undegraded protein (RUP) flow from the rumen is over-predicted and microbial flow is under-predicted by the NRC (2001) model (Roman-Garcia et al., 2016; White et al., 2016; White et al., 2017bc) which leads to biased predictions of AA flow from the rumen. Bateman et al. (2005) and Broderick et al. (2010) observed similar problems. Correcting the prediction problems of microbial and RUP flows, and accounting for both incomplete recovery of AA from acid hydrolysis during analyses and change in hydration associated with peptide bond formation resulted in predictions of AA flow with minimal mean bias and no slope bias (Fleming et al., 2019), however, the errors of prediction for total duodenal flow are relatively high at 25% or greater. A portion of this is likely due to measurement error, but there may be ingredient nutrient profiles that are poorly represented in the feed library.

An additional challenge is associated with predictions of the digested AA supply. Current assumptions suggest the digestibilities of individual AA in the rumen and in the intestine are the same as for the total protein (NRC, 2001; Van Amburgh et al., 2015). Assessment of the available in situ, mobile bag, and *in vitro* data suggests there is heterogeneity in the digestibility of individual AA within and across ingredients, but the data are not strong enough to derive individual measurements for each ingredient (Hvelplund and Hesselholt, 1987; White et al., 2017a).

We have recently adapted an approach used by Maxin et al. (2013) to assess the absorbed supply of each AA from individual dietary ingredients (Estes et al., 2018). The method makes use of a 4 to 8 h constant infusion of a ^{13}C labelled AA mixture derived from enriched algae to assess the plasma entry rate of each AA. Because infusions and sampling are via the jugular vein, measurements can be made with minimal animal preparation. Errors of determination for AA availability from each ingredient are approximately 10% using this method, which is a large improvement over previously used methods (Titgemeyer et al., 1989).

We have used the isotope-dilution method to derive AA bioavailabilities for 9 ingredients typically used in North American dairy rations (maize silage, alfalfa hay, grass hay, maize grain, maize distillers grains, brewers grains, soybean hulls, blood meal, and feather meal). As demonstrated in Figure 2, the bioavailability varies by AA and thus is not equal to the digestibility of the CP in the ingredient as assumed in all models. This work does not identify if the deviations are due to ruminal degradation or intestinal digestibility. Of equal concern was the lack of alignment of CP availability predicted from *in vitro* assays commonly used in the industry with the *in vivo* measures based on the AA observations (Figure 3). Further, laboratory bias was also present for feather meal, but the pattern was different across laboratories indicating that the laboratory methods are not reliable.

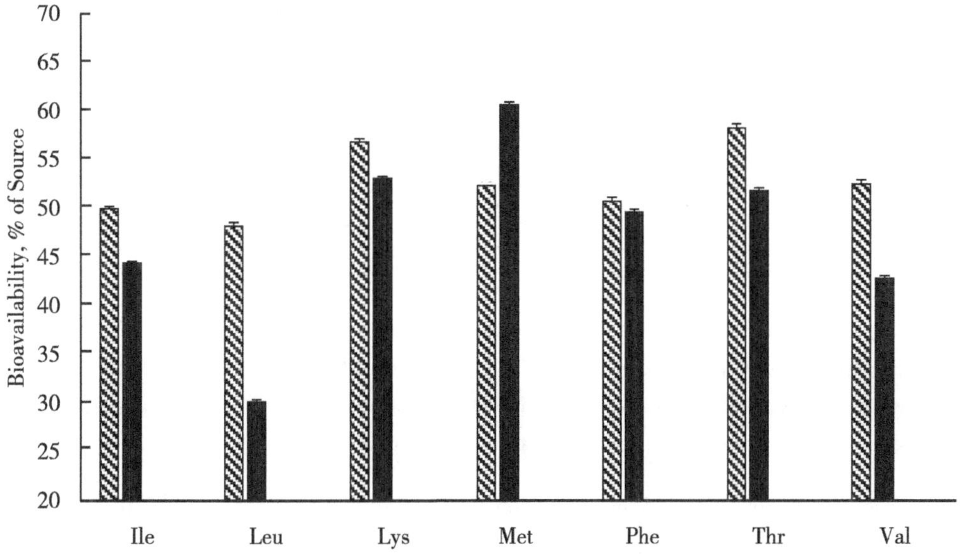

Figure 2 Plasma amino acid appearance for blood meal (stiped) and feather meal (solid) determined using an *in vivo*, isotope - based approach. work was conducted using growing heifers. adapted from Estes et al. (2018)

Further application of this method should allow generation of a table of AA bioavailabilities for all ingredients comparable to energy or CP tables. Such a table would greatly improve our knowledge of AA supply from RUP and avoid the current *in situ* challenges with assigning AA composition to the residue proteins. Additional work is required to identify existing or new *in situ* or *in vitro* methods that can be used to quickly assess the availability of AA from source ingredients that consistently match *in vivo* measures.

Predicting Milk Protein Output from Absorbed AA

The first limiting nutrient and AA concept is based on a hypothesis which has become known as the Law of the Minimum. Sprengel (1828) formulated this concept based on plant growth responses to soil minerals. The original thesis stated that a nutrient can limit plant growth, and when limiting, growth will be proportional to supply. This is strongly supported by volumes of data over the past 175 years. However, Von Liebig (see Paris, 1992 for a translation) sub-

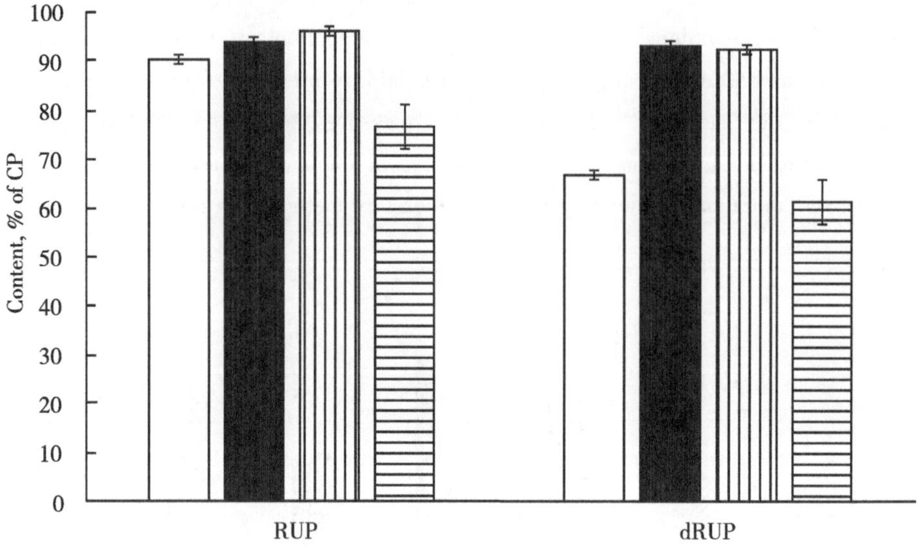

Figure 3 ***In vitro*** **estimates of blood meal ruminally undegraded protein (RUP, % of CP) and digested RUP (dRUP, % of CP) from 3 different commercial laboratories (white, black, and vertically striped bars) as compared to** ***in vivo*** **estimates derived using isotope dilution (horizontally striped bar). adapted from estes et al. (2018)**

sequently restated and expanded the hypothesis indicating that if a nutrient was limiting growth, responses to other nutrients could not occur (von Liebig, 1862).

Mitchell and Block (1946) used von Liebig's extension of Sprengel's thesis to develop the concept of the order of limiting AA, which was described using the analogy of a water barrel with broken staves. Based on this formulation, if any nutrient is limiting milk production, then only the addition of that nutrient to the diet will result in a positive milk yield response, e.g. the single-limiting nutrient paradigm. The ideal protein concept loosely aligns with this framework in that it is assumed there is an ideal AA profile that should be provided to an animal and that profile will remain largely fixed as production levels change.

The ideal protein concept is based on the assumption that there is a fixed, unique set of AA inputs that are required to maximize or optimize production. For that to be true, the efficiencies of use of absorbed AA must remain constant with respect to one another. For example, if the ideal protein target for methionine is half of that for histidine, doubling the need for methionine should exactly double the need for histidine. If this is true, then one can easily determine which nutrient is most limiting by calculating the allowable milk yield from each AA, and this can be extended to energy and other required inputs. If the result of that calculation indicates that inadequate histidine is being provided, then one would predict a response to the addition of histidine, and the same for any other nutrient that is apparently deficient. However, the transfer efficiency of absorbed AA to milk protein is not fixed. Because AA removal from blood is regulated in concert with needs for milk protein synthesis (Bequette et al.,

2000), the efficiency of AA transfer from the gut to milk protein is variable. This complicates application of the ideal protein calculations and undermines the concept of a first-limiting nutrient.

Based on the cell signaling pathways regulating protein synthesis, responses to individual AA and energy should be additive and largely independent (Arriola Apelo et al., 2014a; Castro et al., 2016). AA and energy supplies within the cell are sensed through various mechanisms that impact intracellular signaling pathwayssuch as the mTOR, Akt, and AMPPK pathways. This signaling integrates information regarding the intracellular supply of several key AA (Appuhamy et al., 2011; Appuhamy et al., 2012a), the supply of energy in the cell (Appuhamy et al., 2009), and hormonal signals (such as those from insulin) indicating overall animal status (Appuhamy et al., 2011) and likely IGF-1 as well (Figure 4). These signaling pathways regulate rates of protein synthesis thus ensuring that rates of protein synthesis are matched to substrate supply and energy state in the animal.

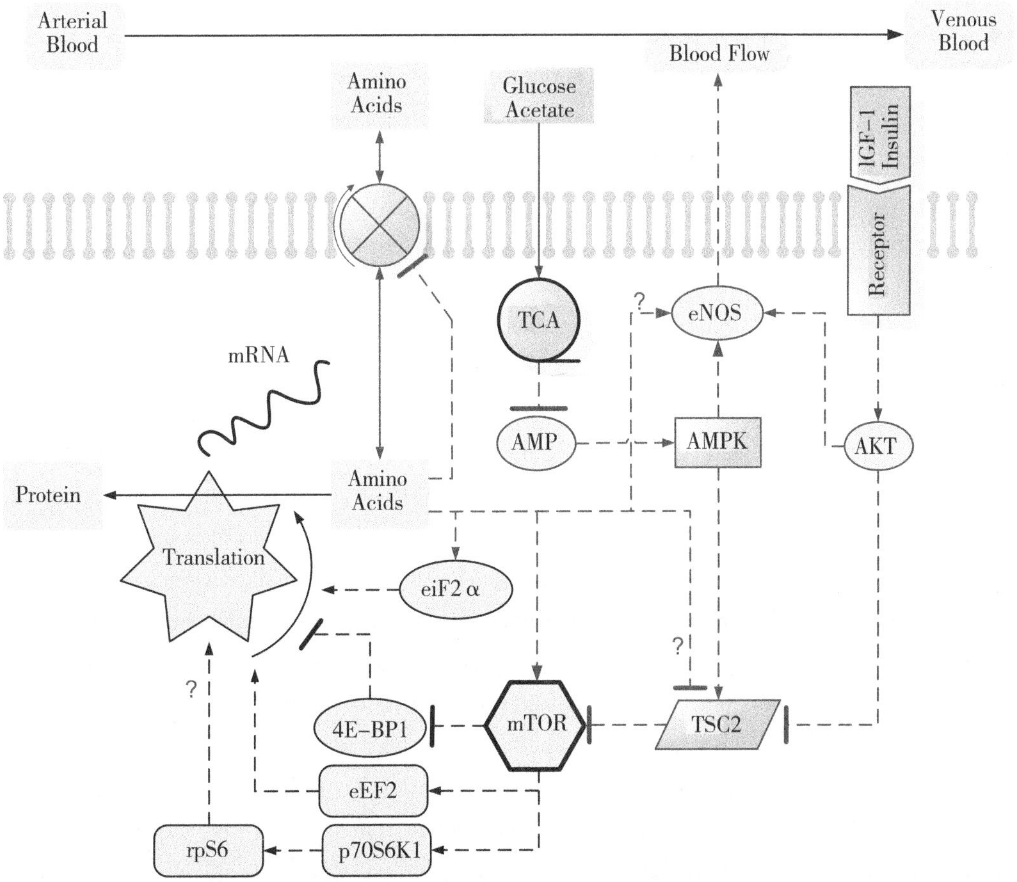

Figure 4 A partial schematic of the regulation of protein synthesis by mammalian target of rapamycin (mTOR) and associated pathways. TCA = tricarboxylic acid cycle

Because milk protein output is also regulated by genetic potential, there must be linkage between cell demands for AA and AA transport (Bequette et al., 2000) to ensure adequate substrate delivery. Regulation of transport activity causes variable efficiency of transfer which undermines the ideal protein concept by creating a range of inputs that can achieve similar efficiency. This also contributes to muted responses to variation in AA supply with declining efficiency as supply increases. If provision of more than one nutrient can offset the loss or deficiency of another, there is almost an infinite number of combinations of AA and energy substrates concentrations that will result in the very same amount of milk protein. Addition of a single AA that can impact cell signaling, while all others are held constant, will push milk protein synthesis higher regardless of which is perceived to be "first-limiting". This has been demonstrated with mammary cells (Clark et al., 1978; Appuhamy et al., 2012b; Appuhamy et al., 2014; Arriola Apelo et al., 2014b), in lactating mice (Liu et al., 2017), and in lactating cows (Rius et al., 2010ab; Yoder et al., 2018). This concept is further supported by meta analyses (Hanigan et al., 2000; Castro et al., 2016; Hanigan et al., 2018). These results clearly demonstrate that the response surface is complex and not consistent with the "Law of the Minimum" when applied to milk protein production.

Mammary Affinity for AA

If we are to match the supply of each AA with animal needs, we must understand the relationships among supply of each and the rate of removal for the major post-absorptive processes which include the portal-drained viscera, the liver, and mammary.

Although the primary driving force for AA uptake may be the balance of AA supply and demand, those forces can be modified by the relative supply of other AA. For example, glutamine concentrations in blood are at least partially determined by metabolism in other tissues, and its transport into mammary tissue is sodium (Na) - dependent (Calvert et al., 1998). Thus, glutamine can be concentrated within the cell and be exchanged for other AA that are not Na-dependent. Glutamine also may play a role in maintenance of cell volume, which has been linked to rates of protein synthesis. The interactive influence of glutamine or other non-essential AA on transport of essential AA could be quite varied. It remains uncertain how large these influences are as AA transport is quite complicated with more than 25 different AA transporters being expressed in epithelial cells and many interactions between AA occur (Calvert and Shennan, 1996; Shennan et al., 1997; Calvert et al., 1998; Bröer, 2008). Consequently, work is needed to characterize transport activity responses for each of the EAA.

Initial *in vivo* work demonstrated that measurements of isotopic enrichment combined with mammary arterivenous difference measurements could be used to derive bi-directional exchange of an AA across the mammary tissue (Hanigan et al., 2009). We have extended this approach to simultaneously assess bi-directional transport rates for all of the AA contained in a labelled algae AA mix that can be purchased from several isotope vendors. This work was initially conducted *in vitro* using mammary cells, and we have recently completed the animal work for an *in vivo* trial. The reduction in transport activity with increasing supply is demonstrated by reduced uptake flux rate constants for leucine and phenylalanine as extracellular AA concentrations increase (Figure 5). Significant additional work is required with such an approach to identify the interactions among AA so that the entire response surface for at least each of the EAA can be worked out. Armed with that information, one could then test a few of the interactions *in vivo* to verify the *in vitro* work.

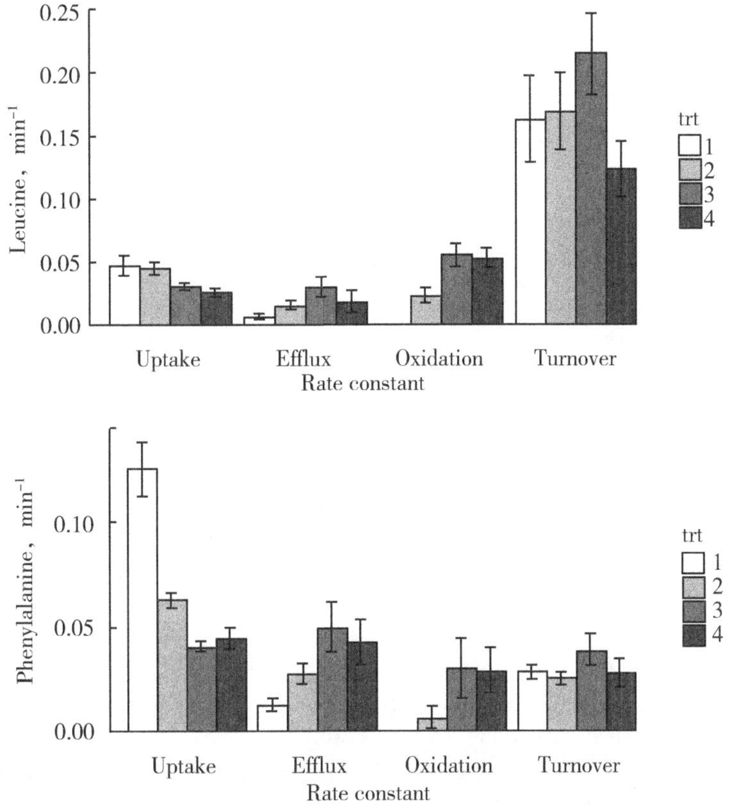

Figure 5 Apparent rate constants (min^{-1}) for leucine and phenylalanine uptake, efflux, catabolism, and sequestration in protein by mammary cells in culture subjected to varying concentrations of total amino acids. treatments were: 1=25%, 2=100%, 3=200%, and 4 = 300% of *in vivo* plasma AA concentrations for high producing lactating dairy cows. from yoder et al. (submitted)

Conclusion

Rations can be balanced at levels well below 15% CP, probably even below 13%, if we are able to reliably match AA supply with true animal needs. However, current models of AA requirements used in field application programs are incompatible with making such predictions. We have devised a new prediction scheme that will be a better representation of the biology, and thus should provide much greater accuracy allowing us to achieve N efficiencies of 35% or greater in lactating cattle. However, additional experimental work is required to better define the supply of AA to the animal and to define the mammary responses to the mix of AA and energy provided from the full range of potential diets. Such information will provide additional gains in prediction precision, allowing even further reductions in dietary CP. Adoption of these systems will improve N efficiency and minimize land use for animal protein production, both of which are essential to feed the future world population.

References

Agle, M., Hristov, A. N., Zaman, S., et al., 2008. Effect of dietary protein level and degradability and energy density on ammonia losses from manure in dairy cows [J]. J. Dairy Sci., 91: 324-324.

Appuhamy, J. A., Knoebel, N. A., Nayananjalie, W. A., et al., 2012a. Isoleucine and leucine independently regulate mTOR signaling and protein synthesis in MAC - T cells and bovine mammary tissue slices [J]. J. Nutr., 142 (3): 484-491.

Appuhamy, J. A. D. R. N., Bell, A. L., Nayananjalie, W. A. D., et al., 2011. Essential amino acids regulate both initiation and elongation of mRNA translation independent of insulin in MAC-T Cells and bovine mammary tissue slices [J]. J. Nutr., 141 (6): 1209-1215.

Appuhamy, J. A. D. R. N., Bray, C. T., Escobar, J., et al., 2009. Effects of acetate and essential amino acids on protein synthesis signaling in bovine mammary epithelial cells in-vitro [J]. J. Dairy Sci., 92 (e-Suppl1): 44.

Appuhamy, J. A. D. R. N., Knoebel, N. A., Nayananjalie, W. A. D., et al., 2012b. Isoleucine and Leucine Independently Regulate mTOR Signaling and Protein Synthesis in MAC-T Cells and Bovine Mammary Tissue Slices [J]. J. Nutr., 142 (3): 484-491.

Appuhamy, J. A. D. R. N., Nayananjalie, W. A., England, E. M., et al., 2014. Effects of AMP-activated protein kinase (AMPK) signaling and essential amino acids on mammalian target of rapamycin (mTOR) signaling and protein synthesis rates in mammary cells [J]. J. Dairy Sci., 97 (1): 419-429.

Arriola Apelo, S. I., Knapp, J. R., Hanigan, M. D., 2014a. Invited review: Current representation and future trends of predicting amino acid utilization in the lactating dairy cow [J]. J. Dairy Sci., 97 (7): 4000-4017.

Arriola Apelo, S. I., Singer, L. M., Ray, W. K., et al., 2014b. Casein synthesis is independently and additively related to individual essential amino acid supply [J]. J. Dairy Sci., 97 (5): 2998-3005.

Bateman, H., Clark, J. H., Murphy, M. R., 2005. Development of a System to Predict Feed Protein Flow to the Small Intestine of Cattle [J]. J. Dairy Sci., 88: 282-295.

Bequette, B. J., Hanigan, M. D., Calder, A. G., et al., 2000. Amino acid exchange by the mammary gland of lactating goats when histidine limits milk production [J]. J. Dairy Sci., 83: 765-775.

Broderick, G., Huhtanen, P., Ahvenjarvi, S., et al., 2010. Quantifying ruminal nitrogen metabolism using the omasal sampling technique in cattle-A meta-analysis [J]. J. Dairy Sci., 93: 3216-3230.

Bröer, S., 2008. Amino acid transport across mammalian intestinal and renal epithelia. Physiol. Rev., 88 (1): 249-286.

Calvert, D. T., Kim, T. G., Choung, J. J., et al., 1998. Characteristics of L-glutamine transport by lactating mammary tissue [J]. J. Dairy Res., 65 (2): 199-208.

Calvert, D. T., Shennan, D. B., 1996. Evidence for an interaction between cationic and neutral amino acids at the blood-facing aspect of the lactating rat mammary epithelium [J]. J. Dairy Res., 63: 25-33.

Caraviello, D. Z., Weigel, K. A., Fricke, P. M., et al., 2006. Survey of Management Practices on Reproductive Performance of Dairy Cattle on Large US Commercial Farms [J]. J. Dairy Sci., 89 (12): 4723-4735.

Castro, J. J., Arriola Apelo, S. I., Appuhamy, J. A., et al., 2016. Development of a model describing regulation of casein synthesis by the mammalian target of rapamycin (mTOR) signaling pathway in response to insulin, amino acids, and acetate [J]. J. Dairy Sci., 99 (8): 6714-6736.

Clark, R. M., Chandler, P. T., Park, C. S., 1978. Limiting amino acids for milk protein synthesis by bovine mammary cells in culture [J]. J. Dairy Sci., 61: 408-413.

Delgado, C. L., 2003. Rising consumption of meat and milk in developing countries has created a new food revolution [J]. J. Nutr., 133 (11 Suppl 2): 3907S-3910S.

Ertl, P., Klocker, H., Hörtenhuber, S., et al., 2015a. The net contribution of dairy production to human food supply: The case of Austrian dairy farms [J]. Agr. Syst., 137: 119-125.

Ertl, P., Zebeli, Q., Zollitsch, W., et al., 2015b. Feeding of by-products completely replaced cereals and pulses in dairy cows and enhanced edible feed conversion ratio [J]. J. Dairy Sci., 98 (2): 1225-1233.

Estes, K. A., White, R. R., Yoder, P. S., et al., 2018. An *in vivo* stable isotope-based approach for assessment of absorbed amino acids from individual feed ingredients within complete diets [J]. J. Dairy Sci., 101 (8): 7040-7060.

Hanigan, M. D., France, J., Crompton, L. A., et al., 2000. Evaluation of a representation of the limiting amino acid theory for milk protein synthesis [M]. CABI, Wallingford.

Hanigan, M. D., France, J., Mabjeesh, S. J., et al., 2009. High Rates of Mammary Tissue Protein Turnover in Lactating Goats Are Energetically Costly [J]. J. Nutr., 139 (6): 1118-1127.

Hanigan, M. D., Lapierre, H., Martineau, R., et al., 2018. Predicting milk protein production from amino acid supply [J]. J. Dairy Sci., 101 (Suppl. 2): 410.

Haque, M. N., Rulquin, H., Andrade, A., et al., 2012. Milk protein synthesis in response to the provision of an "ideal" amino acid profile at 2 levels of metabolizable protein supply in dairy cows [J]. J. Dairy Sci., 95 (10): 5876-5887.

Hristov, A. N., Price, W. J., Shafii, B., 2004. A meta-analysis examining the relationship among dietary factors, dry matter intake, and milk and milk protein yield in dairy cows [J]. J. Dairy Sci., 87 (7): 2184-2196.

Hvelplund, T., Hesselholt, M., 1987. Digestibility of individual amino acids in rumen microbial protein and undegraded dietary protein in the small intestine of sheep [J]. Acta Agriculturae Scandinavica, 37 (4): 469-477.

Lambin, E. F., Meyfroidt, P., 2011. Global land use change, economic globalization, and the looming land scarcity [J]. P. Natl. Acad. Sci., 108 (9): 3465-3472.

Liu, G. M., Hanigan, M. D., Lin, X. Y., et al., 2017. Methionine, leucine, isoleucine, or threonine effects on mammary cell signaling and pup growth in lactating mice [J]. J. Dairy Sci., 100 (5):

4038-4050.

Maxin, G., Ouellet, D. R., Lapierre, H., 2013. Effect of substitution of soybean meal by canola meal or distillers grains in dairy rations on amino acid and glucose availability [J]. J. Dairy Sci., 96 (12): 7806-7817.

Mitchell, H. H., Block, R. J., 1946. Some relationships between the amino acid contents of proteins and their nutritive values for the rat [J]. J. Biol. Chem., 163 (3): 599-620.

NRC, 2001. Nutrient requirements of dairy cattle [M]. 7th rev. ed. National Academy Press, Washington, D. C.

Paris, Q., 1992. The von Liebig hypothesis [J]. Am. J. Agric. Econ., 74 (4): 1019-1028.

Polan, C. E., Cummins, K. A., Sniffen, C. J., et al., 1991. Responses of dairy cows to supplemental rumen-protected forms of methionine and lysine [J]. J. Dairy Sci., 74: 2997-3013.

Rius, A. G., Appuhamy, J. A. D. R. N., Cyriac, J., et al., 2010a. Regulation of protein synthesis in mammary glands of lactating dairy cows by starch and amino acids [J]. J. Dairy Sci., 93 (7): 3114-3127.

Rius, A. G., McGilliard, M. L., Umberger, C. A., et al., 2010b. Interactions of energy and predicted metabolizable protein in determining nitrogen efficiency in the lactating dairy cow [J]. J. Dairy Sci., 93 (5): 2034-2043.

Roman-Garcia, Y., White, R. R., Firkins, J. L., 2016. Meta-analysis of postruminal microbial nitrogen flows in dairy cattle. I. Derivation of equations [J]. J. Dairy Sci., 99 (10): 7918-7931.

Shennan, D. B., Millar, I. D., Calvert, D. T., 1997. Mammary-tissue amino acid transport systems [J]. Proc. Nutr. Soc., 56 (1A): 177-191.

Sprengel, C., 1828. Von den substanzen der ackerkrume und des untergrundes (About the substances in the plow layer and the subsoil) [J]. Journal für Technische und Ökonomische Chemie., 3: 42-99.

St-Pierre, N., 2012. The costs of nutrients, comparison of feedstuffs prices and the current dairy situation. Buckeye News.

Titgemeyer, E. C., Merchen, N. R., Berger, L. L., 1989. Evaluation of soybean meal, corn gluten meal, blood meal and fish meal as sources of nitrogen and amino acids disappearing from the small intestine of steers [J]. J. Anim. Sci., 67 (1): 262-275.

UN, 2017. World population prospects: the 2017 revision. Key findings and advance tables [EB/OL]. ESA/P/WP/248. United Nations, Department of Economic and Social Affairs, Population Division, New York.

Uuml, D., Lling, R., Menzi, H., et al., 2001. Emissions of ammonia, nitrous oxide and methane from different types of dairy manure during storage as affected by dietary protein content [J]. J. Agr. Sci., 137 (02): 235-250.

Van Amburgh, M. E., Collao-Saenz, E. A., Higgs, R. J., et al., 2015. The Cornell Net Carbohydrate and Protein System: Updates to the model and evaluation of version 6. 5 [J]. J. Dairy Sci., 98 (9): 6361-6380.

von Liebig, J., 1862. Die chemie in ihrer anwendung auf agricultur und physiologie [M]. Vol. II. 7. Aufl ed. Friedrich Vieweg, Braunschweig.

Vyas, D., Erdman, R. A., 2009. Meta-analysis of milk protein yield responses to lysine and methionine supplementation [J]. J. Dairy Sci., 92 (10): 5011-5018.

White, R. R., Kononoff, P. J., Firkins, J. L., 2017a. Methodological and feed factors affecting prediction of ruminal degradability and intestinal digestibility of essential amino acids [J]. J. Dairy Sci., 100 (3): 1946-1950.

White, R. R., Roman-Garcia, Y., Firkins, J. L., 2016. Meta-analysis of postruminal microbial nitrogen flows in dairy cattle. II. Approaches to and implications of more mechanistic prediction [J]. J. Dairy Sci., 99 (10): 7932-7944.

White, R. R., Roman-Garcia, Y., Firkins, J. L., et al., 2017b. Evaluation of the National Research Council (2001) dairy model and derivation of new prediction equations. 2. Rumen degradable and undegradable protein [J]. J. Dairy Sci., 100 (5): 3611-3627.

White, R. R., Roman-Garcia, Y., Firkins, J. L., et al., 2017c. Evaluation of the National Research Council (2001) dairy model and derivation of new prediction equations. 1. Digestibility of fiber, fat, protein, and nonfiber carbohydrate [J]. J. Dairy Sci., 100 (5): 3591-3610.

Wilkinson, J. M., 2011. Re-defining efficiency of feed use by livestock [J]. Animal, 5 (7): 1014-1022.

Yoder, P. S., Huang, X., Hanigan, M. D., 2018. Effects of infused leucine and isoleucine or methionine, lysine, and histidine on cow performance [J]. J. Dairy Sci., 101 (Suppl. 2): 407.

Session 2

Advances in Milk Quality and Safety

Effect of Iron in Farm and Processing Water Sources on Milk Quality

Susan E. Duncan[1], Aili Wang[1], and Georgianna Mann[2]
[1]Virginia Polytechnic Institute and State University (Virginia Tech), Blacksburg, Virginia, USA; [2]University of Mississippi, University, MS, USA

Abstract

High mineral concentrations (>0.3 mg/kg Fe and others) may be associated with natural levels in ground water, contaminating sources, drought conditions, or storage systems. Implications of dietary water mineral chemistry on biological fluids are not well described. Water is an important nutrient for dairy cattle, who consume high quantities (90 to 150 L) of water daily. As water is a major component of milk, water source influence on bovine milk synthesis and subsequent milk quality is of interest in understanding implications to cow and calf health, milk processing and stability, and milk nutrient availability for human health. When in excess that exceeds the antioxidant mechanisms in biological systems, iron and copper initiate oxidative reactions that may lead to oxidative stress, initiating biological responses that lead to changes in biological fluids such as saliva and milk. Excess iron in bovine water sources influences the milk proteome, which can affect milk stability as well as influencing sensory and nutritional quality and further processing into dairy products.

Water Use for Dairy Production and Processing

The United States, India, and China produce 29% of the world's milk (14.6%, 8.4%, 6.0%, respectively), but India, with more than 43.6 million cows, and China, with 12.5 million cows, have 21.2% of the global dairy cow population compared to the U.S., at 3.4% (FAOstat, 2012). Water usage on the dairy farm, for bovine drinking source, to clean and cool cows, and irrigate the land, increases water demand to approximately $4.51\times 10^{-9}km^3/$(cow · month) (4,508 L; 1,191 US liquid gallons) (Chase, 2006). Dairy production, compared to beef, consumes a relatively high amount of water (Simmons, 2011). Approximately four gallons of water is needed to produce one gallon of milk.

Water used in processing fluid milk is for heating, cooling, washing, and clean-up. To avoid contamination, potable water used in milking operations on U.S. dairy farms usually cannot be reused, as reclaimed water must fulfill strict regulations to avoid product contamination (Grade "A" Pasteurized Milk Ordinance, 2011). Water use for dairy processing has become more efficient; in 2007, 65 percent less water was used per gallon of milk produced than in 1944 (Raouche et al., 2009; Simmons, 2011). Some dairy processing plants have effectively implemented conservation strategies to reduce the ratio of water to milk (1 : 1).

Water resources are an essential and valuable asset for dairy production and processing. Global water demand for livestock (animals used for food production) is projected in 2025 at 2.4×10^{14} L (235.7 km^3), an increase of over 630% compared to the 37 km^3 (3.7×10^{13} L) utilized in 1995. In the 2005 U. S. Geological Survey, a total of 0.008 km^3/d (2,140 Mgal) of water was withdrawn and used for livestock (livestock, feed lots, dairy operations) and aquaculture, totaling 3 percent of the water withdrawn in the United States. Of this, 60 percent was supplied by groundwater rather than surface water (Barber, 2009). The global dairy industry will be affected by the increased demand for water resources (Rosegrant and Cai, 2002). China, with less than 6% of the total global water resources and approximately one-fifth of the world's population, is challenged with effectively managing water resources for both human and agricultural needs (Yu et al., 2015). In China, more than 60% of all water usage is associated with agriculture. Contaminants from point and non-point contamination cause concerns for potable and non-potable water uses in food-producing agricultural activities. Worldwide iron concentrations in groundwater vary greatly (e.g. 0.04-14.8 mg/L in regions of India; up to 16 mg/L in regions of China) (Yu et al., 2015). A broad range of water-sourced iron, from not detectable to 123 mg/L has been reported across the United States (Mann et al., 2013; Socha et al., 2003).

Water as an Iron Source: Relationship to Dairy Production and Milk Processing

High concentrations of iron and other heavy metals may be associated with natural levels in ground water, run-off from mining or other contaminating sources, drought conditions, or even from the watering systems used for storing water for animal consumption (McNeill, 2006; Bury et al., 2011). Iron concentrations in groundwater sources are variable. In a study of mineral composition of well water, iron concentration ranged between less than 10 μg/kg to greater than 300 μg/kg. Areas of the United States with lower concentrations include the mid-west and the Great Plains. The northern United States tends to have higher concentrations of iron in comparison to the south. However, the greatest concentrations of iron seem to be most prevalent in the mid-Atlantic to north eastern United States (Ayotte et al., 2011). Most dairy production in the United States is located in the west, which has less risk of high iron concentrations in ground water, and in the upper Midwest and northeast where there is greater incidence of high iron-contaminated water (MacDonald et al., 2007).

Understanding possible implications of the mineral characteristics of water sources on cow health and milk quality is imperative (Collignon, 2009). Water provides a significant portion of a lactating cow's mineral intake because of the high volume consumed (90-150 L water per day). Iron can affect palatability of the water provided to cattle, which may cause the animals to consume less water, leading to less milk production (Burlingame et al., 2007; USEPA, National Primary Drinking Water Regulations, 2011; Genther and Beede, 2013). Iron in bovine drinking water should be less than 0.4 mg/L (Socha et al., 2003). Of 2,437 water samples collected across the United States, 41% exceeded this recommendation (Socha et al., 2003). An excess of any heavy metal, particularly iron and copper, may cause adverse effects on milk quality (Hegenauer et al., 1979a). Since water contributes more than 87% of the total weight of milk, water chemistry may deeply impact milk quality. Understanding of the role of excess iron from water sources on bovine health, milk synthesis, and milk and dairy product quality is still developing.

A Study of Iron in Bovine Drinking Water: Implications on to Milk Production and Phosphorus Absorption, Milk Processing, and Milk Proteins

An interdisciplinary team of researchers at Virginia Tech (Blacksburg, VA) conducted a study to determine the effect of varying levels of iron in bovine drinking water on phosphorus absorption in lactating dairy cows and the effects on milk quality, including oxidative stability and the milk proteome. Our faculty and graduate student team included expertise in environmental science and engineering, dairy production, and dairy processing. The interdisciplinary team allowed for expertise relating to mineral and water chemistry, ensuring that the experimental design and handling of the cows' diet and delivery of the water sources was done appropriately, and that milk processing was completed effectively. We were able to complete a broad spectrum of analyses on the water, cow, and milk because of the team cooperation.

Effect of excess iron on lactating dairy cows

We designed the experiment using four ruminally-cannulated early lactation cows on the Virginia Tech dairy farm. Each cow received ferrous lactate solutions by abomasal infusion to approximate iron concentrations in water of 0, 2, 5, or 12.5 mg iron/L, based on an assumed water intake of 100 L/day (Feng et al., 2013). A 4×4 Latin Square design with 14-day periods allowed each cow to receive each treatment, thus serving as their own control. After a one week washout period, cows were infused daily (7 days) with 1 L of appropriate ferrous lactate treatment for 7 days. Cows were fed twice daily, with continuous access to diet and water, and milked twice daily. TMR, feed refusals, blood, and feces were collected each day. On days 11 to 14, milk yield and milk samples were collected; blood samples were collected on days 13 and 14 of each period.

Excess Fe can interfere with absorption of other minerals, notably Cu, Zn. While a negative relationship between Fe consumption and P excretion has been established in rats (Campos et al., 1998), we did not observe any effect on intake and digestibility of total P, inorganic P, or phytate P in lactating dairy cows receiving infusions of ferrous lactate up to 12.5 mg Fe/L in drinking water, based on 100 L of water intake/d (Feng et al., 2013). Dry matter intake, milk yield, and milk composition were not affected by infusion of iron – loaded water; however, digestibility of DM, NDF, and nitrogen decreased with increasing iron infusion (equivalent to 0, 2, 5, 12.5 mg Fe/L of water) (Feng et al., 2013). The chemistry of iron – phosphate interactions is a dominant concept in soil mineralogy and plant nutrition but this relationship does not seem to follow through with respect to dairy cattle. We hypothesized that microbial populations in the large intestine may have been altered with increasing concentrations of ferrous lactate, thus decreasing levels in NDF, nitrogen and DM digestibility. Previous research has suggested that excess iron supplementation decreases microbial activity in the rumen (Hubbert et al., 1958; Martinez and Church, 1970; Harrison et al., 1992). While phosphorus availability has been affected by high iron intake in chicks, there was no evidence of effect in our study on lactating dairy cows. Other studies have suggested that plasma Pi is decreased in calves and steers when fed high iron concentrations in the diet (Koong et al., 1970; Standish et al., 1971). Iron infusion did not affect milk yield, total raw milk protein, milk lactose, while a slight difference in milkfat was observed, the effect was small during this short time-frame study (Feng et al., 2013).

Effect of excess iron on milk quality. Milk from the evening milk on day 13 (day 6 of infusion) was collected and transported to the dairy processing laboratory of the Food Science and Technology Department at Virginia Tech (Mann et al., 2013). Milk from each cow was processed independently. Milk was separated into skim and cream, standardized to 3.2% milkfat, homogenized and vat pasteurized, cooled, and packaged into translucent high density polyethylene terephthalate packages. Processed milk was stored for 11 days at 4℃. Milk quality was evaluated within 72 hours of processing and again 7 days later for gross composition, mineral composition, and indices of oxidation by analytical methods. Sensory evaluation discrimination protocols were used to determine if differences in flavor and odor existed from the cow receiving no iron in the infusate compared to the iron – infused cows for each treatment level.

We reported no effects on iron, copper, phosphorus or calcium in the processed milk overall (Mann et al., 2013); however, the cow × treatment interaction was significant for calcium, copper and iron, suggesting that some cows may be more susceptible to effects, leading to differences in mineral composition within the milk. A 3 – way interaction (treatment×cow×period) was observed for copper and for iron, suggesting that these pro-oxidant minerals may be affected in some cows, potentially increasing susceptibility to oxidation. An increase in processed milk aldehydes, which are used as an indication of oxidation, with increasing concentration of abomasally-infused iron was observed in 3 of the 4 cows, with a low but significant correlation. We suggest that the differences in oxidative stability may relate to changes in antioxidants, perhaps relating to the proteome, or to increased susceptibility associated with changes in mineral composition (Mann et al., 2013). Sensory analyses indicated that milk flavor and odor were different with increasing concentration of infused iron into the abomasum (Mann et al., 2013).

Effect of excess iron on milk proteome

Processed milk from each cow for each treatment was frozen until analyses of the proteome could be completed. Two-dimensional gel electrophoresis was used to separate the proteins (Wang et al., 2016). Protein identification was completed by matrix – assisted laser desorption/ionization time – of – flight (MALDI – TOF) high resolution tandem mass spectroscopy analyses. PDQuest software was used to analyze changes in milk protein composition. All classes of milk proteins were identified in the processed milk from the cows receiving the water treatments, as expected (Wang et al., 2016). Variations in milk proteome were observed among the 4 cows used in our study. In general, decreased expression of caseins may be attributed to oxidative stress induced by excess iron. Oxidized proteins may be a consequence of oxidative stress as well. Whey proteins, especially immune proteins including IgG_1 heavy chain, IgM heavy chain, lactoperoxidase, and lactoferrin were decreased more than 2 – fold in one cow as infused iron concentration increased. Serum albumin proteins were also decreased. In contrast, another cow showed increased intensity of these proteins with higher iron infused concentration. The increase in immune defense proteins may be directly related to a more responsive oxidative stress defense mechanism in some cows, with susceptible cows having lower ability to manage the insult and, potentially, contributing to the sporadic incidence of spontaneous oxidation in milk. Post – translational modification of alpha – lactalbumin and beta-lactoglobulin was observed at some level in most cows.

Implications of Excess Iron on Bovine Health, Milk Proteins, and Milk Quality

Iron can be present in water as insoluble ferric iron, soluble ferrous iron, or as dissolved iron under acidic and neutral oxygen-rich conditions (Lenntech, 2013). Iron in feedstuffs is typically in ferric iron form (National Research Council, 2001). Ferrous iron or iron in the dissolved form, as typically found in groundwater, is more readily absorbed, thus more bioavailable.

An iron deficiency in dairy cattle is rarely observed; effects on animal health are more often observed with excesses and may be attributed to large consumption of water containing high iron concentrations (Feng et al., 2013). When iron exceeds the demand of living cells, there will be an increase in the concentration of reactive oxygen and nitrogen species (such as ·OH, O^{2-}·, and NO·), which is called oxidative stress (Puntarulo, 2005). Oxidative stress is "a disturbance in the pro-oxidant-antioxidant balance in favor of the former, leading to potential damage" (Sies, 1985). The resistance of a biological system to oxidative stress depends on its ability to readily detoxify the reactive oxygen species (ROS) or to repair the resulting damage (Halliwell, 2007). Failure to detoxify ROS will lead to production of peroxides and free radicals that damage the cell, which further induces the disruption of cellular signaling (Sies, 1997; Halliwell, 2007). During the iron-dependent conversion, superoxide anion (O^{2-}) and hydrogen peroxide (H_2O_2) transfer to the extremely reactive hydroxyl radical (·OH)(Haber-Weiss reaction) and release toxic compounds that severely destroy the membranes, proteins and DNA (Halliwell and Gutteridge, 1984). Such damage may lead to the peroxidation of nearby lipids, oxidative damage of DNA and other macromolecules (Papanikolaou and Pantopoulos, 2005), and cell membrane damage and, thus, interrupt several biological reactions (Lobo et al., 2010). Damage to protein structures, such as proteolysis or post-translational modification, results in a decrease of protein bioactivity and a change of protein composition. As a consequence, humans or animals may suffer damage of blood vessels, reduced activity of natural killer cells and lymphatic system (Weinberg, 1990), increased risk of microbial or virus infection (Metwally et al., 2004), formation of tumor sites because of iron deposition (Weinberg, 1990), heart failure caused by the great affinity of myocardial cells with iron (Weinberg, 1990), and other chronic diseases such as chronic liver diseases (Sikorska et al., 2003), type Ⅱ diabetes (McCarty, 2003) and hypertension (Piperno et al., 2002).

One direct consequence of oxidative stress is the accumulation of oxidized protein, which is derived from the oxidation of amino acid residue side chains and protein backbone, formation of protein cross-linkages and aggregation, and generation of protein carbonyl derivatives. Protein oxidation was associated with aging and a number of diseases in humans such as Alzheimer's disease, respiratory distress syndrome, muscular dystrophy, amyotrophic lateral sclerosis and rheumatoid arthritis (Schuessler and Schilling, 1984). When free iron is available, which is determined by the concentrations of iron-binding proteins (such as lactoferrin, transferrin) and iron-responsive factors (control the binding and release of iron from iron-binding proteins), derivatives of protein oxidation such as H_2O_2 can generate the even more toxic ·OH by iron-catalyzed cleavage through the Fenton reaction (Berlett and Stadtman, 1997). On the other hand, free iron also catalyzes the formation of free radicals, which will hasten oxidative stress and further oxidize oxygen-derived radicals (Hentze and

Kühn, 1996). It has been demonstrated that the primary oxidative damage to proteins was metal-catalyzed oxidation (Stadtman and Berlett, 1998; Berlett and Stadtman, 1997).

To minimize the damage (protein aggregation and cross-linking) induced by protein oxidation, a series of antioxidant defense mechanisms are activated within dairy cattle. Mammalian cells serve as the first protective defense to rescue moderately damaged polypeptides. Both prokaryotic and eukaryotic cells have enzymes that can directly repair some covalent modifications to the primary structure of proteins, and restore the functions of defective proteins through reduction of oxidized disulfide bonds, repair function of disulfide bonds and reduction of amino acid side chains (Grune et al., 1997). However, mammalian cells have only limited ability to directly repair oxidized protein. Most oxidized proteins are degraded by proteolytic pathway to minimize protein aggregation and remove potentially toxic protein fragments, which is considered as the secondary antioxidant defense (Berlett and Stadtman, 1997). Though it is known that iron acts as a pro-oxidant, little research has been identified that examines the specific role of iron in the water given to dairy cattle (Sugiarto et al., 2010).

Previous studies reported that dairy cattle can tolerate no more than 1000 mg/kg dietary iron under most conditions (National Research Council, 1956). In drinking water with iron concentrations ranges from 0.1 to 0.91 mg/L, between 1 to 95 mg/d of Fe (calculated) would be delivered per day to a dairy cow; the latter is half of a cow's daily requirement. Therefore, at levels of 2 mg/L or more, may create oxidative stress and other health effects (Wang et al., 2016). Excess Fe has been related to increased incidences of mastitis, retained placenta, and a general decrease in immune function in dairy cows (Socha et al., 2003; Weiss et al., 2010). Observed health effects associated with excess iron in the intestinal tract of bovines includes bacterial infection, reduced weight gain, and oxidative stress (Coup and Campbell, 1964; Standish et al., 1971; Bullen et al., 1978; McGuire et al., 1985; Hansen et al., 2010). Declined health condition of dairy cattle immediately depressed their milk production and altered milk composition, including loss of milk yield (Gröhn et al., 2004), decrease of lactose and fat content (Bansal et al., 2005), and increase of sodium, chloride and electrical conductivity (Bruckmaier et al., 2004).

Excess iron intake may affect milk synthesis, with subsequent effects on milk composition, flavor andoxidation rates. Iron averages 0.5 mg/kg with a range of 0.3 - 0.6 mg/kg in raw milk. Copper, also a pro-oxidant metal, is present at 0.1-0.6 mg/kg in raw milk (Goff and Hill, 1993; Hunt and Nielsen, 2009). Similar concentrations of minerals have been found in whole processing (pasteurized, homogenized) milk (1130 mg/kg Ca, 0.11 mg/kg Cu, 0.3 mg/kg Fe, 910 mg/kg P) (Milk Facts, 2013). About 20 percent of iron is located in the fat fraction of milk, with 30-60 percent bound to transferrin or lactoferrin in the aqueous phase, and another 10 percent bound to the casein.

Iron and phosphorous bind to milk proteins (Jenness, 1974; Walstra et al., 1984; Sims and Sharpley, 2005). About 10 percent of iron in milk is bound to casein proteins, with 20–30 percent bound to iron-binding proteins like lactoferrin and transferrin and is in the ferric (Fe^{3+}) form (Jenness, 1974; Fransson and Lonnerdal, 1983; Jensen, 1995). Iron content of bovine milk tends to vary with location, stage of lactation, time of the year and breed. A 0.4 mg/kg fluctuation of iron content in milk has been recorded during a bovine lactation period. Higher levels of iron are expected in early stages of milk production and highest (2–3 times that of normal milk) in milk containing colostrum (1–2 mg/kg) (Underwood, 1971; Murthy et al., 1972; Jensen, 1995). Iron content of milk is not affected by diet but studies on the effects of drinking water are minimal (Murthy et al., 1972; Fransson and Lonnerdal, 1983).

It is possible that an increase in iron can lead to spontaneous oxidation in the final milk product (Hegenauer et al., 1979b; Mann et al., 2013; Wang et al., 2016). Spontaneous oxidation is defined as the oxidation of milk/milkfat due to a number of factors rather than any recognized cause in particular. Spontaneous oxidation may be due to low concentrations of antioxidants but this notion is not fully explored. This can cause significant problems since it often occurs in herds that tend to be well-managed and often have no other problems. Many times only a few cows will produce milk that readily undergoes spontaneous oxidation rather than the entire herd, furthering difficulties in finding a solution to the problem (Barrefors et al., 1995). This spontaneous oxidation flavor oftendevelops without evidence of addition of other oxidants, like heavy metals, and is seemingly unexplained (Frankel, 1991). The susceptibility of milk to spontaneous oxidation often varies and in some milk can occur very quickly. Many commercial dairies have very few cattle displaying these qualities in their milk ($< 10\%$), but subsequent oxidation of commingled milk in the bulk tank can progress rapidly. This spontaneous oxidized flavor of milk renders it unsuitable for human consumption in many cases (Nicholson and Charmley, 1993; Timmons et al., 2001).

Quality and yield of bovine milk proteins are determined by genetics of dairy cows, hormones, dietary energy, and lactation environment (Bionaz et al., 2012). Casein, which accounts for more than 80% of milk proteins, has a high affinity in binding with Fe ions through the clusters of their phosphoserine residues. Caseins with multiple clusters of phosphoserine residues, especially α_{s1}-, α_{s2}-, and β-casein, present high iron-binding capacity (Horne, 1998). In a study of iron-binding properties of calcium-depleted milk (Mittal et al., 2015), it was found that among milk proteins (casein and whey proteins), iron was bound mostly with casein irrespective of their state of aggregation. Caseinophosphopeptides (CPP) and casein hydrolysates, which are hydrolytic products of casein, are also able to bind with metal ions such as iron through chelation by their phosphoseryl residues. Wang et al. (2016) observed decreased expression of caseins and accumulation of oxidized protein due to oxidative stress attributed to water-sourced iron in dairy cows. They identified large hydrolytic products of casein in milk collected from cows receiving abomasally-infused iron-containing water at low (2 mg/L), medium (5 mg/L) and high (12.5 mg/L) treatments. The increased concentration of hydrolytic products of caseins may function as pro-oxidants along with iron and other free radicals.

Whey proteins of milk possess a variety of nutritional and biological properties, thus have potential value for reducing disease risks such as cancer (Gill and Cross, 2000; de Wit, 1998), inflammation (Clare et al., 2003), chronic stress-induced disease (Ganjam et al., 1997), and HIV (Oona et al., 1997; Micke et al., 2002). Bovine whey proteins have many biological functions including β-lactoglobulin (mediate and transport immunoglobulins during colostrum formation), α-lactalbumin (lactose synthase component and possible antimicrobial/anticancer activity), immunoglobulins (serving as antibodies to protect the mammary gland from infection), serum albumin (anti-mutagenic, anticancer, and immunomodulation activity), lactoferrin (iron-binding, iron transport, antimicrobial/anti-inflammatory/anticancer activities, immune system modulation), and lactoperoxidase (antimicrobial and antioxidant properties) (Alonso-Fauste et al., 2012; Haug et al., 2007; Swaisgood, 1995; Levieux and Ollier, 1999; Loimaranta et al., 1999; Adlerova et al., 2008). Changes in whey protein structure due to oxidation may alter biological function. Wang et al. (2016) reported changes in whey proteins attributed to water-sourced iron, with a high degree of variability among the four cows in the experiment; as low as Fe 2 mg/L caused oxidative stress in the cows.

Lactoferrin is the primary whey protein in milk and has a high affinity for ferric iron (Jenness, 1974; Fransson and Lonnerdal, 1983). Lactoferrin has two binding sites for iron. Manganese and zinc also bind to the same site (Lönnerdal et al., 1985). Lactoferrin levels have been shown to decrease throughout milk production but are higher in cattle with infections, suggesting a role in infection control (Rainard et al., 1982; Satue-Gracia et al., 2000). Antimicrobial activity of lactoferrin has been studied and it may aid in nutritional uptake in infants (Lönnerdal, 2009). Aside from antimicrobial properties, lactoferrin has demonstrated the ability to act as an antioxidant by binding iron (Satue-Gracia et al., 2000). As an iron-binding protein, lactoferrin works as a metal chelator to remove excess iron from living cells by binding and transporting iron ions. Based on its metal chelation property, lactoferrin is able to restrict lipid peroxidation and potential radical-generating reactions through scavenger of lipid byproducts *in vitro* (Sies, 1997). Due to its anti-inflammatory and antioxidant functions, lactoferrin is involved in a wide range of biological activities such as prevention of cell injury and tissue damage, antimicrobial defense, immune modulation, cellular growth and differentiation, and cancer prevention. Lactoferrin concentration in milkmay be affected by additional dietary iron. Lactoferrin concentration increased in human breast milk when mothers were given elevated dietary iron (Zapata et al., 1994).

Oxidation of milk, and consequently, milk products causes unacceptable off-flavors that can result in substantial economic losses. These impacts in profit loss may not only occur in milk, but also in foods using oxidized dairy ingredients. If ignored, this chemical reaction can create an undesirable and unacceptable product from an otherwise sound food source, affecting consumer satisfaction and product integrity.

High quality freshly processed milk has a mild and mellow flavor; therefore oxidative defects can be readily noted and are not easily masked. The "oxidized" flavor can often be detected by consumers, especially in fluid milk (Frankel, 2005). Flavor profiles of milk affected by metal-induced oxidation include cardboard, papery, metallic, painty, cappy, oily, and fishy (Havemose et al., 2006; Alvarez, 2009). Metal-induced flavor is characterized by a rapid taste reaction when the product is placed in the mouth. The flavor also has a tendency to linger even after the sample is expectorated (Bodyfelt et al., 1988; Clark et al., 2009). Metal-induced oxidized flavors in milk also have an astringent (pucker) mouthfeel sensation (Alvarez, 2009). These negative sensory characteristics from the milk subsequently can affect dairy product (cream, butter, yogurt, ice cream, milk powder) quality and shelf-life as well as other food products in which these are incorporated.

Metallic flavors, originally pinpointed to metals used in pipes in dairy processing plants, have caused serious metallic off-flavors with high frequency. Maintenance of product production without metallic off-flavors is a challenge, as water supplies must be constantly controlled against the exposure to copper, iron and manganese. Butter and cheese have a water rinse processing step and water quality must be considered. Hard water can relay metals to the final dairy product. Metallic or oxidized flavors in butter may be initiated by rinsing water; the oxidized flavors are readily detected during the classification process are often given a "below grade" rating (USDA, 1989; Clark et al., 2009). While many off-flavors are attributed to light-oxidized flavors, metals such as copper are also responsible for the formation of off-flavors (Jenq et al., 1988; Cadwallader et al., 2007).

Conclusions

There is still more research needed to help dairy producers and processors recognize the influence of iron in bovine drinking water on the health of the lactating animal, quality of raw milk, and the related dairy products. Such information assists dairy producersand processors in establishing good agricultural practices for bovine drinking water standards applicable for milk quality control. Inter disciplinary studies can better integrate the knowledge and understanding needed for a systems approach for addressing environmental impacts of water on bovine health and implications to milk and dairy quality. The changes in milk proteome attributed to iron in bovine drinking water, as observed in our studies, may be indications that high iron concentrations in bovine drinking water contribute to oxidative stress. Milk quality from cows that are susceptible to oxidative stress may be more susceptible to spontaneous oxidation.

Acknowledgements

This project was funded, in part, by the Virginia Tech College of Agriculture and Life Sciences Pratt Endowment, Virginia Agricultural Experiment Station (Blacksburg), the Hatch Program of the National Institute of Food and Agriculture, US Department of Agriculture (Washington, DC), and the VT Water INTERface Interdisciplinary Graduate Education Program. The authors acknowledge Xin Feng and Katharine Knowlton (Department of Dairy Science, Virginia Tech, Blacksburg) and Andrea Dietrich (Department of Civil and Environmental Engineering, Virginia Tech, Blacksburg) for their contributions the studies described in this paper.

References

Adlerova, L., Bartoskova, A., Faldyna, M., 2008. Lactoferrin: a review [J]. Veterinarni Medicina., 53 (9): 457-468.

Alonso-Fauste, I., Andrés, M., Iturralde, M., et al., 2012. Proteomic characterization by 2-DE in bovine serum and whey from healthy and mastitis affected farm animals [J]. J. Proteomics., 75: 3015-3030.

Alvarez, V. B., 2009. Fluid milk and cream products. Chapter 5 in The Sensory Evaluation of Dairy Products [M]. 2nd ed. S. Clark, M. Costello, M. Drake, and F. Bodyfelt, ed. New York, USA: Springer.

Ayotte, J. D., Gronberg, J. A. M., Apodaca, L. E., 2011. Trace elements and radon in groundwater across the United States, 1992-2003 [R]. In Scientific Investigations Report. U.S. Geological Survey, Reston, VA.

Bansal, B. K., Hamann, J., Grabowski, N. T., et al., 2005. Variation in the composition of selected milk fraction samples from healthy and mastitis quarters, and its significance for mastitis diagnosis [J]. J. Dairy Res., 72: 144-152.

Barber, N. L., 2009. Summary of estimated water use in the United States in 2005 [R]. U.S. D. o. t. Interior, ed. U.S. Geological Survey, Reston, VA.

Barrefors, P., Granelli, K., Appelqvist, L. A., et al., 1995. Chemical characterization of raw milk samples with and without oxidative off-flavor [J]. J. Dairy Sci., 78 (12): 2691-2699.

Berlett, B. S., Stadtman, E. R., 1997. Protein oxidation in aging, disease, and oxidative stress [J]. J. Biol. Chem., 272: 20313-20316.

Bionaz, M., Hurley, W., Loor, J., 2012. Milk protein synthesis in the lactating mammary gland: Insights from transcriptomics analyses [EB/OL]. INTECH Open Access Publisher. http: //creativecommons. org/licenses/by/3. 0.

Bodyfelt, F. W., Tobias, J., Trout, G. M., 1988. The Sensory Evaluation of Dairy Products [M]. New York: Van Nostrand Reinhold.

Bruckmaier, R. M., Weiss, D., Wiedemann, M., et al., 2004. Changes of physicochemical indicators during mastitis and the effects of milk ejection on their sensitivity [J]. J. Dairy Res., 71 (3): 316-321.

Bullen, J. J., Rogers, H., Griffiths, E., 1978. Role of iron in bacterial infection [J]. Curr. Top. Microbiol. Immunol., 80: 1-35.

Burlingame, G., Dietrich, A, Whelton, A., 2007. Understanding the basics of tap water taste [J]. J. Am. Water Works Assoc., 99 (5): 100.

Bury, N. R., Boyle, D., Cooper, C. A., 2011. Iron [M]. Pages 201-251 in Fish Physiology. Vol. 31, Part A. A. P. F. Chris M. Wood and J. B. Colin, ed. Academic Press.

Cadwallader, K. R., Drake, M. A., McGorrin. R. J., 2007. The flavor and flavor stability of skim and whole milk powders [C]. Pages 217-251 in Flavor of Dairy Products. Vol. 971. American Chemical Society, Ann Arbor, MI.

Campos, M. S., Barrionuevo, M., Alferez, M. J. M., et al., 1998. Lisbona, F. Interactions among iron, calcium, phosphorus and magnesium in the nutritionally iron-deficient rat [J]. Exper. Physiol., 83: 771-781.

Chase, L., 2006. How much water do dairy farms use [R]. in AgFocus NWNY Dairy, Livestock and Field Crops Team. Vol. 2012. Cornell University.

Clare, D. A., Catignani, C. L., Swaisgood, H. E., 2003. Biodefense properties of milk, the role of antimicrobial proteins and peptides [J]. Current Pharmaceutical Design, 9: 1239-1255.

Clark, S., M. Costello, M., Drake, M. A., et al., 2009. The Sensory Evaluation of Dairy Products [M]. 2 ed. Springer US, New York.

Collignon, P. J., 2009. Water recycling-forwards or backwards for public health [J]. Med. J. Australia, 191 (4): 238-239.

Coup, M. R., Campbell, A. G., 1964. The effect of excessive iron intake upon the health and production of dairy cows [J]. J. Agric. Res., 7: 624-638.

de Wit, J. N., 1998. Nutritional and functional characteristics of whey proteins in food products [J]. J. Dairy Res., 81: 597-608.

FAOstat [EB/OL]. 2012. http: //www. fao. org/faostat/en/.

Feng, X., Knowlton, K. F., Dietrich, A. M., et al., 2013. Effect of abomasal ferrous lactate infusion on phosphorus absorption in lactating dairy cows [J]. J. Dairy Sci., 96: 4586-4591.

Frankel, E. N., 1991. Recent advances in lipid oxidation. J. Sci. Food Agric., 54 (4): 495-511.

Fransson, G. B., Lönnerdal, B. O., 1983. Distribution of trace elements and minerals in human and cow's milk [J]. Pediatric Res., 17 (11): 912-915.

Ganjam, L. S., Thornton, W. H., Marshall, R. T., et al., 1997. Antiproliferative effects of yoghurt fractions obtained by membrane dialysis on cultured mammalian intestinal cells [J]. J. Dairy Sci., 80: 2325-2339.

Genther, O. N., Beede, D. K., 2013. Preference and drinking behavior of lactating dairy cows offered water with different concentrations, valences, and sources of iron [J]. J. Dairy Sci., 96 (2): 1164-1176.

Gill, H. S., Cross, M. L., 2000. Anticancer properties of bovine milk [J]. British J. Bovine Milk, 84: 161-165.

Goff, H. D., Hill, A. R., 1993. Chemistry and physics [M]. Pages 1-30 in *Dairy Science and Technology Handbook.* Vol. 1. Y. H. Hui, ed. John Wiley & Sons.

Gröhn, Y. T., Wilson, D. J., Gonzǿlez, R. N., et al., 2004. Effect of pathogen-specific clinical mastitis on milk yield in dairy cows [J]. J. Dairy Sci., 87: 3358-3374.

Grune, T., Reinheckel, T., Davies, K. J., 1997. Degradation of oxidized proteins in mammalian cells [J]. The FASEB J. 11: 526-534.

Halliwell, B., Gutteridge, J. M. C., 1984. Oxygen toxicity, oxygen radicals, transition metals and disease. *Biochem* [J]. Biochem. J., 219: 1-14.

Halliwell, B., 2007. Biochemistry of oxidative stress [J]. Biochem. Soc. Trans., 35 (5): 1147-1150.

Hansen, S. L., Ashwell, M. S., Moeser, A. J., et al., 2010. High dietary iron reduces transporters involved in iron and manganese metabolism and increases intestinal permeability in calves [J]. J. Dairy

Sci., 93: 656-665.

Harrison, G. A., Dawson, K. A., Heinken, R. W., 1992. Effects of high iron and sulfate ion concentrations on dry matter digestion and volatile fatty acid production by ruminal microorganisms [J]. J. Anim. Sci., 70: 1188-1194.

Haug, A., Hostmark, A. T., Harstad, O. M., 2007. Bovine milk in human nutrition-a review [J]. Lipids Health Dis., 6: 1-25.

Havemose, M. S., Weisbjerg, M. R., Bredie, W. L. P., et al., 2006. Oxidative stability of milk influenced by fatty acids, antioxidants, and copper derived from feed [J]. J. Dairy Sci., 89 (6): 1970-1980.

Hegenauer, J., Ludwig, D., Saltman, P., 1979a. Effects of supplemental iron and copper on lipid oxidation in milk. 2. Comparison of metal complexes in heated and pasteurized milk [J]. J. Agric. Food Chem., 27 (4): 868-871.

Hegenauer, J., Saltman, P., Ludwig, D., et al., 1979b. Effects of supplemental iron and copper on lipid oxidation in milk. 1. Comparison of metal complexes in emulsified and homogenized milk [J]. J. Agric. Food Chem., 27 (4): 860-867.

Hentze, M. W., Kühn, L. C., 1996. Molecular control of vertebrate iron metabolism: mRNA-based regulatory circuits operated by iron, nitric oxide, and oxidative stress [J]. Proc. National Acad. Sci., 93 (16), 8175-8182.

Horne, D. S., 1998. Casein interactions: Casting light on the black boxes, the structure in dairy products [J]. Int. Dairy J., 8, 171-177.

Hubbert, F. Jr., Cheng, E, Burroughs, W., 1958. Mineral requirement of rumen microorganisms for cellulose digestion *in vitro* [J]. J. Anim. Sci., 17: 1188-1194.

Hunt, C. D., Nielsen, F. H., 2009. Nutritional aspects of minerals in bovine and human milks [M]. Pages 391-456 in Advanced Dairy Chemistry. Vol. 3. 3 ed. Paul McSweeney and P. F. Fox, ed. Springer-Verlag.

Jenness, R., 1974. Biosynthesis and composition of milk [J]. J. Investig. Dermatol., 63 (1): 109-118.

Jenq, W., Bassette, R., Crang, R. E., 1988. Effects of light and copper ions on volatile aldehydes of milk and milk fractions [J]. J. Dairy Sci., 71 (9): 2366-2372.

Jensen, R. G., 1995. Water-soluble vitamins in bovine milk [M]. In *Handbook of milk composition*, Jensen, R. G., Ed., Academic Press, San Diego: 464-467.

Koong, L. J., Wise, M. B., Barrick, E. R., 1970. Effect of elevated dietary levels of iron on the performance and blood constituents of calves [J]. J. Anim. Sci., 31: 422-427.

Lenntech, 2013. Iron and water: reaction mechanisms, environmental impact and health effects [EB/OL]. http: //www. lenntech. com/periodic/water/iron/iron-and-water. htm.

Levieux, D., Ollier, A., 1999. Bovine immunoglobulin G, β-lactoglobulin, α-lactalbumin and serum albumin in colostrum and milk during the early post partum period [J]. J. Dairy Res., 66 (3): 421-430.

Lobo, V., Patil, A., Phatak, A., et al., 2010. Free radicals, antioxidants and functional foods: impact on human health [J]. *Pharmacogn Rev.*, 4 (8): 118-126.

Loimaranta, V., Lain, M., Soèderling, E., et al., 1999. Effects of bovine immune and non - immune whey preparations on the composition and pH response of human dental plaque [J]. Eur. J. Oral Sci., 107 (4): 244-250.

Lönnerdal, B., 2009. Nutritional roles of lactoferrin. Current Opinion Clin [J]. Nutr. Metabolic Care, 12 (3): 293-297.

Lönnerdal, B., Keen, C. L., Hurley, L. S., 1981. Iron, copper, zinc, and manganese in milk [J]. Annu. Rev. Nutr., 1 (1): 149-174.

MacDonald, J. M., O'Donoghue, E. J., McBride, et al., 2007. Profits, costs, and the changing structure of dairy farming [R]. September Page 35 in *USDA Economic Research Report*. Economic Research Service.

Mann, G. R., Duncan, S. E., Knowlton, K. F., et al., 2013. Effects of mineral content of bovine drinking water: Does iron content affect milk quality? [J]. J. Dairy Sci., 96: 7478-7489.

Martinez, A., Church, D. C., 1970. Effect of various mineral elements on *in vitro* rumen cellulose digestion [J]. J. Anim. Sci., 31: 982-990.

McCarty, M. F., 2003. Hyperinsulinemia may boost both hematocrit and iron absorption by up-regulating activity of hypoxia-inducible factor-1α [J]. Med. Hypotheses., 61: 567-573.

McGuire, S. O., Miller, W. J., Gentry, R. P., et al., 1985. Influence of high dietary iron as ferrous carbonate and ferrous sulfate on iron metabolism in young calves [J]. J. Dairy Sci., 68: 2621-2628.

McNeill, L. S., 2006. Water quality factors influencing iron and lead corrosion in drinking water [D]. Dissertation: 102.

Metwally, M. A., Zein, C. O., Zein, N. N., 2004. Clinical significance of hepatic iron deposition and serum iron values in patients with chronic hepatitis C infection [J]. Am. J. Gastroenterol., 99: 286-291.

Micke, P., Beeh, K. M., Buhl, R., 2002. Effects of long-term supplementation with whey proteins on plasma GSH levels of HIV infected patients [J]. European J. Nutr., 41: 12-18.

Mittal, V. A., Ellis, A., Ye, A., et al., 2015. Influence of calcium depletion on iron - binding properties of milk [J]. J. Dairy Sci., 98: 2103-2113.

Murthy, G. K., Rhea, U.S., Peeler, J. T., 1972. Copper, iron, manganese, strontium, and zinc content of market milk [J]. J. Dairy Sci., 55: 1666-1674.

National Research Council (US), Committee on Animal Nutrition, 1956. Nutrient requirements of dairy cattle [S]. National Academies.

Nicholson, J. W. G., Charmley, E., 1993. Injectable alpha - tocopherol for control of oxidized flavor in milk from dairy cows [J]. Can. J. Anim. Sci., 73 (2): 381-392.

Oona, M., Raego, T., Maaroos, H. I., et al., 1997. Helicobacter pylori in children with abdominal complaints: has immune bovine colostrum some influence on gastritis? [J]. Bacteriology Abstracts (Microbiology B)., 6: 49-57.

Papanikolaou, G., Pantopoulos, K., 2005. Iron metabolism and toxicity [J]. Toxicol Appl Pharmacol., 202: 199-211.

Piperno, A., Trombini, P., Gelosa, M., et al., 2002. Increased serum ferritin is common in men with essential hypertension [J]. J. Hypertens., 20: 1513-1518.

Puntarulo, S., 2005. Iron, oxidative stress and human health [J]. Mol. Aspects Med., 26: 299-312.

Rainard, P., Poutrel, B., Caffin, J. P., 1982. Lactoferrin and transferrin in bovine milk in relation to certain physiological and pathological factors [J]. Ann. Vet. Res., 13 (4): 321-328.

Raouche, S., Naille, S., Dobenesque, M., et al., 2009. Iron fortification of skim milk: Minerals and 57Fe Mössbauer study [J]. Int. Dairy J., 19 (1): 56-63.

Rosegrant, M. W., Cai, X. M., 2002. Global water demand and supply projections Part-2. Results and prospects to 2025 [J]. Water International, 27 (2): 170-182.

Satue-Gracia, M. T., Frankel, E. N., Rangavajhyala, N., et al., 2000. Lactoferrin in infant formulas: Effect on oxidation. J. Agric [J]. Food Chem., 48 (10): 4984-4990.

Schuessler, H., Schilling, K., 1984. Oxygen effect in the radiolysis of proteins [J]. Part 2 bovine serum albumin [J].Int. J. Radiat. Biol., 45: 267-281.

Sies, H., 1985. Oxidative stress: introductory remarks [M]. In Oxidative Stress, Sies, H., Ed., Academic Press, London: 1-8.

Sies, H., 1997. Oxidative stress: oxidants and antioxidants [J]. Exp. Physiol., 82: 291-295.

Sikorska, K., Stalke, P., Lakomy E. A., et al., 2003. Disturbances of iron metabolism in chronic liver diseases [J]. Med. Sci. Monit., 3: 64-67.

Simmons, J., 2011. Making safe, affordable and abundant food a global reality [EB/OL]. Page 11 in Elanco Animal Health. Http: //www. ncbiotech. org/sites/default/files/.../Three-Rights-White-Paper-Revised. pdf. Accessed April 1, 2017.

Sims, J. T., Sharpley, A. N., 2005. Phosphorus: Agriculture and the environment [R]. American Society of Agronomy, Madison, Wisconsin.

Socha, M. T., Ensley, S. M., Tomlinson, D. J., et al., 2003. Variability of water composition and potential impact on animal performance [C]. Pages 85-96 in Proc. Intermountain Nutr. Conf., Utah State University, Logan.

Stadtman, E. R., Berlett, B. S., 1998. Reactive oxygen-mediated protein oxidation in aging and disease [J]. Drug Metab. Rev., 30: 225-243.

Standish, J. F., Ammerman, C. B., Palmer, A. Z., et al., 1971, Influence of dietary iron and phosphorus on performance, tissue mineral composition and mineral absorption in steers [J]. J. Anim. Sci., 33: 171-178.

Sugiarto, M., Ye, A., Taylor, M. W., et al., 2010. Milk protein-iron complexes: Inhibition of lipid oxidation in an emulsion [J]. Dairy Sci. Technol., 90 (1): 87-98.

Swaisgood, H. E., 1995. Protein and amino acid composition of bovine milk [M]. Handbook of milk composition., 1: 464-468.

Timmons, J. S., Weiss, W. P., Palmquist, D. L., et al., 2001. Relationships among dietary roasted soybeans, milk componenets and spontaneous oxidized flavor of milk [J]. J. Dairy Sci., 84 (11): 2440-2449.

Underwood, E. J., 1971. Trace Elements in Human and Animal Nutrition [M]. No. xiv. Academic Press, New York.

United States Environmental Protection Agency, 2011. National Primary Drinking Water Regulations [S]. 42 CFR Part 141-142. Fed. Regist.

U.S. Department of Health and Human Services, Food and Drug Administration, 2011. Grade "A" Pasteurized Milk Ordinance [EB/OL]. Http: //www.fda.gov/downloads/Food/GuidanceRegulation/UCM209789. pdfAccessed April 1, 2017.

Walstra, P., Jenness, R., Badings, H. T., 1984. Dairy Chemistry and Physics [M]. New York: Wiley.

Wang, A., Duncan, S. E., Knowlton, K. F., et al., 2016. Milk protein composition and stability changes affected by iron in water sources [J]. J. Dairy Sci., 99 (6): 4206-4219.

Weinberg, E. D., 1990. Cellular iron metabolism in health and disease. Drug Metab. Rev, 22: 531-579.

Weiss, W. P., Pinos-Rodrigues, J. M., Socha, M. T., 2010. Effects of feeding supplemental organic iron to late gestation and early lactation dairy cows [J]. J. Dairy Sci., 93: 2153-2160.

Yu, X., Geng, Y., Heck, P., et al., 2015. A review of China's rural water management [J]. Sustainability, 7: 5773-5792.

Zapata, C. V., Donangelo, C. M., Trugo, N. M. F., 1994. Effect of iron supplementation during lactation on human milk composition [J]. J. Nutr. Biochem., 5: 331-337.

Impact of Milk Hauling on Raw Milk Quality

Emily Darchuk, Eva Kuhn, Joy Waite-Cusic, Lisbeth Goddik
Department of Food Science & Technology,
Oregon State University, Corvallis, OR 97331, USA

Abstract

Milk tanker trucks are cleaned following predetermined Sanitation Standard Operating Procedures (SSOP) to assure quality and safety of the milk being transported and because the law requires it. Traditionally tankers were cleaned after every load of raw milk. Today, dairy processors strive to be efficient and control unnecessary cost, therefore truck cleaning has decreased in frequency to satisfy the regulatory limit of CIP cleaning required within 24 h of first use. The implications are important. If a tanker is not cleaned sufficiently, there are risks of decreased raw milk quality and safety risks. In contrast, if the tanker is cleaner more than needed, there is a waste of water, chemicals, energy, time, and efficiency. Thus dairy companies must strive to understand the rate and extend of fouling within the tankers to develop data based SSOP for their tankers. The objective of this study was to evaluate current raw milk hauling practices and determine if cleaning of tankers is appropriate. All research presented in this report were conducted in dairy manufacturing settings and involved assistance from milk hauling companies, truck drivers, dairy receiving bay and QA employees, and laboratory technicians. All assisted the OSU graduate students as they conducted experiments "in-situ." The five dairy manufacturing facilities are located in California, Washington, and Oregon. The facilities are all large and receive from 80-300 tanker trucks with raw milk each day. The studies ranged from evaluating existing data on raw milk quality from 23,000 milk hauls to studies focusing on specific tankers and their cleaning following short and high frequency hauls to long and low frequency hauls. Cleaning treatments were introduced that included cleaning after each load to cleaning after 10 loads. In addition, a worst case scenario was examined involved having a truck haul low quality milk in the morning, then stand dirty over the warm summer day, followed by hauling high quality milk. Finally receiving hoses and pumps in the receiving bays were examined. The data demonstrate that current milk hauling procedures are adequate to properly clean trailers and prevent contamination of subsequent loads. However, it also became evident that different CIP systems are not equally efficient. Therefore the dairy industry should continue to clean as required by law at least once every 24 h but care should be taken to monitor CIP efficiency and perform preventive maintenance of systems as required to maintain proper function.

Introduction

Milk hauling overview

Hauling is an important link in the milk supply chain which involves the transfer of milk from the producing farms to a manufacturer or cooperative. The transportation of milk occurs within tanker compartments towed by a truck. Although this is a highly regulated process, variations in procedures and equipment do occur based on scale, region and hauling company. The process outlined within this thesis is representative of the conditions and processes that the milk experienced within this study and is illustrative of the industrialized milk supply chain within the United States. It is also important to note that all dimensions quoted within this thesis should be assumed to be realistic estimates that are either based on measurements taken of an actual tanker or from information provided by hauling industry contacts.

Milk industry background

The US dairy industry has undergone drastic changes over the past few decades. Milk was previously produced on small farms but consolidation has been influenced by an economy of scale that has benefited some larger dairy operations. Although this trend has led to shifts towards larger herd size, the makeup of the dairy industry is varied and can be regionally dependent. Dairy farms now can contain over 15,000 cows, but the majority of operations still contain fewer than 100 head. The largest percentage of production come from farms with over 2,000 head which is also the farm size that has seen the fastest growth (MacDonald et al., 2007). In 2014 the average herd size was 204 cows which was nearly double the average herd size just 10 years prior (Progressive Dairyman, 2015).

This variability and change in herd size has had an impact on milk hauling. As herd sizes grow the volume of milk that can be collected from any given farm increases, allowing for fewer farm pickups per load. The spread of small and large farms has created situations where milk tankers either travel extended distances to pick up a large farm load or collect milk from multiple smaller farms prior to delivery. For large farms, multiple pickups can be required daily.

Hauling companies must work closely with both farmers and manufacturers to manage both the producer and processor milk supply all while scheduling the most efficient tanker routes to reduce resource usage. Although milk can remain in a farm bulk tank for up to 72 h (Food and Drug Administration, 2013) more frequent pick-ups benefit the farmer as tank capacity is limited and premiums are often paid based on producer milk quality at the time of delivery. Concurrently, the receiving bay of dairy manufacturing plants can be a bottle neck in production and deliveries must be staggered throughout the day to balance production, silo capacity, and truck resources. This creates a dynamic environment for the milk hauler, where schedules need to be constantly updated to account for both producer and manufacturer's needs. As haulers are typically contracted, it is in their best interest to identify efficiencies in the process which create situations where a tanker is only cleaned when required by the Pasteurized Milk Ordinance (PMO).

Hauling regulations

As defined by the PMO, a bulk milk tanker is a vehicle and associated equipment (tanks, pumps, hoses) used by a hauler to transport raw milk from a dairy farm to a milk plant (Food and Drug Administration, 2013). Milk tanker trucks are regulated by the PMO and gross vehicle weight (GVW) regulations outlined by each state's department of transportation which are based on bridge laws. The PMO regulates the materials and procedures for the hauling process where as the GVW outlines regulations based on safety and road maintenance concerns. Aside from regulation, configurations, usage and engineering of milk tankers can vary greatly depending on manufacturer and hauling company.

Within the Pacific Northwest (Oregon and Washington), both single and double trailer configured trucks are used to haul milk. Within industry, single compartment trailers are called tanker tubes whereas the double trailers are referred to as farm transfer systems (FTS) or double - bottoms (Karpoff and Webster, 1984). Both Oregon and Washington have the same GVW regulations, limiting a permitted vehicles' maximum weight to be no more than 50,000kg (FHWA, 2000; ODOT). Aside from the GVW, the legal operating weight of a truck is determined by the tire size, wheelbase, and number of axles, which all impact the manufacturer's design to maximize load efficiency through reducing truck weight (ODOT; Sharma and Mahoney, 1983; FHWA, 2000). As haulers are paid partly by how much weight they can haul the design of milk trucks has been carefully considered to protect the product while maximizing efficiency.

Tanker truck design

Although the PMO allows some flexibility in the type of material used in the design of trucks, most tankers are constructed from 300 series AISI stainless steel (Food and Drug Administration, 2013). Stainless steel is an alloy produced when chromium is added to iron and carbon to protect the steel from corrosion and oxidation through the development of a passive layer (Lo et al., 2009; ISSF, 2010). Within dairy tankers, basic grade 304 stainless steel is commonly used, this material is also referred to as Austenitic Cr - Ni stainless steel, with a composition of 18%Cr and 9%Ni (ISSF, 2010).

Milk tankers are designed like a thermos with a 10 gage metal end cap, 12 gage metal interior tank and an 18 gage exterior shell. The tanker consists of two cylinders fabricated from stainless steel metal sheets welded around a 1.5 inch polystyrene core which acts both as support and as an insulator between the internal and external diameter of the tank. Polystyrene is an extruded foam in which air is entrapped within the cell structure which provides an insulating effect (Dow Plastics, 2014). For its weight and price, the combination of polystyrene foam and thin gauge 304 stainless steel is very strong, and offers a smooth internal surface that allows for high efficacy cleaning utilizing clean in place (CIP) systems (Figure 1).

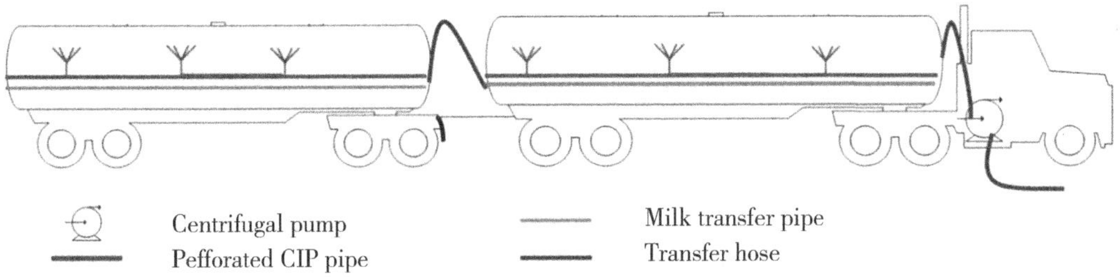

Figure 1 Diagram of a tanker truck; although all pipes are cleaned during a CIP wash the red pipes are specific to the CIP cycle, blue pipes are used for milk loading and the purple lines are shared and consist of the flexible transfer hoses

Tanker engineering

Tanker trucks are designed to protect milk quality as well as transport large volumes of milk efficiently from a farm to a plant. Understanding how a tanker is designed is critical to understanding how hauling can impact milk quality. Two areas of tanker engineering that are critical to milk quality are (i) rate of heat transfer between the cold milk and warmer outside temperatures (ii) the quantity of residual milk that can remain in a truck following delivery.

Tanker insulation efficiency background

To understand the efficacy of an insulated truck, a heat transfer formula can be used to determine the theoretical temperature change that can occur during two conditions the truck can encounter while on a route giving insight into the rate of heat transfer while a truck is stationary and in motion.

When using either of these models we are assuming that the tanker compartment is full of milk, creating a negligible head space. This is a condition common in the front compartment as this is filled to capacity prior to transferring milk into the back trailer. Due to this there is often residual head space in the back compartment, the volume of which will vary depending on load number and farm size. As the volume of head space is variable in the back compartment, all calculations were done to estimate heat transfer of milk in the front compartment. In evaluation of these calculations, it is important to note the key assumptions made and the understanding that these are extreme examples of the situations that milk tankers undergo, providing a worst case scenario estimate into the expected rate of heat transfer.

Assumption one

Milk remains in a tanker continuously for 24 h. Due to industry pressure to maximize use of equipment, hauls typically occur consecutively, leaving little time for the tanker to sit empty before picking up the next load. Based on this our calculations assume a tanker is full for the entire 24 h period.

Assumption two

Tankers are continuously in motion or still. Tankers in motion are representative of longer haul situations, during which trucks travel long distances between deliveries. Trucks in motion experience greater temperature changes as compared to stationary trucks which would be more representative of shorter haul situations.

Tanker insulation efficiency calculations

Although some temperature change occurs over time in insulated tankers; these trucks, even when exposed to very warm temperatures (35℃), experience little change to the receiving raw milk temperature as compared to the temperature it was pumped into the truck on the farm. Based on the Churchill-Bernstein Equation (Perry et al., 1997), a tanker at constant motion (60 mph) filled with cold milk (5℃) will gain less than 2℃ over a 24 h period. This same tanker in stationary conditions will gain less than 1℃. Per the PMO, all grade A milk must arrive at the plant under 7.2℃, allowing for tanker trucks to be used for extended periods without issue as long as milk is loaded at cold enough temperatures at the farm (Food and Drug Administration, 2013).

It should be noted that there are also non-insulated areas of the truck such as the transfer pump hoses located in between the tanks. As there is no insulation, this area will see elevated temperatures very quickly. Although this is an area of potential risk for microbial growth the risk is mitigated through purging the hose with air after pumping to reduce the amount of residual milk remaining in the line. These results show that regardless of the hauling situation, milk can remain within refrigerated temperatures over a 24 h use period, limiting bacterial growth.

Tanker load out efficiency

The other aspect of tanker engineering is understanding how much residual milk can remain in a truck following delivery. Calculation of the internal surface area provides understanding into the volume of residual milk that can build up on the walls of the tanker between washes and cleans. It is this milk which harbors bacteria that could directly contaminate future loads or create long term issues through the formation of biofilms. Milk tankers are weighed coming into and leaving the plant so tracking the residual milk left in the tanker is achievable. As milk is pumped from the tanker into the plant, the only milk remaining in the truck is within a foam which coats the inside surface of the walls. The formation of this foam can occur from movement of under filled loads or as a result of seal issues in receiving pumps or hoses. This foam creates a thin layer across the surface of the tank which later collapses back into milk upon transport. Once foam is formed within a tank the only way to remove it is with a water or chemical rinse. Typically the weight of the residual foam is negligible and even at worst case scenario only a few gallons of foam remain in the truck (hauling contact, 2014). Based on industry data, typical shrinkage of a load is less than 0.02% the total weight of a tanker (Industry Sponsor, 2015). After pumping out a 34,000 kg load of milk there will be less than 10 kg of milk remaining in a truck. It is this remaining milk that can grow bacteria or form biofilms, so minimizing the residual milk through a highly effective pumping systems helps to prevent quality issues within the truck and in downstream product.

Tanker sanitation concerns

How a tanker is utilized impacts how favorable the conditions can be for biofilm formation and thus the potential for quality defects. Milk tankers provide an opportunistic environment for biofilms to grow due to the surface interface with the milk and tanker walls, extended periods of time the truck is empty but not clean and the varying internal surface temperature that can occur when a truck is empty (Donlan, 2002). Biofilms are created when a community of bacteria create an exopolysacharide shell which can protect them from harsh conditions such as CIP treatments. Once biofilms form they can be difficult to remove and have the potential to enter the milk plant where they can thrive (Marchand et al., 2012). Thermo-resistant, enzyme and biofilm forming bacteria have also be isolated from the internal surface of a dairy tanker, suggesting hauling could be a potential cause of milk quality issues. Although biofilm formation within a tanker is a concern, the risk of their development is less likely as compared to other areas of the plant due to the low temperatures, low shear, and smooth surface area that the raw milk is exposed to within a tanker truck (Marchand et al., 2012). The risk of development is also managed through cleaning treatments, but the overall tanker sanitation is only as good as the cleaning treatments and preventative maintenance that it obtains.

Tanker cleaning

Milk tanker trucks are required to undergo a CIP treatment after every 24 h of use but are allowed to be used for multiple loads between washes. Washes are loosely regulated by the PMO and trucks must display a wash tag on the exterior of each tank documenting the last time it was cleaned. Within the 24 h period, a truck can be used as needed to haul the milk from the farm to the plant which may involve long hauls, frequent use or extended waiting periods during which the tanker is soiled but empty. The PMO only mandates minimum temperatures and frequency of cleans allowing manufacturers a great deal of flexibility in their choice of chemicals, pressures and frequency beyond regulated 24 h CIP treatments (Food and Drug Administration, 2013). Manufacturers typically work with chemical companies to design a sanitation regime that meet their quality, cost and efficiency goals. This flexibility allows for plant to plant variability in cleaning efficacy which can impact day to day sanitation within tanker trucks that deliver to multiple facilities.

Although the CIP process is regulated, the chemicals used, temperatures met and pressures achieved vary from plant to plant. To begin the clean, the truck pulls into the receiving bay and the receiving hose is connected to the plant water supply. During this set up, flow diversions are created so that the water and chemicals will utilize the perforated CIP pipe. This pipe runs the length of the tank and is designed similarly to a sprinkler system to create pressurized spray reaching all areas of the tanker. Using a power take off (PTO) to power the pump from the motor of the truck, water and chemical are pumped through the receiving hose into the first tanker, washing both the milk transfer pipe as well as the internal surface of the tank. The same solution travels through the transfer pump hose into the back compartment simultaneously cleaning both the front and back compartments at the same pressure. Although a specific CIP procedure is not detailed for tanker trucks in the PMO, it is a process which is documented and evaluated during inspections from state regulators. Typical CIP processes involve a water rinse, detergent, and water rinse followed by a sanitizer treatment. Although tanks and trucks are typically cleaned as a unit it is important to note that they are three independent pieces of equipment and thus may have differing conditions based on previous use. CIP temperatures reach upward of 170°F, making cleaning a very resource intensive step of the manufacturing process. CIP treatments within a facility can make up half of a dairy plants' energy usage (DMI, 2010) so it is important for companies to find a balance between cleaning frequently enough to maintain milk quality while also managing resource usage.

Tanker design summary

Evaluation of the design and industry use of the tanker trucks is critical to understanding the results of our study. The cold conditions maintained by the insulated tanker helps to substantially slow the bacterial growth in the milk maintaining quality during transportation. This is further aided by the small amount of residual milk left in the tanker following load out reducing impact on future loads in between cleans. Impact of residual milk is also reduced through conducting a CIP wash following every 24 h of use, although variability in CIP practices can create sanitation issues that vary based on how an individual facility uses and maintains their equipment.

Methodology

Study overview

This research covered five individual studies that were performed within the standard operations of a commercial dairy manufacturing plant. Samples were analyzed using common quality metrics to ensure that the study was representative of industry practices.

- Study 1: Evaluation of existing data for 23,000 raw milk hauls obtained during a 24 month period.
- Study 2: High frequency short distance milk hauls with up to 10 trips per 24 h. Different cleaning treatments from once after each load to once per 10 loads were evaluated. The study was repeated summer and winter to determine seasonal impacts.
- Study 3: Low frequency long distance milk hauls with 2 long distance trips per 24 h. Introduction of a rinse step between the 2 hauls was investigated.
- Study 4: Worst case scenario investigating the practice of hauling milk in the morning, leaving the truck during warm day time hours and then reusing the dirty truck was investigated.
- Study 5: Swabbing of flexible receiving hoses and pumps in receiving bays to examine possible build-up of contaminants in between CIP.

Tanker trucks

Milk was hauled within one double trailer tanker truck with a flexible transfer hose to connect the two compartments. These trailers were transported by a truck that carried the transfer pump and hose which loaded the milk from the farm bulk tank into the trailer compartment. Prior to the study, all equipment had passed regulatory inspection.

Cleaning treatments

Study 2 & 3 investigated the addition of cleaning treatments incremental to the standard operating procedure of a 24 h CIP which served as a standard use variable (control). Two trucks underwent different cleaning treatments each day, creating four replicated days for each of the four cleaning treatments over the eight day study. Cleaning treatments were partial stages of the full CIP cycle and utilized existing chemicals and equipment. All water rinses conducted were 2-3 minutes in duration and utilized ambient temperature water. Water samples were analyzed from the CIP system daily to ensure that rinse water was not a source of contamination. All cleaning treatments, including CIP, were conducted in the receiving bay of the plant immediately after unloading milk and prior to continuing on to the next load.

Tanker routes

Study 2: Each truck was assigned to a route which determined what farm milk each truck would pick up within a 24 h period. Routes were selected based on their ability to be repeated daily and were specific to each study location. Within each route there were up to 9 loads scheduled. Each load was either filled from a single farm or was commingled and contained multiple farms within the same truck. A load was completed when a full truck delivered milk to the manufacturing plant.

Study 3: To reduce variability in producer milk quality, one route was repeated for the duration of this study. This route consisted of milk from one farm which was collected twice daily. Prior to the first load, the truck underwent a CIP treatment at the manufacturing plant. Following CIP, the tanker would travel to the farm which was located approximately five hours away. All milk was loaded from a single bulk tank, filling both trailer compartments of the truck. Once loaded, the truck would return to deliver the milk to the same manufacturing plant. Following delivery of the first load, the truck would either return to the same farm without any cleaning treatment (standard use-SU) or a water rinse followed by a sanitizing spray (RS) would occur prior to the second farm pick up. Regardless of treatment, following the delivery of the second load the truck would undergo a CIP treatment and a new treatment day would begin. All cleaning treatments, including CIP, were conducted in the receiving bay of the plant immediately after unloading milk and prior to continuing on to the next load.

Study 4: Two dairy farms were identified with historically high and low microbiological counts. One truck was utilized for the study. The truck picked up milk from the historically high count producer in the morning. After unloading, the truck would either immediately pick up a load of raw milk from the historically low count producer (control treatment) or the truck would stand empty and dirty for 6 h before picking up raw milk from the low count producer.

Sampling

Samples were collected for each load at both the farm and plant. Prior to the study, training of both the receivers and haulers was conducted to ensure that sampling and cleaning procedures were consistent throughout the study. All samples were kept below 7℃ during storage and transport and were tested within 36 h of sampling at a corporate laboratory. Haulers followed PMO regulations when collecting producer samples from the farm bulk tank (Food and Drug Administration, 2013). Receivers took tanker samples using a sanitized stainless steel dipper from the top hatch of the front and back tanker trailer. A different dipper was used for each compartment of the truck to avoid cross contamination.

Surface swabs

Sponge-stick swabs moistened with Letheen broth (3M US, St. Paul, MN) were used after unloading milk to measure residual bacteria left on the internal surface of tank. For every load, a 900 cm^2 area (30 cm×30 cm) was swabbed per manufacturer's instructions. Following CIP or RS treatment, a second swab was taken to measure the efficacy of the clean. Receivers were trained to rotate the area of the ceiling swabbed with each incoming load and before and after cleaning treatments.

Microbiological analysis

All milk samples were analyzed for individual bacteria count (IBC), thermophilic spore count (TSC), and preliminary incubation (PI) most probable number (MPN). Individual bacteria counts of all milk samples were conducted using a Bactoscan FC (FOSS, Hillerød, Denmark). Thermophilic spores were quantified using the method described by Wehr and Frank (2004). Preliminary incubation was conducted by adding a diluted samples to a TEMPO Total Viable Count (TVC) vial (bioMérieux; Marcy l'Etoile, France). The TVC vials were incubated at (13±1)℃ for 18 h followed by (32±1)℃ for 48 h and enumerated using TEMPO reader following manufacturer instructions.

Both the rinse water from CIP systems and sponge swabs were evaluated for aerobic plate count (APC) using Petrifilm (3M US) incubated at (32±1)℃ for 48 h. Petrifilms were enumerated using an automated counter (3M Petrifilm reader).

Results

Study 1: Negative impact of milk hauling was defined as when milk sampled directly from the tanker truck had a higher microbiological count that the corresponding milk load sampled from the bulk tank at the farm. The data from 23,000 hauls revealed that the incidence of negative impact was higher during winter months than in the summer. There was also a trend that when there was a negative impact of milk hauling, it occurred with raw milk from historically high count milk farms. In conclusion, it is rare that microbiological counts increases during transport and when it does occur, it is likely with poor quality raw milk.

Study 2: We could not document any negative impact on raw milk quality from milk hauling when trucks were used up to 10 times within a 24 h period without cleaning. Since there was no negative impact for the control that was cleaned by CIP once per 24 h, there was no improvement gained from addition of extra cleaning cycles between loads. The same results were obtained for summer and winter experiments. We did demonstrate that the effectiveness of tanker CIP varied among different plants. Swabbing the interior of cleaned tankers demonstrated this differences and we recommend companies do so on a routine basis.

Study 3: The question we examined was if the introduction of a brief rinse after the first load could prevent any negative impact on quality of milk in the subsequent load when milk was transported over long distances. We did not detect any negative impact on milk quality and therefore there is no apparent justification for adding the extra rinse step.

Study 4: Initial benchtop research had demonstrated that dirty milk cans left empty for extended periods of time can contribute to microbial quality of pasteurized milk subsequently filled into these milk cans. When conducting this experiment in industry setting using milk trucks, we could not detect any negative impact on milk quality in subsequent loads. Perhaps because we used high quality raw milk in the second load instead of pasteurized milk. Thus any potential impact was more difficult to determine.

Study 5: The swabbing and analysis of soft hoses and pumps within dairy factory receiving bays is currently on-going and results may be available shortly.

Conclusions

It appears that current milk hauling procedures are adequate for preventing contamination of raw milk when tankers are cleaned by CIP every 24 h. When all procedures are followed properly, it is not possible to identify a negative impact. When negative impact was identified, it was associated with transport of poor quality milk primarily during winter. Warmer summer temperatures did not appear to negatively impact transportation; this is likely due to the excellent insulation of the tankers. Different CIP systems do not function with similar efficiency. Interior swabbing of clean trucks demonstrated significant differences between dairy plants.

Acknowledgements

We thank Washington Dairy Products Commission and Dairy Management Inc., for sponsoring this research. We thank our industry partners for allowing us access to their milk receiving stations and Q. A. data.

References

Darchuk, E. M., Meunier-Goddik, L., Waite-Cusic, J., 2015. Microbial quality of raw milk following commercial long distance hauling [J]. J. Dairy Sci., 98: 8572-6.

Darchuk, E. M., Waite-Cusic, J., Meunier-Goddik, L., 2015. Impact of commercial hauling practices and tanker cleaning treatments on raw milk microbiological quality [J]. J. Dairy Sci., 98: 7384-7393.

DMI., 2010. U.S. Dairy Sustainability Commitment Progress Report [R].

Donlan, R. M., 2002. Biofilms: Microbial Life on Surfaces [J]. Emerg. Infect. Dis., 8: 881-890.

Food and Drug Administration, 2013. Grade "A" Pasteurized Milk Ordinance [R]. U.S Department of Health and Human Services Public Health Service, Washington, DC.

Karpoff, E., Webster. F. C., 1984. Innovative methods of milk transport: double-bottoms and TOFC [J]. J. Northeast. Agric. Econ. Counc., 13: 1-60.

MacDonald, J. M., O'Donoghue, E. J., McBride, W. D., et al., 2007. Profits, costs, and the changing structure of dairy farming [R]. US Department of Agriculture, Economic Research Service.

Marchand, S., De Block, J., De Jonghe, V., et al., 2012. Biofilm Formation in Milk Production and Processing Environments; Influence on Milk Quality and Safety [J]. Compr. Rev. Food Sci. Food Saf., 11: 133-147.

Paez, R., Taverna, M., Charlon, V., et al., 2013. Application of ATP-Bioluminescence Technique for Assessing Cleanliness of Milking Equipment, Bulk Tank and Milk Transport Tankers [J]. Food Prot. Trends., 23: 308-314.

Progressive Dairyman, 2015. 2014 Dairy Stats [R]. Progressive Dairyman.

Sharma, J., Mahoney, J., 1983. Evaluation of present legislation and regulation on tire sizes, configurations and load limits [D]. University of Washington.

Teh, K. H., Flint, S., Palmer, J., et al., 2012. Proteolysis produced within biofilms of bacterial isolates from raw milk tankers [J]. Int. J. Food Microbiol., 157: 28-34.

Teh, K. H., Flint, S., Palmer, J., et al., 2014. Biofilm-an unrecognised source of spoilage enzymes in dairy products? [J]. Int. Dairy J., 34: 32-40.

Teh, K. H., Flint, S., Palmer, J., et al., 2011. Thermo-resistant enzyme-producing bacteria isolated from the internal surfaces of raw milk tankers [J]. Int. Dairy J., 21: 742-747.

Teh, K. H., Lindsay, D., Palmer, J., et al., 2013. Lipolysis within single culture and co-culture biofilms of dairy origin [J]. Int. J. Food Microbiol., 163: 129-135.

Wehr, H. M., Frank, J. F., 2004. Standard methods for the examination of dairy products [M]. 17th ed. American Public Health Association Inc., Washington, DC.

Virulence, UHT Survival, and Biofilms in *Bacillus cereus* and Other *Bacillus* spp. in Milk

J. L. McKillip, Ball State University,
Department of Biology, Muncie, IN USA

Abstract

We isolated a *Bacillus amyloliquefaciens* strain present in ultra-high temperature (UHT) pasteurized organic whole milk in order to ascertain virulence determinants present in this species, and potential for biofilm formation compared to the mesophilic type strain *Bacillus cereus* ATCC14579. The overall goal of this project was to genotypically and phenotypically characterize thermoduric *Bacillus amyloliquefaciens* virulence and biofilm potential, including the presence of the global regulator effector PlcR, when cultures are sublethally-stressed by growth in subinhibitory concentrations of carvacrol, an antimicrobial extract from essential oil of oregano that has been shown to increase virulence of *Bacillus* spp. in other contexts. Recovery of bacteria from milk necessitated a nonselective enrichment in Brain Heart Infusion (BHI) broth, incubated aerobically (shaking) for >24 h, after which time samples were serially diluted and spread-plated onto Tryptic Soy Agar (TSA) plates to recover *Bacillus* spp. Resulting colonies were streak plated onto TSA to ensure purity of culture before catalase testing, Gram and spore-staining to presumptively identify to the genus level. Pure cultures were biochemically identified to the species level using the Microgen Bacillus ID system (Hardy Diagnostics), and validated further using fatty acid profiling and 16S rDNA sequencing. In order to confirm presence of the target virulence and regulator genes in each UHT milk isolate, DNA was extracted from TSB-grown pure cultures during late log phase. Quantified DNA template was used in real-time (SYBR Green-based) PCR with primers specific for each of the target genes: *plcR* (encoding a pleiotropic extracellular virulence factor regulator), *codY* (encoding another global effector protein), *nheA* and *hblC* (encoding enterotoxins), and the16S rRNA gene is a control. PCR indicated that *plcR*, *codY*, *nheA*, *hblC* genes were all present in the *Bacillus amyloliquefaciens* isolate. This was determined by comparing mean melting temperatures (*Tm*) for each PCR product. Both *B. cereus*

ATCC14579 and *B.amyloliquefaciens* produce biofilms, with the former producing 10% more relative biofilm material, with a slight (but not significantly different) decrease upon exposure of cultures to subinhibitory carvacrol. The significance of this project will be to determine if parameters regarding shipment, storage and shelf life of UHT organic milk should be revisited, in order to ensure quality before consumption of product that may harbor thermoduric toxigenic *Bacillus* spp.

Introduction

B.cereus-general background, ubiquity and general features

Foodborne illness from a variety of microorganisms effects on average 76 million individuals in the U.S. each year resulting in some 5,000 deaths (Mead et al., 1999). Worldwide statistics on *Bacillus cereus* foodborne illness are underestimated due to a variety of factors, including emetic symptoms similar to *Staphylococcus aureus* intoxication and diarrheal symptoms similar to those elicited by *Clostridium perfringens* type A. Most affected individuals do not seek medical attention due to the short duration of signs and symptoms. *B.cereus* seems to account for between 1.4% - 12% of foodborne illness outbreaks worldwide (Stenfors et al., 2008).

B.cereus is a large (1.0 - 1.2 μm by 3.0 - 5.0 μm) Gram - positive aerobic - to - facultative spore-forming rod-shaped bacterium (Figure 1). The word bacillus in Latin translates to small rod, while cereus translates to wax - like. The genus *Bacillus* can be split into two groups: *B.subtilis* and *B.cereus*. The *B.cereus* group consists of *B.cereus*, *B.thuringiensis*, *B.mycoides*, *B.anthracis*, and *B.weihenstephanensis*. The members of this group produce lecithinase, but do not produce acid from mannitol, distinguishing them from other *Bacillus* species. As the flagship pathogen of this group, *B.cereus* is ubiquitous in soil and freshwater environments in all temperate zones of the world (Gilbert and Kramer, 1986; Kotiranta et al., 2000; Kramer and Gilbert, 1989; Schoeni and Wong, 2005). This bacterial genus is capable of contaminating a wide range of food products, including rice, chicken, vegetables, spices, and dairy products. Contamination in the dairy industry may occur when *B.cereus* spores come in contact with the udders of cows (Andersson et al., 1995), if the spores colonize feed or bedding, or if the spores survive pasteurization (Claus and Berkley, 1986; Sneath, 1986). This is a serious problem in the food industry because *B.cereus* endospores are in many instances partially resistant to the heat of pasteurization, dehydration, gamma radiation, and other physical stresses. This resistance is due to the ultrastructure of the endospore of course, but also in part to the hydrophobic nature of the spores that allows them to adhere strongly to surfaces and develop biofilm-like properties (Mattson et al., 2000; Ronner et al., 1990). For example, an irradiation dose of 1.25-4 kGy needs to be administered to reduce spores by 90% (De Lara et al., 2002). Also, pasteurization may result in the activation and germination of spores (Hanson et al., 2005). In addition, *B.cereus* endospores germinate in response to particular nutrients such as glycine or in response to physical stress such as temperature (spore germination can occur over 5 - 50℃ in cooked rice) (Granum, 1994) and high pressures (i. e. 500 MPa) (Black et al., 2007). Thus foods need to be cooked at least at a temperature of 100℃ or above to kill most of the endospores (Griffiths and Shraft, 2002).

B.cereus toxin production

B.cereus produces several types of toxins, including four hemolysins (Granum, 1994), three distinct phospholipases, a heat/acid stable emetic toxin called cereulide (a plasmid encoded cyclic peptide) that causes vomiting in infected individuals, and several heat-labile enterotoxins [hemolysin BL (Hbl), nonhemolytic enterotoxin (Nhe), and cytotoxin K] (Lund et al., 2000) that all cause diarrhea. Cereulide has an incubation period of 0.5 - 6 h while the total duration of the emetic syndrome is 6-24 h (Ehling-Schulz et al., 2004). The incubation time for the Hbl and/or Nhe-mediated gastroenteritis is on average 12 h and the duration of signs and symptoms is between 12-24 h (Kramer and Gilbert, 1989). An infectious dose ranging between 105 - 108 viable cells or spores is necessary to elicit symptoms, while the concentration of cereulide necessary to elicit disease, has not yet been conclusively determined (Gilbert and Kramer, 1986). While total duration of the emetic syndrome is less than 24 h and is usually self-limiting, two rare cases in children have been documented where the cereulide toxin was responsible for inhibiting hepatic mitochondrial fatty-acid oxidation which lead to liver failure and resulted in the death of both children (Dierick et al., 2005; Mahler et al., 1997).

The cereulide toxin is encoded by the cereulide synthetase (*ces*) gene located on a 208-kb megaplasmid (Ehling-Schulz et al., 2006). Cereulide exerts its toxic effects by binding to 5-HT3 receptors on the vagus afferent nerve, which induces an imbalance of cellular potassium leading to mitochondrial swelling (Agata et al., 1994; Mikkola et al., 1999; Sakurai et al., 1994). A 2nM concentration of cereulide causes inhibition of RNA synthesis and the above mentioned cellular cytoxicity events. At high doses of cereulide, massive degeneration of hepatocytes was observed (Yokoyama et al., 1999).

The hemolysin BL (Hbl) consists of three proteins termed B, L1, and L2 (Beecher and Wong, 1994, 1997; Ryan et al., 1997). These toxins are produced from the Hbl operon that codes for HblC, HblD, and HblA toxins, which bind to the membrane of eukaryotes where they oligomerize to form pores allowing fluid accumulation into the cell. In addition, the B and L1 components of the Hbl enterotoxin complex produce a unique discontinuous beta-hemolysis pattern on blood agar (Beecher and MacMillan, 1990). The nonhemolytic enterotoxin (Nhe) is also composed of three protein components (39 kDa, 45 kDa, and

105 kDa) that demonstrate homology with each other and the Hbl protein components (Schoeni and Wong, 2005). These proteins produced from the Nhe operon are called NheA, NheB, and NheC. The exact mode of action of how all three proteins work together to act as pore-forming cytotoxins remains to be fully elucidated. All three proteins seem to be required to achieve a cytotoxic effect on their host cell (Lindback et al., 2004). While the function of NheC is not yet understood, it is theorized to possibly act as a catalyst to cause NheA to bind to NheB that has attached itself to the host cell membrane leading to cell lysis. Thus due to similar structural and functional properties, both the Hbl and Nhe toxins are believed to belong to a superfamily of pore-forming cytotoxins (Fagerlund et al., 2008). There have been limited studies on the exact mode of action for *Bacillus* enterotoxins. These toxins, such as Nhe, form pores in lipid bilayers and that a reverse absorption of fluid, Na^+, and Cl^- by epithelial cells, causing a malabsorption of glucose and amino acids. This causes mucosal damage leading to necrosis. Adenylate cyclase is believed to contribute to the process of reverse absorption of fluid in epithelial cells (Kramer and Gilbert, 1989).

While a great deal of information on virulence gene presence and expression is known in *B.cereus*, very little has been done to explore the virulence potential of thermoduric spore-formers that may be found in UHT milk. This study addresses this need by detection and characterization of several virulence determinants relevant to the dairy industry. Owing to the conserved nature of many of these genes, we hypothesize that *Bacillus* spp. found in UHT milk would harbor and express these toxin and regulator genes under conditions similar to those previously described for mesophilic and psychrotrophic *Bacillus* spp.

Bacillus spp. biofilms

Adherence of microbial biofilms to dairy production surfaces makes sanitization more difficult, and increases cost *via* labor and chemical usage along with lost production time. FDA involvement and subsequent product recalls can also occur causing further financial problems for dairies. Araújo et al. (2009) have proposed a basic mechanism for biofilm adhesion based on six general stages. First, the biofilm surface must be primed for adhesion with the existence of food deposits. The biofilm-producing microorganism must then come into contact with the primed surface. Positive and negative biochemical forces including van der Waals forces and other electrostatic forces then allow the biofilm to make a non-permanent attachment to the surface when microorganism are between 20 and 50 nm away. Irreversible adhesion results within 1.5 nm when extracellular polysaccharide production, ionic bonds, and hydrophobic forces occur. The fourth stage is described by the multiplication of bacterial cells and an increase in secreted polysaccharides and the fifth stage involves strong metabolism in the biofilm. Lastly, microorganisms begin to be released from the biofilm during the sixth stage, shedding bacteria to generate new biofilms elsewhere.

Several authors have identified a variety of mesophilic *Bacillus* subspecies capable of surviving ultra-high temperature pasteurization *via* endospore formation (Araújo et al., 2009; Lindsay et al., 2002; Scheldeman et al., 2006; Sutyak et al., 2008). Using bacterial cultures sampled from dairies, 16s rRNA, and PCR amplification some of the most prevalent and potentially problematic species, in regards to biofilm production, have been characterized. These species include *B.cereus*, *B. amyloliquefaciens*, and several others.

The level of virulence activity in *B.cereus* cells is due to a number of different environmental factors, including temperature, pH, oxygen tension, glucose concentrations, and specific antimicrobial chemical compounds (Glatz and Goepfeort, 1976; Sutherland and Limond, 1993). Biofilm production is understood to be under similar regulation as toxins and other extracellular virulence determinants, which suggests that subinhibitory stress may have great influence on overall potential for *Bacillus* spp. to become problematic in dairy microbiology settings.

Carvacrol (essential oil of oregano) has been shown to inhibit *B. cereus* by damaging cell membranes and changing K^+/H^+ concentration gradients in the cell and can deplete ATP levels in 7 min at 2 mM carvacrol concentration (Ultee, et al., 1999). We plan to treat *B.cereus* and the UHT isolate cultures with carvacrol and quantify levels of *hblC*, *nheA* virulence gene expression, along with the global regulator genes *codY* and *plcR*. Subinhibitory concentrations of carvacrol will also be used to ascertain relative effects on biofilm formation in the UHT isolate compared to the *B.cereus* type strain. We hypothesize that sublethal levels of carvacrol will significantly increase enterotoxin production, biofilm formation, and global regulator expression in UHT *Bacillus* spp. to levels that are directly correlated to one another. This would confirm that the PlcR and CodY global effectors directly mediate the induction of all of these extracellular factors during chemical stress by carvacrol, an otherwise attractive GRAS antimicrobial control strategy.

Materials and Methods

Recovery and identification of Bacillus spp. from UHT-pasteurized organic milk

Organic dairy (white) milk was obtained at local retail. Recovery of bacteria from milk necessitated a nonselective enrichment in Brain Heart Infusion (BHI) broth, incubated aerobically (shaking) for >24 h, after which time samples were serially diluted and spread plated onto Tryptic Soy Agar (TSA) plates to recover *Bacillus* spp. Resulting colonies were streak plated onto TSA and Blood Agar to ensure purity of culture before catalase testing, Gram and spore-staining to presumptively identify to the genus level. Pure cultures were biochemically identified to the species level using the Microgen Bacillus ID system (Hardy Diagnostics, Springboro, OH USA), and validated further using fatty acid profiling and 16S rDNA sequencing (MIDI Labs Inc., Newark DE, USA).

PCR confirmation of plcR, codY, nheA, hblC and 16S genes

In order to confirm presence of the target virulence and regulator genes in the UHT milk isolate, DNA was extracted from TSB-grown pure cultures during late log phase using the method of Phelps and McKillip (2002). Quantified DNA template was used in real-time (SYBR Green-based) PCR with primers specific for each of the target genes (Table 1) and theMasterAmp™ High Fidelity PCR Kit (Epicentre Biotechnologies, Madison, WI), which included 12.5 μl MasterAmp 2× PCR premix, 0.5 μl MasterAmp TAQurate DNA Polymerase Mix, 100 pmol each primer, and 0.5 μg template DNA per reaction. All reactions were completed at least in triplicate and the mean Ct and Tm values were recorded for analyses. With respect to targets, *PlcR* is a pleiotrophic extracellular virulence factor regulator, *CodY* is a flagellar repressor, *NheA* and *HblC* are primary enterotoxins (Nimmer et al., 2014), and the 16S gene is a conserved housekeeping control target.

Table 1 List of primers used for real-time PCR

5′→3′ Sequence	Target Gene	Amplicon Size (bp)	Reference
GCTCTATGAACTAGCAGGAAAC	nheA (forward)	561	27
GCTACTTACTTGATCTTCAACG	nheA (reverse)		
ATGAAAACTAAAATAATGACAG	hblc (forward)	300	27
ATCCTTTACTTTTTGAATTTAA	hblc (reverse)		
ACTAGGATCCATGCAAGCAGAGAAATTAG	plcR (forward)	860	Phelps and McKillip, 2002
ACTAAGGTCCTTATCTGCTGATTTTATTTAC	plcR (reverse)		
AGCGAGACTCAATTACACCA	codY (forward)	669	
AATGCGTTATTACAGAGCGC	codY (reverse)		
AAGTCGAGCGAATGGATTGA	16S (forward)	1469	
TCCGATACGGCTACCTTGTT	16S (reverse)		

Growth cuwe of each isolate in a sterile milk system

In order to ascertain growth rate over time of this natural isolate, a growth curve in a sterile organic milk system was completed. Pure cultures of the *Bacillus* spp. were inoculated (10^2 CFU/mL) into sterilized organic milk and incubated at ambient temperature (23℃) for 72 hours. Every two hours, samples were removed for Standard Plate Count (SPC) -based density determination and optical density measurements at A_{600}.

RNA isolation and NASBA (RNA amplification) to assess relative gene expression during log/stationary phase

Aliquots were taken beyond mid - log phase (~ 10^5 CFU/mL as determined by growth curve) for RNA isolation using TRIzol® (and DNase - I treated), and mRNA for each gene target noted above will be amplified with real - time NASBA using a bioMerieux kit and transcript-specific primers.

Immunoassay confirmation of CodY, NHE, and HBL

Enzyme-linked immunosorbent assay (ELISA) will be used to confirm NHE and CodY and Latex Agglutination will be used to confirm HBL. This work is ongoing (A. Grutsch, data not shown).

Biofilm assay

Cultures of the UHT isolate and an the reference type strain *Bacillus cereus* ATCC14579 were grown overnight in TSB and subsequently placed into a 96-well microtiter plate for quantitiative biofilm measurement using the procedure of O'Toole (2011). This crystal violet dye-binding assay allows for relative quantification of biofilm-formation using absorbance at A_{550} as a means to presumptively identify biofilms and to monitor their development. Cultures were prepared alone or while growing in the presence of subinhibitory concentrations (SIC) of carvacrol (Sigma-Aldrich, St. Louis, MO USA) (1 mmol), previously used in our lab as a sublethal stressor and key inducer of virulence in *B.cereus* (Figure 1).

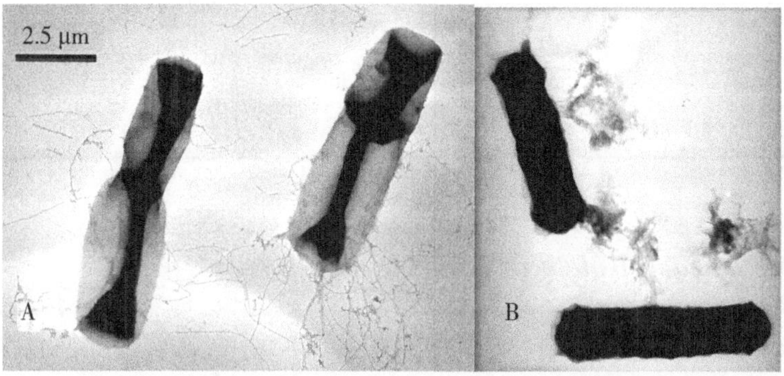

Figure 1 ***Bacillus cereus*** **(ATCC14579) during mid-log phase growth (A) and at the same stage when grown in the presence of 1 mmol of carvacrol (B, P. Nimmer, this study)**

Results

Recovery and identification of Bacillus spp. from UHT-pasteurized organic milk

Nonselective enrichment of UHT organic milk samples in Brain Heart Infusion (BHI) Broth yielded a single pure culture bacterial isolate that was a Gram-positive spore-forming rod, obligately aerobic and catalase-positive. Presumptive ID as a *Bacillus* spp. allowed for further characterization using the *Bacillus* Microgen ID system, with results matching most closely (90%) with *Bacillus firmus* based on biochemical reactions on 19 unique substrates, including cellobiose, arabinose, inositol, and others. Upon fatty acid profiling using MALD-TOF, this isolate was identified as *B. amyloliquefaciens*.

PCR confirmation of plcR, codY, nheA, hblC and 16S genes

All target genes were consistently detectable in the *B. amyloliquefaciens* UHT strain when compared to our reference type strain *B. cereus* ATCC14579 (Figure 2). Ongoing work in our lab will soon ascertain relative levels of gene expression in *B. amyliquefaciens* in broth and organic milk at the transcriptional and translational levels (A. Grutsch, data not shown).

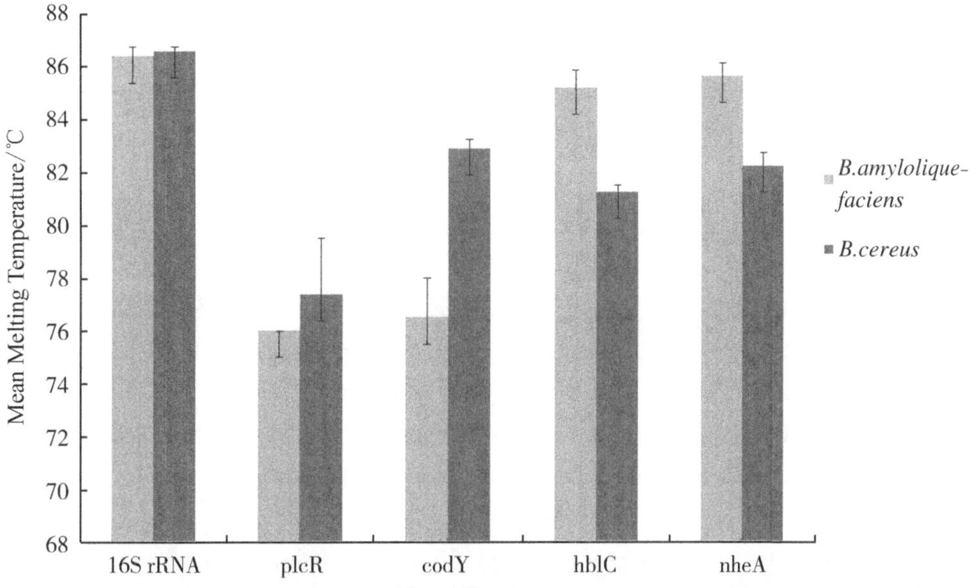

Figure 2 Mean tm value differences-*B.cereus* ATCC14579 and *B.amyloliquefaciens* (UHT isolate)

Biofilm Assay

Each *Bacillus*species was compared in their abilities to product measureable biofilm on a preliminary microtiter-based dye binding assay. Both *B.cereus* ATCC14579 and *B.amyloliquefaciens* produce biofilms, with the UHT isolate producing approximately 10% less when cultivated for 48 h at ambient temperature (23℃) (Figure 3). When grown in the presence of subinhibitory carvacrol, each *Bacillus* spp. produced slightly less biofilm material overall, although not significantly less than nonstressed cultures.

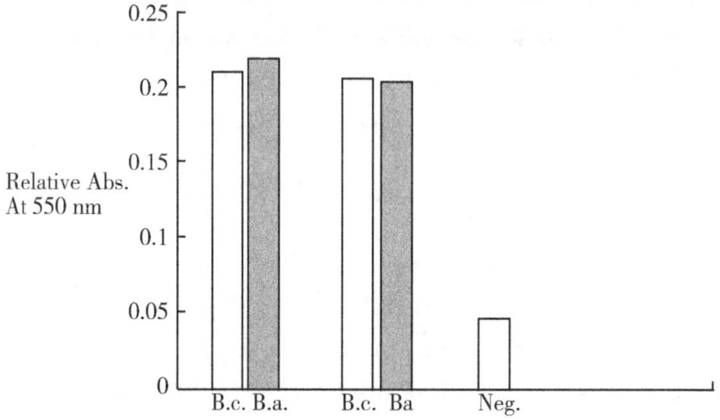

Figure 3 Relative biofilm production abilities of *B.cereus* ATCC14579 (B. c) and *B.amyloliquefaciens* (B. a.) using the crystal violet dye-binding assay described in Materials and Methods. Underscored bars represent cultures grown in the presence of subinhibitory levels of carvacrol

Discussion

Although work in our laboratory is ongoing, these data provide strong indications that ultra-high temperature pasteurization of milk leaves a window of opportunity for potentially enterotoxigenic *Bacillus* spp. to pose a threat to human health, and certainly can limit conditions needed for proper and safe shipment and storage of such products. Relative few reports have investigated the virulence potential of UHT isolates, with most studies focusing on general species characterization and growth properties (Araújo et al., 2009; Lindsay et al., 2002; Scheldeman et al., 2006; Sutyak et al., 2008). *B.amyloliquefaciens* seems to be a common isolate from UHT milk products, and our isolate here is most closely aligned with this species based on fatty acid profiling. Practically speaking, assigning a species name to the isolate is secondary in importance to characterizing the strain with regard to virulence potential, including biofilm-production ability. Based on results of this study, the UHT isolate harbors both of the key regulator genes encoding CodY and PlcR, two well-conserved and heavily-studied global effectors of virulence regulation in *Bacillus* spp. (Nimmer et al., 2014). The *B. amyliquefaciens* strain identified herein also encodes the two most well-characterized enterotoxins in *Bacillaceae*, NHE and HBL, which are responsible for foodborne illness in a wide variety of foods, including fluid and powdered milk products. To our knowledge, this is the first report that a UHT milk isolate of *Bacillus* spp. encodes these virulence determinants, leaving tremendous opportunity to further investigate the effect of physical and chemical stressors on expression of these proteins during processing, shipment, and storage conditions.

Biofilm-production ability in the *B. amyliquefaciens* was also preliminarily ascertained here using a conventional crystal violet dye-binding assay. Our results confirm that this UHT isolate does in fact product a substantial biofilm (also observed qualitatively in the broth and sterile milk systems in which the species has been cultivated), and current work in our laboratory is involved with more definitively quantifying rate and amount of biofilm material on glass, stainless steel, and wax paperboard substrates during incubation in milk systems. The specific protein and carbohydrate composition of the biofilm material will be determined as well, during incubation of the culture alone or in the presence of the subinhibitory carvacrol concentrations. These data will offer insight on potential control strategies for biofilms in the dairy industry.

References

Agata, N., Mori, M., Ohta, M., et al., 1994. A novel dodecadepsipeptide, cereulide, isolated from *Bacillus cereus* causes vacuole formation in Hep-2 cells [J]. FEMS Microbiol. Lett., 121: 31-34.

Andersson, A., Ronner, U., Granum, P. E., 1995. What problems does the food industry have with the spore-forming pathogens *Bacillus cereus* and *Clostridium perfringens* [J]. Int. J. Food Microbiol., 28: 145-155.

Araújo, E. A., Bernardes, P. C., Andrade, N. J., et al., 2009. Gibbs free energy of adhesion of *Bacillus cereus* isolated from dairy plants on different food processing surfaces evaluated by the hydrophobicity [J]. Int. J. Food Sci. Tech., 44: 2519-2525.

Beecher, D. J., Macmillan, J. D., 1990. A novel bicomponent hemolysin from *Bacillus cereus* [J]. Infect. Immun., 58: 2220-2227.

Beecher, D. J., Wong, A. C. L., 1994. Improved purification and characterization of hemolysin BL, a hemolytic dermonecrotic vascular permeability factor from *Bacillus cereus* [J]. Infect. Immun., 62: 980-986.

Beecher, D. J., Wong, A. C. L., 1997. Tripartite hemolysin BL from *Bacillus cereus*: hemolytic analysis of component interaction and model for its characteristic paradoxical zone phenomenon [J]. J. Biol. Chem., 272: 233-239.

Black, E. P., et al., 2007. Response of spores to high pressure [J]. Comp. Rev. Food Sci. Food Safety, 6: 103-119.

Claus, D., Berkley, R. C. W., 1986. Genus *Bacillus* Cohn 1872, 174 [M]. Pages 1105-1137 in Bergey's Manual of Systematic Bacteriology. P. H. A Sneath, N. S. Mair, M. E. Sharpe, and J. G. Holt, ed., Williams and Wilkins, Baltimore.

De Lara, J., Fernandez, P. S., Periago, P. M., et al., 2002. Irradiation of spores of *Bacillus cereus* and *Bacillus subtilis* with electron beams [J]. Innov. Food Sci. Emerg. Technol., 3: 379-384.

Dierick, K., Van Coillie, E., Swiecicka, I., et al., 2005 Fatal family outbreak of *Bacillus cereus*-associated food poisoning [J]. J. Clin. Microbiol., 43: 4277-4279.

Ehling-Schulz, M., Fricker, M., Scherer, S., 2004. *Bacillus cereus*, the causative agent of an emetic type of food-borne illness [J]. Mol. Nutr. Food Res., 48: 479-487.

Ehling-Schulz, M., Fricker, M., Grallert, H., et al., 2006. Cereulide synthetase gene cluster from emetic *Bacillus cereus*: structure and location on a mega virulence plasmid related to *Bacillus anthracis* toxin plasmid pXO1 [J]. BMC Microbiol., 6: 20.

Fagerlund, A., Lindback, T., Storest, A. K., et al., 2008. *Bacillus cereus* Nhe is a pore-forming toxin with structural and functional properties similar to the ClyA (HlyE, SheA) family of haemolysins, able to induce osmotic lysis in epithelia [J]. Microbiology, 154: 693-704.

Gilbert, R. J., Kramer, J. M., 1986. *Bacillus cereus* food poisoning [M]. Pages 85-93, in Progress in Food Safety (Proceedings of Symposium). Food Research Institute, D. C. Cliver & B. A. Cochrane, ed. University of Wisconsin-Madison, Madison, WI.

Glatz, B. A., Goepfert, J. M., 1976. Defined conditions for synthesis of *Bacillus cereus* enterotoxin by fermenter-grown cultures [J]. Appl. Environ. Microbiol., 32: 400-404.

Granum, P. E., 1994. *Bacillus cereus* and its toxins [J]. J. Appl. Bacteriol. Symp. Suppl., 76: 615-665.

Griffiths, M. W., Schraft, H., 2002. *Bacillus cereus* food poisoning [M]. Pages 261-270, in Foodborne Diseases, 2nd ed. D. Cliver, ed. Elsevier Science, Ltd., London, England.

Hanson, M. L., Wendorff, W. L., Houck, K. B., 2005. Effect of heat treatment of milk on activation of *Bacillus* spores [J]. J. Food Prot., 68: 1484-1486.

Kotiranta, A., Lounatmaa, K., Haapasalo, M., 2000. Epidemiology and pathogenesis of *Bacillus cereus* infections [J]. Microbes Infect., 2: 189-198.

Kramer, J. M., Gilbert, R. J., 1989. *Bacillus cereus* and other *Bacillus* species. Pages 21 - 70 in Foodborne Bacterial Pathogens [J]. M. P. Doyle, ed. Marcel Dekker, New York.

Lindback, T., Fagerlund, A., Rodland, M. S., et al., 2004. Characterization of the *Bacillus cereus* Nhe enterotoxin [J]. Microbiology, 150: 3959-3967.

Lindsay, D., Brözel, V., Mostert, J., et al., 2002. Differential efficacy of a chlorine dioxide-containing sanitizer against single species and binary biofilms of a dairy-associated *Bacillus cereus* and a *Pseudomonas fluorescens* isolate [J]. J. Appl. Microbiol., 92: 352-361.

Lund, T., De Buyser, M. L., Granum, P. E., 2000. A new cytotoxin from *Bacillus cereus* that may cause necrotic enteritis [J]. Mol. Microbiol., 38: 254-261.

Mahler, H., Pasi, A., Kramer, J. M., et al., 1997. Fulminant liver failure in association with the emetic toxin of *Bacillus cereus* [J]. New Engl. J. Med., 336: 1142-1148.

Mattson, M. P., Culmsee, C., Fu, Z., et al., 2000. Roles of nuclear factor κB in neuronal survival and plasticity [J]. J. Neurochem., 74: 443-456.

Mead, P. S., Slutsker, L., Dietz, V., et al., 1999. Food-related illness and death in the United States [J]. Emerg. Infec. Dis., 5: 607-625.

Mikkola, R., Saris, N. E. L., Grigoriev, P. A., et al., 1999. Ionophoretic properties and mitochondrial effects of cereulide: the emetic toxin of *Bacillus cereus* [J]. Eur. J. Biochem., 263: 112-117.

Nimmer, P. S., Beer, M. R., McKillip, J. L., 2014. *Bacillus cereus*: a bacterial species of environmental and clinical significance [J]. J. Liberal Arts Sci., 18: 21-32.

Phelps, R. J., McKillip, J. L., 2002. Enterotoxin Production in Natural Isolates of *Bacillaceae* outside the *Bacillus cereus* Group [J]. Appl. Environ. Microbiol., 68: 3147-3151.

Ronner, U., Husmark, U., Henrikson, A., 1990. Adhesion of *Bacillus cereus* spores in relation to hydrophobicity [J]. J. Appl. Bacteriol., 69: 550-556.

Ryan, P. A., MacMillan, J. D., Zilinskas, B. A., 1997. Molecular cloning and characterization of the genes encoding the L1 and L2 components of hemolysin BL from *Bacillus cereus* [J]. J. Bacteriol., 179: 2551-2556.

Sakurai, N., Kolke, K., Irie, Y., et al., 1994. The rice culture filtrate of *Bacillus cereus* isolated from emetic type food poisoning causes mitochondrial swelling in a HEp-2 cell [J]. Microbiol. Immunol., 38: 337-343.

Scheldeman, P., Herman, L., Foster, S., et al., 2006. *Bacillus sporothermodurans* and other highly heat-resistant spore formers in milk [J]. J. Appl. Microbiol., 101: 542-555.

Schoeni, J. L., Wong, A. C., 2005. *Bacillus cereus* food poisoning and its toxins [J]. J. Food Prot., 68: 636-648.

Sneath, P. H. A., 1986. Endospore-forming Gram-positive rods and cocci [M]. Pages 1104-1105 in Bergey's Manual of Systematic Bacteriology. P. H. A. Sneath, N. S. Mair, M. E. Sharpe and J. G. Holt, ed. Williams and Wilkins, Baltimore, MD.

StenforsArnesen, L. P., Fagerlund, A., Granum, P. E., 2008. From soil to gut: *Bacillus cereus* and its food poisoning toxins [J]. FEMS Microbiol. Rev., 32: 579-606.

Sutyak, K. E., Wirawan, R. E., Aroutcheva, A. A., et al., 2008. Isolation of the *Bacillus subtilis* antimicrobial peptide subtilosin from the dairy product-derived *Bacillus amyloliquefaciens* [J]. J. Appl. Microbiol., 104: 1067-1074.

Sutherland, A. D., Limond, A. M., 1993. Influence of pH and sugars on the growth and production of diarrhoeagenic toxin by *Bacillus cereus* [J]. J. Dairy Res., 60: 575-580.

Ultee, A., Kets, E. P. W., Smid, E. J., 1999. Mechanisms of action of carvacrol on the food-borne pathogen *Bacillus cereus* [J]. Appl. Environ. Microbiol., 10: 4606-4610.

The Bovine Milk Microbiome: Origins and Potential Health Implications for Cows and Calves

Mark A. McGuire[1], Janet E. Williams[1], and Michelle K. McGuire[2]
[1]University of Idaho, Moscow, ID, USA
[2]Washington State University, Pullman, WA, USA

Introduction

Milk is a complex fluid which contains nutrients (protein, fat, carbohydrate, vitamins, minerals, and water) and myriad other biological substances important for the growth and development of the neonate (McGuire and McGuire, 2015, 2017). Typically, the quality of milk is judged by how well the nutrient content matches the nutrient requirement of the offspring. In addition, milk contains "living" cells whose roles in the neonate are unclear. For instance, milk contains several types of host cells including immune cells (lymphocytes, macrophages, and polymorphonuclear leukocytes) and those from epithelial lineages (Berry et al., 2007). Progenitor/stem cells have been isolated and propagated from human milk (Berry et al., 2007). Again, specific roles for the neonate are not clear. In addition, non-host cells, such as bacteria, have long been identified in milk, particularly that produced during mastitis. In fact, long-standing dogma suggests that the presence of bacteria in milk represents infectious mastitis; otherwise milk is sterile. This is, however, clearly not true. When a variety of culture media are used, research now shows that bovine milk contains a significant population of lactic acid bacteria including *Lactococcus*, *Streptococcus*, *Lactobacillus*, *Leuconostoc*, and *Enterococcus* sp. Others, such as the psychotrophs, can be found after cold storage (Quigley et al., 2011, 2013). The presence of viable bacteria in human milk using culture-dependent methods has also been demonstrated (e.g., Heikkilä and Saris, 2003; Hunt et al., 2012; Martín et al., 2003); in all these studies the researchers were limited by which bacterial taxa they were prepared to identify—often driven by specific interests of a research group. In summary, bacteria are commonplace in milk and may have important roles in both mammary gland and neonatal health (Jeurink et al., 2013; McGuire and McGuire, 2015, 2017; Quigley et al., 2013). The advent of culture-independent methodsfor bacterial community analysis, coupled with broad culturing approaches, are changing the perception of the complexity and role of bacteria in milk.

Bacteria in Mastitis

Bovine mastitis, an inflammatory condition of the mammary gland, is a tremendously important disease for the dairy industry, affecting up to 30% of dairy cows (Barkema et al., 1998; Olde Riekerink et al, 2008). Bovine mastitis is generally considered to be caused by *Escherichia coli*, environmental streptococci, *Klebsiella* spp., and coagulase-negative staphylococcus species (Oliveira et al., 2013), although *Staphylococcus aureus* has been found to be the most commonly isolated organism from clinical cases in smaller herds characteristic of Canada and the Netherlands (Barkema et al., 1998). Bovine mastitis risk is greatest in the first few weeks postpartum (Pyörälä, 2008), with effect of parity uncertain in early lactation (Barkema et al., 1998; Green et al., 2007; McDougall et al., 2009). To help prevent mastitis, dairymen are encouraged to follow proper milking procedures, use post-milking teat disinfectants, disallow cross-suckling of calves, and segregate or cull chronically infected animals (Hogan et al., 1999). In addition to being a serious economic hardship for the dairy industry due to loss of milk sales, mastitis prevention and treatment results in the majority of antibiotic use in dairy herds (Saini et al., 2012), which may represent a concern in terms of increasing antibiotic resistance in both animals and humans (National Research Council, 1999; Oliver et al., 2011).

Culture-dependent Analysis of Mastitic Milk

Currently the "gold standard" for mastitis pathogen identification is culture based (Hogan et al., 1999) but limitations in this approach have been identified. For instance, several studies have shown that culture-dependent methods do not detect microbial growth in at least 20%-30% of milk samples produced from quarters with clinical evidence of inflammation (Bradley et al., 2007; Hogan et al., 1999). Lack of bacterial growth in these types of milk samples can occur for many reasons: bacterial growth inhibitors in milk, low numbers of bacteria in milk, intermittent shedding of the pathogen, bacteria eliminated by the immune system in the presence of continued inflammation, and inadequate culture conditions (Britten, 2012; Koskinen et al., 2009, 2010; Taponen et al., 2009). Use of genomic-based (culture-independent) bacterial detection methods, such as polymerase chain reaction (PCR) using primers specific for mastitis pathogens, may better identify organisms of concern (Koskinen et al., 2009, 2010; Taponen et al., 2009) and, therefore, may be a more sensitive approach than use of culture-dependent methods.

Culture-independent Analysis of Mastitic Milk

Culture-independent assessment of bacteria is becoming a popular method to understand complex bacterial communities, and was used extensively in the Human Microbiome Project (2012). High - throughput sequencing allows for deeper evaluation of the bacteria present through the use of "universal" primers targeting the 16S rRNA gene found in all bacteria (Pace, 1997; Woese, 1987; Woese and Fox, 1977). Using universal primers for the V1-V2 region of the 16S rRNA gene, Oikonomou et al. (2012) confirmed that mastitis pathogens identified in bovine milk by culture were usually detected as the most abundant organism by culture-independent sequencing. Further, they discovered a diverse bacterial community present in culture-negative samples. Kuehn et al. (2013) specifically examined culture-negative milk samples from cows with clinical mastitis. Again, assessment of the V1-V2 region of the 16S rRNA gene allowed identification of a diverse bacterial community present in the culture-negative milk samples. These studies (Kuehn et al., 2013; Oikonomou et al., 2012) demonstrate that culture-independent analysis of a bacterial community provides a different view of the bacteria present than culture-dependent methods. In particular, not surprisingly, the relative abundance of anaerobic bacteria was substantial. Recently, Lima and coworkers (2017) sequenced the V4 hypervariable region of the 16S rRNA gene in bacterial DNA collected from colostrum. They reported greater richness in colostrum samples from primiparous cows compared to multiparous cows and differences in the overall taxonomic structure between the two groups. Additionally, bacterial communities in colostrum were less diverse from primiparous cows that were diagnosed with clinical mastitis during the first 30 d postpartum than primiparous cows that were not. Thus, use of culture-independent analysis may offer new possible pathogens to consider, how the microbial ecology of colostrum influences risk of mastitis, and the possibility of a polymicrobial disease origin in some cases.

Culture-independent Analysis of Healthy Milk

Both studies (Kuehn et al., 2013; Oikonomou et al., 2012) also compared mastitic milk samples to "healthy" milk from cows with no clinical signs of inflammation and low somatic cell counts (SCC). The bacterial community present in healthy bovine milk wasreported to be both rich and diverse. Kuehn et al. (2013) found the most abundant genera in milk from healthy quarters (no cultivable growth) to be *Ralstonia*, *Pseudomonas*, *Sphingomonas*, *Stenotropomonas*, *Psychrobacter*, *Bradyrhizobium*, *Corynebacterium*, *Pelomonas*, and *Staphylococcus* while Oikonomou et al. (2012) detected these and *Streptococcus*, *Lactobacillus*, and *Enterococcus* as well. In another study, Oikonomou et al. (2014) found that *Faecalibacterium*, unclassified *Lachnospiraceae*, *Propionibacterium*, and *Aeribacillus*were present in every milk sample produced by a healthy quarter (no cultivable growth and SCC<20,000 cells/mL). The bacterial community in this milk was easily discriminated from milk with greater SCC. Diversity indices were greater in the milk with < 20, 000 cells/mL than milk with greater number of cells. In preliminary results from 103 cows (Brooker, Williams, McGuire and McGuire, unpublished; Figure 1), bacterial communities in milk produced bylow - SCC quarters (< 200, 000 cells/mL) was more diverse and rich than milk from quarters with greater SCC. The microbiome of bulk tank milk samples from 19 dairy farms also demonstrated an inverse relationship between diversity and bacterial presence even when SCC was<250,000 cells/mL (Rodrigues et al., 2017).

Individual Variability in the Bovine Milk Microbiome

It is worth noting that the generamost prevalent in bovine milk are generallyalso the most abundant genera found in milk produced by healthy women (Hunt et al., 2011; Jeurink et al., 2013). In fact, more is currently known about the human milk microbiome than the bovine milk microbiome, so comparison between species may be useful. In our work, the microbial community in human milk as assessed using primers for the V1-V2 hypervariable region of the 16S rRNA gene was found to be relatively consistent and unique to a woman over a period of 5 weeks. Approximately half of the bacterial community in all milk samples from these 16 women included members of the genera *Staphylococcus*, *Streptococcus*, *Serratia*, *Pseudomonas*, *Corynebacterium*, *Ralstonia*, *Propionibacterium*, *Sphingomonas*, and *Bradyrhizobium.* The other half of the microbial community was more individualized. This personalized "fingerprint" could be driven by genetic, environment, or a combination of factors. Differences among cows within and between herds have not been directly addressed, but preliminary results from our group (Brooker, Williams, McGuire and McGuire, unpublished; Figure 1) suggest a difference in the overall milk microbiome between 2 dairy herds. In support of a potential difference among herds, the microbiome of bulk-tank milk from 19 dairies substantially differed in the relative abundance of bacteria even at the phylum taxonomic level (Rodrigues et al., 2017). Such discrimination suggests environment may be a significant factor in shaping the bovine milk microbiome.

This is important because variation in milk microbiome among farms and bulk tanks may impact processing characteristics. Significant complexity in the raw milk sampled from tanker trucks ($n=899$) arriving at two dairy processors in California was detected (Kable et al., 2016). Even with seasonal variability, a core microbiome of 29 taxonomic groups was identified using universal primers to the V4 region of the 16S rRNA gene. *Streptococcus*, *Staphylococcus*, unidentified members of *Clostridiales*, *Corynebacterium*, *Turicibacter*, and *Acinetobacter* were the most abundant of these groups. Thus, there appears to be herd differences in the microbiome of raw milk (as detected among samples from bulk tanks and tanker trucks) most likely originating with the variation in bacterial diversity of milk fromacow.

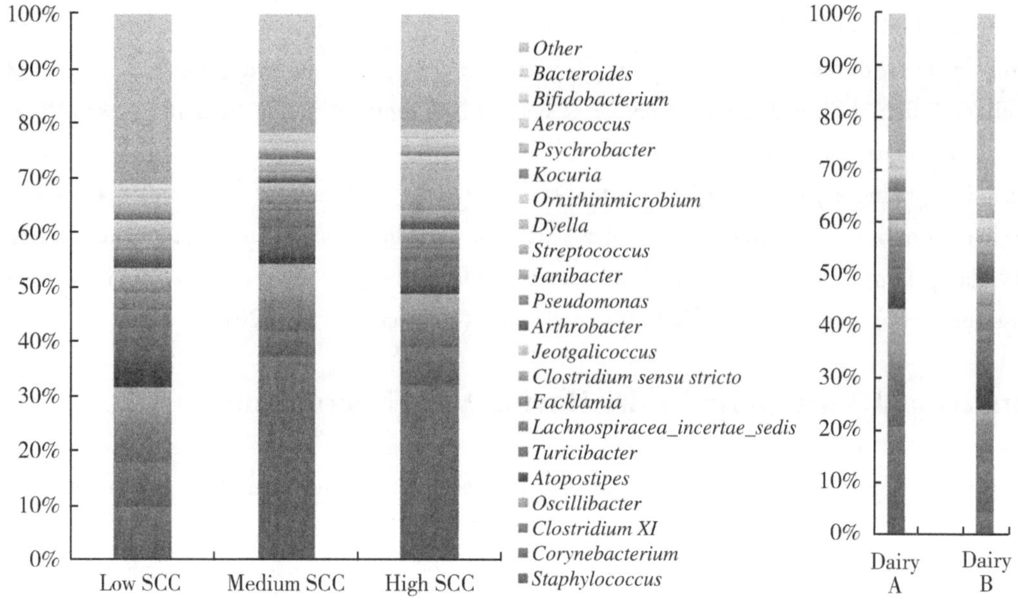

Figure 1 The top 20 bacterial genera found in milk produced by cows sampled by quarter on two dairies (*n* = 53 for dairy A, *n* = 50 for dairy B). Milk from individual quarters across farm was stratified by somatic cell count (SCC) (< 200, 000 cells/mL = Low SCC; 200, 000 to 400,000 cells/mL=Medium SCC; >400,000 cells/mL=High SCC) with 349, 26, and 35 samples, respectively (Brooker, Williams, McGuire and McGuire, unpublished)

Origin of the Milk Microbiome

The origin of the milk microbiome is also an area of active debate, but a growing literature suggests that its genesis is likely complex (Fernández et al., 2013; Rodríguez, 2014). Again, most research on this topic has been conducted in women. Where the bacteria in human milk come from is not entirely known, but two likely overarching sources are external (via the mammary ducts) and internal via an entero - mammary [gastrointestinal (GI) tract to mammary tissue] pathways. In humans, there is substantial retrograde flow of milk from the infant's mouth into the mammary gland during suckling (Ramsay et al., 2004). As such, anything that is present in the infant's mouth is likely to be present (at least for a short period of time) in the mammary ductal system. Indeed, both human milk and saliva are rich in *Streptococcus* spp. (Nasidze et al., 2009), suggesting the possibility that one might shape the other. Similarly, some bacterial taxa commonly isolated from skin (e.g., *Staphylococcus* and *Corynebacterium*) are also found in human milk (Gao et al., 2010). However, correlation does not infer causality, and delineating the potential directionality of this relationship (if, indeed, a causal relationship exists) will require controlled intervention studies.

Studies also point to the possibility of selective translocation of bacteria from the mother's GI tract to the mammary gland. One of the first studies to carefully examine this was conducted by Perez and colleagues (2007) who aseptically collected milk and peripheral blood from healthy breastfeeding women and feces from their infants. They also studied bacterial translocation to extra-intestinal tissues in mice. Their results suggested that intestinally-derived bacteria might be transported to the lactating mammary gland within mononuclear cells. In support of such transfer, Jiménez et al. (2008) demonstrated that *Lactobacillus* strains naïve to the breastfeeding woman appeared in her milk following oral administration of the strains. The bacteria were identified not only by culture - dependent but also culture - independent methods to confirm the specific strains consumed.

A series of interesting findings have also established that viable bacteria reside in mammary tissue of women who have never breastfed (Chan et al., 2016; LaTuga et al., 2014; Urbaniak et al., 2014; Xuan et al., 2014). These findings strongly suggest that one source of bacteria for milk includes the mammary gland itself. Thus, although additional work is needed, there is mounting evidence that the bacteria in human milk originate from a variety of sources. Creative intervention studies utilizing unique and/or labeled bacterial strains delivered orally to the mother and/or infant will be needed to untangle this complex web of potential origins of the milk microbiome.

Although limited, ahandful of researchers have examined the origin (s) of the bovine milk microbiome. For example, Zhang et al. (2014) compared the microbiome of milk produced by cows undergoing a subacute ruminal acidosis challenge through the feeding of a high- (70%), relative to a low- (40%), concentrate diet. Forty-eight operational taxonomic units (OTU; similar to bacterial species) were found to be affected by diet. The OTUswhich increased with the high-concentrate diet may exacerbate risk of mastitis and reduce the flavor and taste of the milk. The changes detected in the bovine milk microbiome support a possible entero-mammary path in cows. It is noteworthy, however, that the genera detected in the milk across diets was different from those in the studies previously mentioned (Kuehn et al., 2013; Oikonomou et al., 2012, 2014). Zhang et al. (2014) attributed those differences to the bacteria found on the skin of the teat (Verdier-Metz et al., 2012). To assess the extent of skin contamination or source of milk bacteria, our research group (Reynolds, Yahvah, Williams, Fox, McGuire, and McGuire, unpublished) biopsied bovine milk by needle aspiration through the wall of the teat after a surgical scrub of the teat wall. This work revealed that the bacterial community structure in the needle aspirate was similar to that of milk removed and collected via hand stripping with a sterile glove into a sterile tube immediately prior to the aspirate (Figure 2). This suggests that teat skin contribution to the bovine milk microbiome is minimal, and contamination during sampling can be avoided through aseptic methods recommended by the National Mastitis Council (Hogan et al.,

1999).

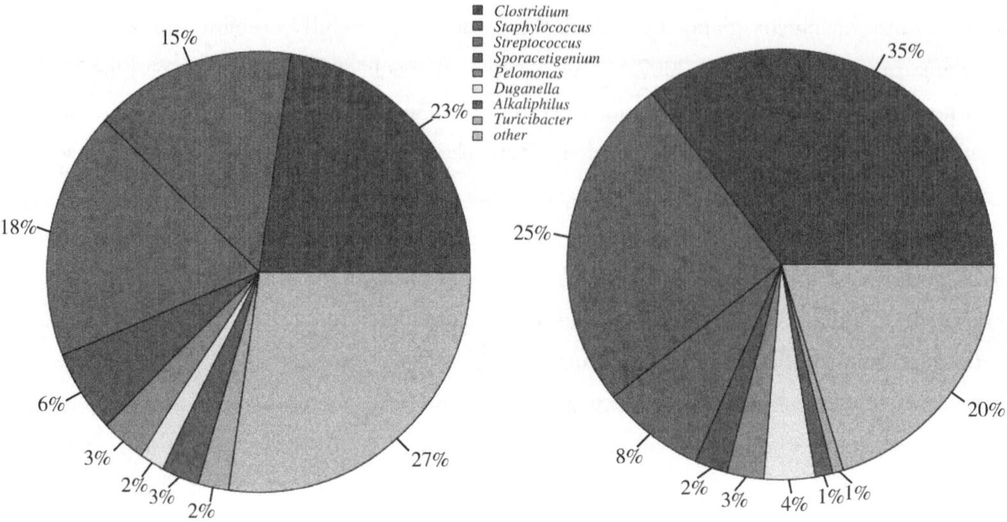

Figure 2 Bovine milk microbiome ($n=30$) in a hand strip sample after cleaning (left pie chart) and via aspiration by needle (right pie chart) through the teat wall (unpublished data from Reynolds, Yahvah, Williams, Fox, McGuire, and McGuire, unpublished)

Effect of Milk-borne Bacteria on the Nursing Neonate

The presence of bacteria in milk appears to be important for the assembly and colonization of the GI tract of the suckling neonate. There is compelling evidence that the infant's GI bacterial ecology is influenced by delivery mode (Dominguez-Bello et al., 2010), early feeding choices (Bezirtzoglou et al., 2011; Thompson et al., 2015), and antibiotic use (Arboleya et al., 2015). For instance, feces of breast-fed infants have a different bacterial community compared to feces from formula-fed infants (Penders et al., 2005). There are many factors involved with different feeding modes, including microbiota between human milk and formula feeding, that may affect the fecal bacterial community. Relatively new to this story is the likely intimate collective connections among maternal GI, milk, and infant GI microbiomes. Martínet al. (2006) showed that *Lactobacillus salivarius* CECT 5713 met the definition of a probiotic. The strain was detected in maternal feces and milk, and infant feces during supplementation through both culture-dependent and-independent methods. This is fairly convincing evidence of a entero-mammary transfer of a bacterium from mother to the nursing infant through milk to inoculate the infant's GI tract. A similar story was described by Albesharat et al. (2011) with the consumption of fermented foods enriched in lactic acid bacteria by lactating women. Thus, it is very likely that bacteria in milk impact the bacterial community in the infant's GI tract.

Direct evidence for the impact of bacteria in milk on a calf's GI tract is lacking, although a number of factors have been identified such as age, maternal microbiota, birth process, diet, and antibiotics (Malmuthuge et al., 2015; Yeoman and White, 2014). The addition of bifidobacteria and lactic acid bacteria to the milk replacer of calves did decrease the frequency of diarrhea, but fecal microbiome was not assessed (Abe et al., 1995). Thus, the potential for milk-borne bacteria to impact the GI tract microbiome is possible. A recent study by Deng et al. (2017) did demonstrate significant differences in the microbiome along the GI tract from calves fed acidified waste milk, untreated waste milk, pasteurized waste milk, or untreated bulk tank milk. Because these treatments should have differed significantly in their living bacteria (and therefore, the bacteria consumed by the recipient calves), these data indirectly support the potential for the milk microbiome to, at least in part, establish the offspring's GI microbiome.

Conclusions

Culture-independent analysis of bacterial communities in milk indicates the presence of a rich and diverse population—in both milk produced by healthy and mastitic animals. This paradigm shift may help researchers and clinicians prevent and treat mastitis and aid in our understanding in the development of the GI microbiome of the suckling neonate. Indeed, milk may be viewed as a probiotic food source capable of transferring certain bacteria from mother to infant. Although much research remains to be done to understand the physiologic mechanisms involved, milk-borne bacteria likelyplay roles in reducing the risk of certain diseases associated with GI problems, as well as development of allergies and asthma. A greater understanding of these important components of milk is warranted in both cows and other species, particularly women.

References

Abe, F., Ishibashi, N., Shimamura, S., 1995. Effect of administration of Bifidobacteria and lactic acid bacteria to newborn calves and piglets [J]. J. Dairy Sci., 78: 2838-2846.

Albesharat, R., Ehrmann, M. A., Korakli, M., et al., 2011. Phenotypic and genotypic analyses of lactic acid bacteria in local fermented food, breast milk, and faeces of mothers and their babies [J]. Syst. Appl.Microbiol., 34: 148-155.

Arboleya, S., Sanchez, B., Milani, C., et al., 2015. Intestinal microbiota development in preterm neonates and effect of perinatal antibiotics [J]. J. Pediatr., 166: 538-544.

Barkema, H. W., Schukken, Y. H., Lam, T. J. G. M., et al., 1998. Incidence of clinical mastitis in dairy herds grouped in three categories by bulk milk somatic cell counts [J]. J. Dairy Sci., 81: 411-419.

Berry, C. A., Thomas, E. C., Piper, K. M. E., et al., 2007. The histology and cytology of the human mammary gland and breastmilk [M]. In: Textbook of Human Lactation. Hale TW, Hartmann PE

(Ed) Hale Publishing LP Amarillo TX pp. 35-47.

Bezirtzoglou, E., Tsiotsias, A., Welling, G. W., 2011. Microbiota profile in feces of breast-and formula-fed newborns by fluorescence in situ hybridization (FISH) [J]. Anaerobe, 17: 478-482.

Bradley, A. J., Leach, K. A., Breen, J. E., et al., 2007. Survey of the incidence and aetiology of mastitis on dairy farms in England and Wales [J]. Vet. Rec., 160: 253-258.

Britten, A. M., 2012. The role of diagnostic microbiology in mastitis control programs [J]. Vet. Clin. North America Food Anim. Pract., 28: 187-202.

Chan, A. A, Bashir, M., Rivas, M. N., et al., 2016. Characterization of the microbiome of nipple aspirate fluid of breast cancer survivors [J]. Sci.Rep., 6: 28061.

Deng, Y.F., Wang, Y.J., Zou, Y., et al., 2017. Influence of dairy by-product waste milk on the microbiomes of different gastrointestinal tract components in pre-weaned dairy calves [J]. Sci. Rep., 7: 42689.

Dominguez-Bello, M. G., Costello, E. K., Contreras, M., et al., 2010. Delivery mode shapes the acquisition and structure of the initial microbiota across multiple body habitats in newborns [J]. Proc. Natl.Acad.Sci.USA, 107: 11971-11975.

Fernández, L., Langa, S., Martín, V., et al., 2013. The human milk microbiota: origin and potential roles in health and disease [J]. Pharmacol. Res., 69: 1-10.

Gao, Z., Perez-Perez, G.I., Chen, Y., et al., 2010. Quantitation of major human cutaneous bacterial and fungal populations [J]. J. Clin. Microbiol., 48: 3575-3581.

Green, M. J., Bradley, A. J., Medley, G. F., et al., 2007. Cow, farm, and management factors during the dry period that determine the rate of clinical mastitis after calving [J]. J. Dairy Sci., 90: 3764-3776.

Heikkilä, M. P., Saris, P. E. J., 2003. Inhibition of *Staphylococcus aureus* by the commensal bacteria of human milk [J]. J.Appl.Microbiol., 95: 471-478.

Hogan, J.S., Gonzalez, R.N., Harmon, S.C., et al., 1999. Laboratory Handbook on Bovine Mastitis [M]. 1st ed. National Mastitis Council, Madison, WI.

Human Microbiome Project Consortium., 2012. Structure, function and diversity of the healthy human microbiome [J]. Nature, 486: 207-214.

Hunt, K. M., Foster, J. A., Forney, L. J., et al., 2011. Characterization of the diversity and temporal stability of bacterial communities in human milk [J]. PLoS One, 6: e21313.

Hunt, K. M., Preuss, J, Nissan, C, et al., 2012. Human milk oligosaccharides promote the growth of staphylococci [J]. Appl.Environ. Microbiol., 78: 4763-4770.

Jeurink, P.V., van Bergenhenegouwen, J., Jiménez, E., et al., 2013. Human milk: a source of more life than we imagine [J]. Benef.Microbes, 4: 17-30.

Jiménez, E., Fernández, L., Maldonado, A., et al., 2008. Oral administration of Lactobacillus strains isolated from breast milk as an alternative for the treatment of infectious mastitis during lactation [J]. Appl. Environ. Microbiol., 74: 4650-4655.

Kable, M. E., Srisengfa, Y., Laird, M., et al., 2016. The core and seasonal microbiota of raw bovine milk in tanker trucks and the impact of transfer to a milk processing facility [J]. mBio., 7: e00836-16.

Kuehn, J. S., Gorden, P. J., Munro, D., et al., 2013. Bacterial community profiling of milk samples as a means to understand culture-negative bovine clinical mastitis [J]. PLoS One, 8: e61959.

Koskinen, M. T., Holopainen, J., Pyörälä, S., et al., 2009. Analytical specificity and sensitivity of a real-time polymerase chain reaction assay for identification of bovine mastitis pathogens [J]. J. Dairy Sci., 92: 952-959.

Koskinen, M. T., Wellenberg, G. J., Sampimon, O. C., et al., 2010. Field comparison of real-time polymerase chain reaction and bacterial culture for identification of bovine mastitis bacteria [J]. J. Dairy

Sci., 93: 5707-5715.

LaTuga, M.S., Stuebe, A., Seed, P.C., 2014. A review of the source and function of microbiota in breast milk [J]. Semin. Reprod. Med., 32: 68-73.

Lima, S. F., Teixeira, A. G. V., Lima, F. S., et al., 2017. The bovine colostrum microbiome and its association with clinical mastitis [J]. J. Dairy Sci., 100: 3031-3042.

Malmuthuge, N., Griebel, P. J., Guan, L. L., 2015. The gut microbiome and its potential role in the development and function of newborn calf gastrointestinal tract [J]. Front. Vet. Sci., 2: 36.

Martín, R., Langa, S., Reviriego, C., et al., 2003. Human milk is a source of lactic acid bacteria for the infant gut [J]. J.Pediatr., 143: 754-758.

Martín, R., Jiménez, E., Olivares, M., et al., 2006. *Lactobacillus salivarius* CECT 5713, a potential probiotic strain isolated from infant feces and breast milk of a mother-child pair [J]. Int.J.Food Microbiol., 112: 35-43.

McDougall, S., Parker, K. I., Heuer, C., et al., 2009. A review of prevention and control of heifer mastitis via non-antibiotic strategies [J]. Vet Microbiol., 134: 177-185.

McGuire, M. K., McGuire, M. A., 2015. Human milk: Mother Nature's prototypical probiotic food? [J]. Adv. Nutr., 6: 112-123.

McGuire, M.K., McGuire, M.A., 2017. Isn't milk sterile? A historical perspective on milk microbes [M]. In: Prebiotics and Probiotics in Human Milk. McGuire M. K., McGuire M. A., Bode L (ed.) pp. 297-314. Elsevier. San Diego CA.

Nasidze, I., Li, J., Quinque, D., et al., 2009. Global diversity in the human salivary microbiome [J]. Genome Res., 19: 636-643.

National Research Council (US) Committee on Drug Use in Food Animals., 1999. The Use of Drugs in Food Animals: Benefits and Risks [M]. Washington (DC): National Academies Press (US).

Oikonomou, G., Machado, V.S., Santisteban, C., et al., 2012. Microbial diversity of bovine mastitic milk as described by pyrosequencing of metagenomics 16s rDNA [J]. PLoS One, 7: e47471.

Oikonomou, G., Bicalho, M. L., Meira, E., et al., 2014. Microbiota of cow's milk; distinguishing healthy, sub-clinically and clinically diseased quarters [J]. PLoS One, 9: e85904.

Olde Riekerink, R. G., Barkema, H. W., Kelton, D. F., et al., 2008. Incidence rate of clinical mastitis on Canadian dairy farms [J]. J. Dairy Sci., 91: 1366-1377.

Oliveira, L., Hulland, C., Ruegg, P. L., 2013. Characterization of clinical mastitis occurring in cows on 50 large dairy herds in Wisconsin [J]. J. Dairy Sci., 96: 7538-7549.

Oliver, S. P., Murinda, S. E., Jayarao, B. M., 2011. Impact of antibiotic use in adult dairy cows on antimicrobial resistance of veterinary and human pathogens: a comprehensive review [J]. Foodborne Pathog. Dis., 8: 337-355.

Pace, N. R., 1997. A molecular view of microbial diversity and the biosphere [J]. Science, 276: 734-740.

Penders, J., Vink, C., Driessen, C., et al., 2005. Quantification of Bifidobacterium spp., Escherichia coli and Clostridium difficile in faecal samples of breast-fed and formula-fed infants by real-time PCR [J]. FEMS Microbiol. Lett., 243: 141-147.

Perez, P.F., Doré, J, Leclerc, M., et al., 2007. Bacterial imprinting of the neonatal immune system: lessons from maternal cells? [J]. Pediatrics, 119: e724-732.

Quigley, L., O'Sullivan, O., Beresford, T. P., et al., 2011. Molecular approaches to analyzing the microbial composition of raw milk and raw milk cheese [J]. Int J Food Microbiol., 150: 81-94.

Quigley, L., O'Sullivan, O., Stanton, C., et al., 2013. The complex microbiota of raw milk [J]. FEMS Microbiol. Rev., 37: 664-698.

Ramsay, D. T., Kent, J. C., Owens, R. A., et al., 2004. Ultrasound imaging of milk ejection in the breast of lactating women [J]. Pediatrics, 113: 361-367.

Rodrigues, M. X., Lima, S. F., Canniatti - Brazaca, S. G., et al., 2017. The microbiome of bulk tank milk: characterization and associations with somatic cell count and bacterial count [J]. J. Dairy Sci., 100: 1-17.

Rodríguez, J. M., 2014. The origin of human milk bacteria: is there a bacterial entero-mammary pathway during late pregnancy and lactation? [J]. Adv.Nutr., 5: 779-784.

Saini, V., McClure, J. T., Léger, D., et al., 2012. Antimicrobial use on Canadian dairy farms [J]. J. Dairy Sci., 95: 1209-1221.

Taponen, S., Salmikivi, L., Simojoki, H., et al., 2009. Real-time polymerase chain reaction-based identification of bacteria in milk samples from bovine clinical mastitis with no growth in conventional culturing [J]. J. Dairy Sci., 92: 2610-2617.

Thompson, A. L., Monteagudo-Mera, A., Cadenas, M. D., et al., 2015. Milk-and solid-feeding practices and daycare attendance are associated with differences in bacterial diversity, predominant communities, and metabolic and immune function of the infant gut microbiome [J]. Front. Cell Infect.Microbiol., 5: 3.

Urbaniak, C., Cummins, J., Brackstone, M., et al., 2014. Microbiota of human breast tissue [J]. Appl. Environ. Microbiol., 80: 3007-3014.

Verdier-Metz, I., Gagne, G., Bornes, S., et al., 2012. Cow teat skin, a potential source of diverse microbial populations for cheese production [J]. Appl. Environ. Microbiol., 78: 326-333.

Woese, C.R., 1987. Bacterial evolution [J]. Microbiol. Rev., 51: 221-271.

Woese, C.R., Fox, G.E., 1977. Phylogenetic structure of the prokaryotic domain: the primary kingdoms [J]. Proc. Natl. Acad. Sci. US, 74: 5088-5090.

Xuan, C., Shamonki, J.M., Chung, A., et al., 2014. Microbial dysbiosis is associated with human breast cancer [J]. PLoS One, 9: e83744.

Yoeman, C.J., White, B.A., 2014. Gastrointestinal tract microbiota and probiotics in production animals [J]. Annu. Rev. Anim. Biosci., 2: 269-286.

Zhang, R., Huo, W., Zhu, W., et al., 2014. Characterization of bacterial community of raw milk from dairy cows during subacute ruminal acidosis challenge by high-throughput sequencing [J]. J. Sci. Food Agric., 95: 1072-1079.

Recent Advances in Milk Ingredient Development: the Role of MFGM and Phospholipids in the Biological Activity of Milk

Rafael Jiménez-Flores

The Ohio State University, Department of Food Science and Technology.
Columbus Ohio, USA

Abstract

This paper presents a personal view on the potential applications of the milk by-products or underutilized fractions with a special focus on the milk fat globule membrane as an ingredient in the processed foods area. Several factors are of importance for this presentation: the reasons for having undervalued components in milk, with added information on individual components and their relationship with health and wellness. We hope to give a glimpse of the reasons on why it is a good idea to use more efficiently the components of milk and the milk fat globule membrane. In addition, we consider current advances in the scientific evaluation of milk components as important players in the food industry and alternate fields. We present relevant information on the role of this components in digestion and nutrition, especially under the new methodology offered by 'metagenomics'.

Keywords: Milk fat globule membrane (MFGM), isolation, analytical tools, biological activities, innovation

Introduction

This paper presents some relevant information on the potential of milk components as ingredients for the food industry, and the fat globule membrane (MFGM) as an ingredient in our food supply. In today's world where milk fat has an image problem due to the load of saturated fat and the role of triglycerides in the modern American diet, it is important to emphasize that there are very important components in milk that have not been fully considered. Those are the minor lipids, glycolipids and glycoproteins that compose the MFGM. Since the industrialization of butter, these components have taken a second seat to primary commodity ingredients such as skim milk powder and whey protein concentrates. The lipids contained in buttermilk powder are not used to their full potential in the processed foods industry. In foods, these lipids exhibit superior emulsifying properties, and help significantly in thickening formulations due to water distribution properties, and aid where foaming is important such as in ice cream. However, the popular current view of buttermilk is of a milk-derived powder with an important liability; it goes rancid in about 6 months due to lipid oxidation.

This work starts by an introduction of the milk components that have great potential in the future because in today's world they are undervalued. These components are undervalued under the concepts of low market price, low understanding of their potential in biological function, they may not be of interest under current commercial priorities, and lack of long term vision of some commercial entities and the food ingredient environment.

The most undervalued components of milk and dairy products are: lactose, mineral components of whey and whey by-products and most importantly in my view, the lipids and phospholipids in milk. A case can be made for each of this components that are at this point in the same commercial and technological position as whey proteins were 30 years ago. The fact that today whey and whey proteins are used in so many foods, isevidence that the intrinsic good properties of the components in milk can be exploited for a commercial gain, and furthermore, for a health and wellness benefit. Also important would be to mention that some cases can also help in protecting the environment, as will be exposed below.

Lactose

From the biological point of view, is important to realize that lactose is the common sugar produced by all mammals in the planet (with exception of Pinnipeds). Perhaps a better understanding on the biological reasons or fundamentals that nature selected lactose as the sugar for mammals is an important scientific quest with potential revolutionary answers. Some evidence that lactose intolerance has been grossly exaggerated, as stated by Mattar et al. (2012), "Symptoms of lactose intolerance might have been exaggerated, such that up to 12 g of lactose is possibly well tolerated by lactase non-persistence individuals, which negates the need for restrictions on lactose-hydrolyzed milk, as well as fermented and matured milk products, preventing any subsequent effects on bone mass density". Moreover, the NIH has consistently been focusing on the microbiome in individuals' intestines as a major determinant on food digestion and nutrition. Another aspect that hinders better commercialization of lactose has been addressed by Wong and Hartel (2014), where the inadequate control of lactose crystallization yields great part of lactose crystals of very large size distribution, where the small crystals hinder many of the proper function of the lactose in commercial opperations.

Looking towards the future, it seems likely that biotechnological applications of lactose promise great advances. Lactose itself or as an aide, are being used in the production of bio-gas, bio-hyddrogen, exopolysaccharides, bacteriocins (from lactic acid bacteria), organic acids, and single cell organisms for biomass such as algae (Modler, 1985; Rosa, Santos et al., 2014; Gomes, Rosa et al., 2015; Seo, Yun et al., 2015; Huppertz and Gazi, 2016; Remon et al., 2016).

Milk Salts

There has not been much progress on the utilization of milk salts in the last 10 years (Pyne 1962). Milk salts, and in particular calcium phosphate is a very interesting substance that also has unique animal origin. Some recent examples of use of calcium phosphate are in the specific areas of: bone healing and surgery (Bohner 2001; Hu et al., 2014), in other biologically active materials still under study and research (Schumacher et al., 2007) and also in other industrial uses such as ink and printing materials (Richard et al., 2014). Another established practice is the use of calcium phosphate in aquaculture, although not widely documented, the sales of calcium phosphate for this purpose is increasing, perhaps due to its higher bio-availability for the fish (Richard et al., 2014).

Also, regarding the future use and consideration for milk salts, is their role in aids as protein functionality agents. One example is the use of the different calcium phosphates in protein solubility in different processed foods (Guinee, 2009). However, it is also evident that milk calcium plays an important role in milk powder functionality (Eshpari et al., 2015).

Milk Lipids as Triglycerides

Current view of milk lipids is rapidly changing, especially in their nutritional and functional attributes and links to human health. First it is important to consider that the spectrum of nutrients in milk and dairy products is unique and with most acceptable flavors and textures, and this is in great part due to the unique composition of milk fat. In addition, it is unclear whether the saturated fat in milk is unfavorable or detrimental in any way to cardiovascular health or with respect to nutritional physiology. There is currently no convincing evidence that moderate ingestion of saturated fatty acids from milk increases the risk of cardiovascular diseases (Arnold and Jahreis, 2011; Adamska and Rutkowska, 2014). However, we should continue to pay attention to current research on the ignored facts of the health benefits of a balanced diet that includes milk fat. Perhaps a very balanced view of milk fat is presented by the review of Markey et al.(2015) where they conclude the following: "Despite the significant saturated fatty acid profile of milk fat, epidemiological evidence and experimental data challenge this perception. Indeed, these data suggest that milk may afford some protection from cardiovascular events, but the evidence is less substantialfor other dairy products. There is a need to carry out well-designed, human dietary intervention studies to determine whether consumption of modified milk and dairy products, with unsaturated fatty acids enhanced profile, may provide more protection from cardiovascular disease in both healthy and at-risk populations" (Markey et al., 2015).

On the topic of milk phospholipids, there is much to be discussed, and it is a topic of particular interest in our research group. Milk phospholipids, also called glycerophospholipids, and sphingolipids and ceramides are quantitatively the most abundant phospholipids (PLs) in milk. They are found most prominently on the milk fat globule membrane (MFGM) and in other vescicles in skim milk (they can't be separated in cream due to their high density). They include, phosphatidylethanolamine (approx. 25% by weight), phosphatidylcholine (approx. 25% by weight), phosphatidylinositol and phosphatidylserine, while sphingomyelin (approx.25% by weight) is the dominant species of sphingolipids There is considerable evidence that PLs have beneficial health effects, such as cellular membrane health, principally intestinal cell membrane, regulation of many inflammatory reactions, significant preventive and therapeutic biological activity on some types of cancer, and inhibition of the cholesterol absorption. Technologically, phospholipids from milk show excellent emulsifying properties and can be used as a delivery system for biologically active compounds and lipid-soluble constituents. Due to the amphiphilic characteristics of these molecules, their extraction, separation and detection are critical points in the isolation and processing approach. While the analytical procedures can be based on extraction by using chloroform and methanol, followed by the determination by high pressure liquid chromatography (HPLC), coupled with Charged Aerosol Detectors (CAD) or mass detector (MS), these procedures are only valuable in the analytical world (Contarini and Povolo, 2013, Cilla, Quintaes et al., 2016). An economically and technological efficient process to obtain milk phospholipids in quantities relevant for the food industry is still to be defined, although there are several commercial ingredients available at high prices. Recently we found that the presence of milk phospholipids greatly influences the way that proteins precicpitate and aggregate under thermal treatment, which is fundamentally important to manipulate texture in processed foods high in protein (Saffon et al., 2015).

Milk Phopsholipids

We have worked for the last 20 years on the milk phospholipids, and there are several worthy reviews on the topic. For this particular review we can quote from our previous chapter on Food and Structure.

Lipids in MFGM

Polar lipids, such as glycerophospholipids and sphingolipids, constitute less than 1%

of the milk total lipids; nevertheless, they function as intracellular signaling molecules, provide a framework structure, and have nutritional and functional properties such as emulsification (Jiménez - Flores and Brisson, 2008; Lopez et al., 2008). Indeed, research has also found a relationship between polar lipid consumption and enhanced health (Spitsberg, 2005). The phospholipids on the MFGM are distributed heterogeneously and present in different percentages (Dewettinck et al., 2008; Jiménez-Flores and Brisson, 2008; Lopez et al., 2008). These phospholipids contain high levels of long - chain fatty acids (FA) such as palmitic, stearic, and tricosanoic acids (Singh, 2006; Sánchez-Juanes et al., 2009). No difference has been observed in the FA composition of MFGM phospholipids regardless of season (Fauquant et al., 2005). Other lipids, such as triglycerides (TG), diglycerides, monoglycerides, sterols, and sterol esters, are mainly present in the milk fat globule core as seen in Table 1. Among these, triglycerides represent about 60% of the neutral lipids in the milk fat globules and cholesterol accounts for 90% of the total sterols in milk fat globules (Mather, 2000; Keenan and Mather, 2006; Rombaut et al., 2006a; Dewettinck et al., 2008).

While phospholipids and cholesterol are very interesting components of membranes, and they function as excellent emulsifiers, their properties have been extensively studied (Krog, 1991; Xiong and Kinsella, 1991; McClements et al., 1993; Goff, 1997; Singh and Dalgleish, 1998; Einhorn - Stoll et al., 2002; Waninge et al., 2005). Sphingomyelin in milk has been studied as a bioactive lipid, and only animalsources of fat contain this class of polar lipid (Prostenik and Gospocic, 1966; Morrison, 1969; Schmelz et al., 1996; Vesper et al., 1999; Nyberg et al., 2000; Motouri et al., 2003; Waninge et al., 2003, Noh and Koo, 2004; Waninge et al., 2005; Rombaut et al., 2007). Sphingolipids constitute up to one-third of the MFGM polar lipid fraction (Dewettinck et al., 2008; Jiménez-Flores and Brisson, 2008). They are characterized by a sphingoid base, which is a long-chain aliphatic amine (12-22 carbon atoms) (Rombaut et al., 2007). Attachment of a fatty acid via a hydroxyl group forms a ceramide that is the basic unit for the formation of sphingo - phospholipids, such as sphingomyelin (SM). The latter has a high degree of saturation that facilitates complex formation with cholesterol, known as a lipid raft. These are rigid domains implicated in cellular processes, like signal transduction, endocytosis, and cholesterol trafficking (Dewettinck et al., 2008; Lopez et al., 2008, 2010). Sphingomyelin is primarily located in the outer membrane of the MFGM, and is present in higher percentages in milk derived from cows fed with unsaturated fatty acids due to the presence of smaller globules and consequently more surface area (Lopez et al., 2008). An analysis of sphingolipids in the most commonly consumed dairy products found similar sphingolipid concentrations in nonfat dry milk and fermented products, indicating that starter cultures do not contribute to SM in dairy products (Jiménez - Flores and Higuera - Ciapara, 2009); however, we can manipulate the composition of this essential polar lipid by changing the diet of cows (Lopez et al., 2008). Furthermore, SM has been shown to reduce aberrant crypt foci and the appearance of colon adenocarcinoma, to modulate immune responses, among other functions (Dewettinck et al., 2008; Jiménez-Flores and Higuera-Ciapara, 2009). Sphingomyelin metabolites also provide health benefits (Schmelz, Dillehay et al. 1996, Thompson and Singh 2006).

Long term feeding of sphingolipids (1%) to laboratory rats significantly decreased total blood cholesterol levels and elevates HDL, suggesting that sphingolipids represent a "functional" constituent of food.

Proteins in MFGM

MFGM proteins constitute 1% - 2% of the total bovine milk proteins, and depending on the milk source and how it is processed, 25%-70% of the MFGM may be polypeptides, ranging in molecular weight (MW) from 15,000 to 240,000 Da (Ye et al., 2002, 2004; Dewettinck et al., 2008; Jimenez-Flores and Brisson, 2008; Jiménez-Flores and Higuera-Ciapara, 2009). Indeed, Murgiano et al. (2009) detected differences between Chianina and Holstein cattle in the amount of proteins associated with mammary gland development, lipid droplets formation, and host defense mechanisms. Mather (2000) presents a detailed description and suggested nomenclature of the known and major MFGM proteins.

It is very important that some consideration has been given to MFGM protein composition from different sources. Affolter and his group have applied proteomic techniques to accurately quantify seven MFGM proteins from buttermilk or whey protein concentrate (Affolter et al., 2010). They used sophisticated laboratory analytical techniques, based in isotope dilution mass spectrometry, that describe in detail the quantitative differences.

From our own research we have learned that in addition to the large number and diverse functions of the major MFGM proteins, they appear to have complex interactions with each other, other proteins, lipids, and other molecules such as hydrophobic and hydrophilic interactions, and the formation of disulfide bonds (Kim and Jimenez-Flores, 1995; Morin et al., 2006; Morin et al., 2007; Morin et al., 2008). It has recently been shown that XDH/XO and BTN form a high molecular weight aggregate through the formation of intermolecular disulfide bonds after a heat treatment at 60℃ for 10 minutes (Ye et al., 2002), while previous studies have shown that with higher temperature heat treatments MFGM proteins form similar high molecular weight complexes with milk proteins. Kim and Jiménez - Flores (1995) reported that with heat treatment of 87℃, b - lactoglobulin (BLG) and other milk serum proteins interact, forming high molecular weight complexes with the MFGM proteins PAS6/7. The mechanism of formation was undefined, but did not appear to be solely due to the formation of disulfide bonds (Kim and Jiménez-Flores, 1995).

MFGM health and well-being

Progress in the knowledge of the composition and role of milk fat globule membrane components has led to the realization that some possess biological properties beyond their nutritional significance. Spitsberg (Spitsberg, 2005), in his review, refers to the MFGM as a nutraceutical with nutritional and pharmaceutical potential. In fact, the MFGM has antimicrobial proteins, illness suppressors, and micronutrient-binding proteins that bind compounds, such as iron, zinc, copper, folate, and vitamin B_1, and other components, with anticarcinogenic properties, (Snow et al., 2010, 2011). Lonnerdal et al. (2006) complemented infant food with bovine MFGM and micronutrients to evaluate its effect on children's growth, nutrition, and morbidity. In a double-blind study with 6-to 11-month-old infants (n =550), fortification of food was beneficial in the copper and vitamin B_{12} status among the subjects, which enhance growth and normal function of the brain and nervous system, respectively (Lonnerdal et al., 2006). Wat et al. (2009) supplemented bovine PL to mice fed a high-fat diet resulting in a reduction in liver weight, total liver lipid, and serum lipid levels, which might be of therapeutic benefit in humans with nonalcoholic fatty liver disease, especially obese and diabetic patients. Choline has recently been identified as an essential component of the human diet. Nutrition studies have revealed that choline-containing phospholipids play a central role in the process of signal transduction within the body and that alterations in this process may lead to abnormalities such as cancer and Alzheimer's disease (Zeisel et al., 1986; Fang and Dalgleish, 1995; Nyberg, et al., 1998; Parodi, 2003; Miyaji et al., 2005; Parodi, 2005). Based on this, we know first-hand that MFGM is now regarded as a valuable source of these essential components of our diet.

We now have some indication of the beneficial properties of MFGM as part of a dairy ingredient. Ward and coworkers have recently shown that an ingredient rich in bovine MFGM components from whey cream prevented intestinal cancer in rats (Snow et al., 2010). Using a similar component in the diet of rats, the same group demonstrated that a MFGM-rich ingredient inhibited gastrointestinal leakiness in mice treated with lipopolysaccharide (Snow, et al., 2011).

Physiological function

The overall physiological function of the MFGM, as we understand it today, is to allow for secretion of the milkfat into the milk. The nature of lipid droplet synthesis and secretion follows a biochemical pathway that has been described recently by Bionaz and Loor (2006). In this work, they describe the action of 45 genes associated with lipid synthesis, and have postulated the interaction of these genes in the achievement of milk synthesis and secretion.

Proteomic analysis for the MFGM has been studied by several groups (Cavaletto et al.,

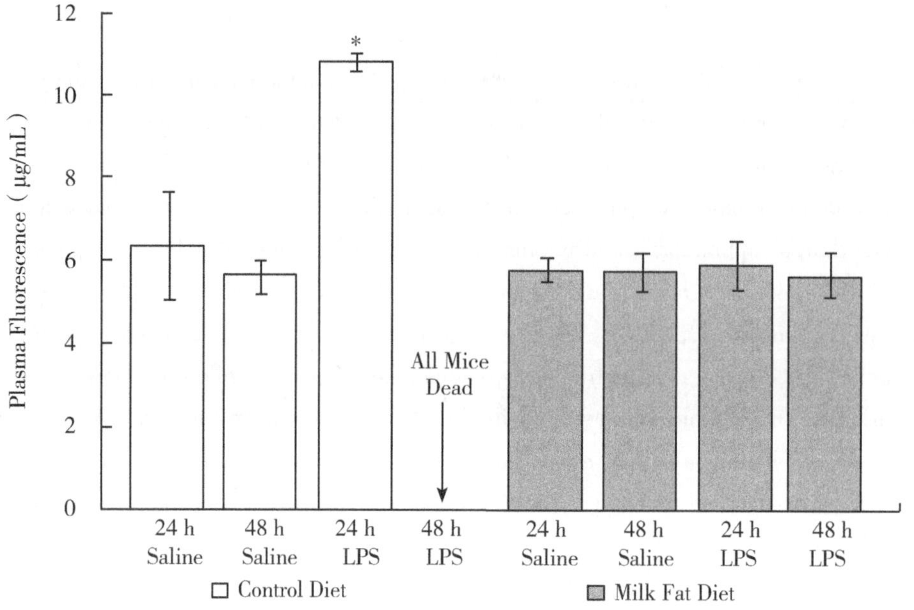

Figure 1 Results from ' leaky gut ' experiment. the four bars (clear) represent the two groups of mice that were fed a normal diet. the first two bars are the mice that were treated with saline solution and the bars indicate the plasma fluorescence in the blood stream (taken as normal levels of fluorescence). the third and fourth bar represent the control group after LPS treatment. The mice showed significantly increased fluorescence, which indicates a grave level of LPS in their blood stream, and the fourth bar could not be recorded because all the mice died. In contrast the filled bars (dark) indicate both groups of mice fed MFGM, and they showed that their blood levels of fluorescence does not change with LPS treatment. It was concluded that MFGM gives protection to the mice to LPS induced inflammation and thus MFGM protects the mice against LPS via preventing ' leaky gut syndrome'

2002; Yamada et al., 2002; Fortunato et al., 2003; Cavaletto et al., 2004; O'Donnell et al., 2004; Hamdan and Righetti, 2005; Manso et al., 2005; Wang et al., 2006; Westermeier, 2006; Beddek et al., 2008; Reinhardt and Lippolis, 2008; Wilson et al., 2008; Gagnaire et al., 2009; Murgiano et al., 2009; Affolter et al., 2010), and there are important and significant similarities among studies and across the studied species. The most abundant proteins in the MFGM, butyrophylin and xanthine oxido/reductase, indicates the importance of these proteins in the secretory process, and very likely in the mechanism of attachment of the MFGM to the surface of the fat globule. Other similarities are found on the high proportion of the proteins in the MFGM that form part of lipid metabolism, lipid transport and trafficking, immunity and defense, and membrane structure.

In summary, the genomic analysis of the origin of the MFGM indicates a highly conserved series of proteins that are essential for the synthesis and secretion of the fat into milk, and the group of proteins that communicate chemically important growth and immunological signals to the neonate, are the main functional proteins of the MFGM. This is organized and structured in a phospholipid matrix that allows their biological function in the secretory mammary cell, and one can only hypothesize at this point, that it also has the elements for transmitting the nutrients and biologically active components and presenting them to theneonate in a functional manner.

Conclusion

The path ahead for the use of dairy ingredients will focus on adding value to those milk components that today have low value due to different technical or economic reasons. Lactose, as the unique sugar it is will either find a technological innovation that could make it an essential component on a functional ingredient, or it is also likely that a non-food application be a factor to increase its demand. Either way, it would be very useful to find the reason and advantage of lactose over any other sugar that could be the basis for mammalian evolution.

Milk minerals have great potential in the health and wellness arena; from nutrition source of calcium and bone health, to surgery and dentistry applications. However, milk minerals have also a promising path in the industrial world where bone replacers, inks and 3D printing media are making a tremendous improvement in our lives.

Finally, milk lipids and phospholipids are finally getting the recognition they have long been lacking. The milk triglicerides are subject to many biochemical, nutritional and epidemiological studies that in general conclude that they are not the culprit of cardio-vascular disease, and moreover, many positive health effects are being discovered.

Milk phospholipids are gaining approval and market due to their various health and wellness potential. Perhaps some technological hurdles to provide an economical and efficient process to isolate or enrich all phospholipid concentration or a process to fractionate them and produce ingredients high in phospholipid content will open profitable markets for these undervalued milk components.

References

Adamska, A., Rutkowska, J., 2014. Odd-and branched-chain fatty acids in milk fat-characteristic and health properties [J]. Postepy Higieny I Medycyny Doswiadczalnej, 68: 957-966.

Affolter, M., Grass, L., Vanrobaeys, F., et al., 2010. Qualitative and quantitative profiling of the bovine milk fat globule membrane proteome [J]. J. Proteomics, 73 (6): 1079-1088.

Arnold, C., Jahreis, G., 2011. Milk Fat and Health [J]. Ernahrungs Umschau, 58 (4): 177-181.

Beddek, A. J., Rawson, P., Peng, L. F., et al., 2008. Profiling the metabolic proteome of bovine mammary tissue [J]. Proteomics, 8 (7): 1502-1515.

Bohner, M., 2001. Physical and chemical aspects of calcium phosphates used in spinal surgery [J]. Eur. Spine J., 10: S114-S121.

Cavaletto, M., Giuffrida, M. G., Conti, A., 2004. The proteomic approach to analysis of human milk fat globule membrane [J]. Clinica Chimica Acta, 347 (1-2): 41-48.

Cavaletto, M., Giuffrida, M. G., Fortunato, D., et al., 2002. A proteomic approach to evaluate the butyrophilin gene family expression in human milk fat globule membrane [J]. Proteomics, 2 (7): 850-856.

Cilla, A., Quintaes, K. D., Barbera, R., et al., 2016. Phospholipids in Human Milk and Infant Formulas: Benefits and Needs for Correct Infant Nutrition [J]. Crit. Rev. Food Sci., 56 (11): 1880-1892.

Contarini, G., Povolo, M., 2013. Phospholipids in Milk Fat: Composition, Biological and Technological Significance, and Analytical Strategies [J]. Int. J. Mol. Sci., 14 (2): 2808-2831.

Dewettinck, K., Rombaut, R., Thienpont, N., et al., 2008. Nutritional and technological aspects of milk fat globule membrane material [J]. Int. Dairy J., 18 (5): 436-457.

Einhorn-Stoll, U., Weiss, M., Kunzek, H., 2002. Influence of the emulsion components and preparation method on the laboratory-scale preparation of o/w emulsions containing different types of dispersed phases and/or emulsifiers [J]. Nahrung-Food, 46 (4): 294-301.

Eshpari, H., Jimenez-Flores R., Tong, P. S., et al., 2015. Partial calcium depletion during membrane filtration affects gelation of reconstituted milk protein concentrates [J]. J. Dairy Sci., 98 (12): 8454-8463.

Fang, Y., Dalgleish, D. G., 1995. Studies on Interactions between Phosphatidylcholine and Casein [J]. Langmuir, 11 (1): 75-79.

Fauquant, C., Briard, V., Leconte, N., et al., 2005. Differently sized native milk fat globules separated by microfiltration: fatty acid composition of the milk fat globule membrane and triglyceride core [J]. Eur. J. Lip. Sci. Tech., 107 (2): 80-86.

Fortunato, D., Giuffrida, M. G., Cavaletto, M., et al., 2003. Structural proteome of human colostral fat globule membrane proteins [J]. Proteomics, 3 (6): 897-905.

Gagnaire, V., Jardin, J., Jan, G., et al., 2009. Invited review: Proteomics of milk and bacteria used in fermented dairy products: From qualitative to quantitative advances [J]. J. Dairy Sci., 92 (3): 811-825.

Goff, H. D., 1997. Colloidal aspects of ice cream-A review [J]. Int. Dairy J., 7 (6-7): 363-373.

Gomes, B. C., Rosa, P. R. F., Etchebehere, C., et al., 2015. Role of homo-and heterofermentative lactic acid bacteria on hydrogen-producing reactors operated with cheese whey wastewater [J]. Int. J. Hydrogen Energ., 40 (28): 8650-8660.

Guinee, T. P., 2009. The role of dairy ingredients in processed cheese products [J]. Dairy-Derived Ingredients: 507-538.

Hamdan, M., Righetti, P. G., 2005. Proteomics today: protein assessment and biomarkers using mass spectrometry, 2D electrophoresis, and microarray technology [M]. New York: John Wiley & Sons.

Hu, M. H., Lee, P. Y., Chen, W. C., et al., 2014. Comparison of three calcium phosphate bone graft substitutes from biomechanical, histological, and crystallographic perspectives using a rat posterolateral lumbar fusion model [J]. Mat. Sci. Eng. C-Mate., 45: 82-88.

Huppertz, T., Gazi, I., 2016. Lactose in dairy ingredients: Effect on processing and storage stability [J]. J. Dairy Sci., 99 (8): 6842-6851.

Jimenez-Flores, R., Brisson, G., 2008. The milk fat globule membrane as an ingredient: why, how, when? [J]. Dairy Sci. Technol., 88 (1): 5-18.

Jiménez-Flores, R., Higuera-Ciapara, I., 2009. Beverages based on milk fat globule memebrane (MFGM) and other novel concepts for dairy-based functional beverages. Functional and speciality beverage technology [M]. P. P. PhD. Boca Raton, New York, Washington DC., CRC Press, Woodhead Publishing Limited. 1: 281-296.

Kim, H. Y., Jimenez-Flores, R., 1995. Heat-Induced Interactions Between the Proteins of Milk Fat Globule Membrane and Skim Milk [J]. J. Dairy Sci., 78 (1): 24-35.

Krog, N., 1991. Thermodynamics of Interfacial Films in Food Emulsions [J]. ACS Symposium Series, 448: 138-145.

Manso, M. A., LÈonil, J., Jan, G., et al., 2005. Application of proteomics to the characterisation of milk and dairy products [J]. Int. Dairy J., 15 (6-9): 845-855.

Markey, O., Hobbs, D. A., Givens, D. I., 2015. Public health implications of milk fats: the current evidence base and future directions [J]. Clin. Lipidol., 10 (1): 5-8.

McClements, D. J., Monahan, F. J., Kinsella, J. E., 1993. Effect of Emulsion Droplets on the Rheology of Whey-Protein Isolate Gels [J]. J. Texture Stud., 24 (4): 411-422.

Miyaji, M., Jin, Z. X., Yamaoka, S., et al., 2005. Umehara H Affiliation: Department of Rheumatology and Clinical Immunology. Role of membrane sphingomyelin and ceramide in platform formation for Fas-mediated apoptosis [J]. J. Exp. Med., 202 (2): 249-259.

Modler, H. W., 1985. Functional-properties of nonfat dairy ingredients-a review-modification of lactose and products containing whey proteins [J]. J. Dairy Sci., 68 (9): 2206-2214.

Morin, P., Britten, M., Jimenez - Flores, R., et al., 2007. Microfiltration of buttermilk and washed cream buttermilk for concentration of milk fat globule membrane components [J]. J. Dairy Sci., 90 (5): 2132-2140.

Morin, P., Pouliot, Y., Britten, M., 2008. Effect of Buttermilk Made from Creams with Different Heat Treatment Histories on Properties of Rennet Gels and Model Cheeses [J]. J. Dairy Sci., 91 (3): 871-882.

Morin, P., Pouliot, Y., Jimenez-Flores, R., 2006. A comparative study of the fractionation of regular buttermilk and whey buttermilk by microfiltration [J]. J. Food Eng., 77 (3): 521-528.

Morrison, W. R., 1969. Polar lipids in bovine milk. I. Long - chain bases in sphingomyelin [J]. Biochemica et Biophysica Acta, 176 (3): 537-546.

Motouri, M., Matsuyama, H., Yamamura, J. - i., et al., 2003. Milk sphingomyelin accelerates enzymatic and morphological maturation of the intestine in artificially reared rats [J]. J. Pediatr. Gastr. Nutr., 36 (2): 241-247.

Murgiano, L., Timperio, A. M., Zolla, L., et al., 2009. Comparison of Milk Fat Globule Membrane (MFGM) Proteins of Chianina and Holstein Cattle Breed Milk Samples Through Proteomics Methods [J]. Nutrients, 1 (2): 302-315.

Noh, S. K., Koo, S. I., 2004. Milk sphingomyelin is more effective than egg sphingomyelin in inhibiting intestinal absorption of cholesterol and fat in rats [J]. J. Nutr., 134 (10): 2611-2616.

Nyberg, L., Duan, R. -D., Nilsson, K., 2000. A mutual inhibitory effect on absorption of sphingomyelin and cholesterol [J]. J. Nutr. Biochem., 11 (5): 244-249.

Nyberg, L., Farooqi, A., Blackberg, L., et al., 1998. Digestion of ceramide by human milk bile salt-stimulated lipase [J]. J. Pediatr. Gastroenterol Nutr., 27 (5): 560-567.

O'Donnell, R., Holland, J. W., Deeth, H. C., et al., 2004. Milk proteomics [J]. Int. Dairy J., 14 (12): 1013-1023.

Parodi, P. W., 2003. Anti-cancer agents in milkfat [J]. Aust. J. Dairy Technol., 58 (2): 114-118.

Parodi, P. W., 2005. Dairy product consumption and the risk of breast cancer [J]. J. Am. Coll. Nutr., 24 (6): 556S-568S.

Prostenik, M., Gospocic, L., 1966. Sphingomyelin from cow's milk [J]. Naturwissenschaften, 53 (16): 407.

Pyne, G. T., 1962. REVIEWS OF PROGRESS OF DAIRY SCIENCE-SOME ASPECTS OF PHYSICAL

CHEMISTRY OF SALTS OF MILK [J]. J. Dairy Res., 29 (1): 101-126.

Reinhardt, T. A., Lippolis, J. D., 2008. Developmental Changes in the Milk Fat Globule Membrane Proteome During the Transition from Colostrum to Milk [J]. J. Dairy Sci., 91 (6): 2307-2318.

Remon, J., Laseca, M., Garcia, L., et al., 2016. Hydrogen production from cheese whey by catalytic steam reforming: Preliminary study using lactose as a model compound [J]. Energy Conversion and Management, 114: 122-141.

Richard, R. C., Oliveira, R. N., Soares, G. D. A., et al., 2014. Direct-write assembly of 3D scaffolds using colloidal calcium phosphates inks [J]. Materia-Rio De Janeiro, 19 (1): 61-67.

Rombaut, R., Dewettinck, K., Van Camp, J., 2007. Phospho-and sphingolipid content of selected dairy products as determined by HPLC coupled to an evaporative light scattering detector (HPLC-ELSD) [J]. J. Food Compos. Anal., 20 (3-4): 308-312.

Rosa, P. R. F., Santos, S. C., Sakamoto, I. K., et al., 2014. Hydrogen production from cheese whey with ethanol-type fermentation: Effect of hydraulic retention time on the microbial community composition [J]. Bioresour. Technol., 161: 10-19.

Saffon, M., Jimenez-Flores, R., Britten, M., et al., 2015. Effect of heating whey proteins in the presence of milk fat globule membrane extract or phospholipids from buttermilk [J]. Int. Dairy J., 48: 60-65.

Sánchez-Juanes, F., Alonso, J. M., Zancada, L., et al., 2009. Distribution and fatty acid content of phospholipids from bovine milk and bovine milk fat globule membranes [J]. Int. Dairy J., 19 (5): 273-278.

Schmelz, E. M., Dillehay, D. L., Webb, S. K., et al., 1996. Sphingomyelin Consumption Suppresses Aberrant Colonic Crypt Foci and Increases the Proportion of Adenomas versus Adenocarcinomas in CF1 Mice Treated with 1, 2-Dimethylhydrazine: Implications for Dietary Sphingolipids and Colon Carcinogenesis [J]. Cancer Res., 56 (21): 4936-4941.

Schumacher, G. E., Antonucci, J. M., O'Donnell, J. N. R., et al., 2007. The use of amorphous calcium phosphate composites as bioactive basing materials-Their effect on the strength of the composite/adhesive/dentin bond [J]. J. Am. Dent. Assoc., 138 (11): 1476-1484.

Seo, Y. H., Yun, Y. M., Lee, H., et al., 2015. Pretreatment of cheese whey for hydrogen production using a simple hydrodynamic cavitation system under alkaline condition [J]. Fuel, 150: 202-207.

Singh, A. M., Dalgleish, D. G., 1998. The emulsifying properties of hydrolyzates of whey proteins [J]. J. Dairy Sci., 81 (4): 918-924.

Singh, H., 2006. The milk fat globule membrane—A biophysical system for food applications [J]. Curr. Opin. Colloid In., 11 (2-3): 154-163.

Snow, D. R., Ward, R. E., Olsen, A., et al., 2011. Membrane-rich milk fat diet provides protection against gastrointestinal leakiness in mice treated with lipopolysaccharide [J]. J. Dairy Sci., 94 (5): 2201-2212.

Spitsberg, V. L., 2005. Invited Review: Bovine Milk Fat Globule Membrane as a Potential Nutraceutical [J]. J. Dairy Sci., 88 (7): 2289-2294.

Thompson, A. K., Singh, H., 2006. Preparation of liposomes from milk fat globule membrane phospholipids using a Microfluidizer [J]. J. Dairy Sci., 89 (2): 410-419.

Vesper, H., Schmelz, E. M., Nikolova-Karakashian, M. N., et al., 1999. Sphingolipids in Food and the Emerging Importance of Sphingolipids to Nutrition [J]. J. Nutr., 129 (7): 1239-1250.

Wang, J. J., Li, D. F., Dangott L. J., et al., 2006. Proteomics and its role in nutrition research [J]. J. Nutr., 136 (7): 1759-1762.

Waninge, R., Nylander, T., Paulsson, M., et al., 2003. Phase equilibria of model milk membrane lipid systems [J]. Chem. Phys. Lipids, 125 (1): 59-68.

Waninge, R., Walstra, P., Bastiaans, J., et al., 2005. Competitive adsorption between beta-casein or beta-lactoglobulin and model milk membrane lipids at oil-water interfaces [J]. J. Agr. Food chem., 53 (3): 716-724.

Westermeier, R., 2006. Sensitive, Quantitative, and Fast Modifications for Coomassie Blue Staining of Polyacrylamide Gels [J]. Proteomics, 6 (S2): 61-64.

Wilson, N. L., Robinson, L. J., Donnet, A., et al., 2008. Glycoproteomics of milk: differences in sugar epitopes on human and bovine milk fat globule membranes [J]. J. Proteome Res., 7 (9): 3687-3696.

Xiong, Y. L., Kinsella, J. E., 1991. Influence of fat globule-membrane composition and fat type on the rheological properties of milk based composite gels [J]. Milchwissenschaft, 46 (4): 207-212.

Yamada, M., Murakami, K., Wallingford, J. C., et al., 2002. Identification of low-abundance proteins of bovine colostral and mature milk using two-dimensional electrophoresis followed by microsequencing and mass spectrometry [J]. Electrophoresis, 23 (7-8): 1153-1160.

Ye, A., Singh, H., Taylor, M. W., et al., 2002. Characterization of protein components of natural and heat-treated milk fat globule membranes [J]. Int. Dairy J., 12 (4): 393-402.

Ye, A., Singh, H., Taylor, M. W., et al., 2004. Interactions of fat globule surface proteins during concentration of whole milk in a pilot-scale multiple-effect evaporator [J]. J. Dairy Res., 71 (4): 471-479.

Zeisel, S. H., Char, D., Sheard, N. F., 1986. Choline, phosphatidylcholine and sphingomyelin in human and bovine milk and infant formulas [J]. J. Nutr., 116 (1): 50-58.

Salmonella Remains a Threat in the Production of Powdered Infant Formula

Patrick Wall

Centre for Food Safety, University College Dublin

Introduction

Powder infant formula (PIF) is the premium destination for the output of many dairy farms and negative publicity associated with illness in infants associated with contaminated product damages the entire dairy production chain. Primary producers of milk, dairy processors and infant formula manufacturers need to be aware of the challenges they are facing and the latest tools available to detect problem bacteria and to assist in risk mitigation.

Infant formula and other powder based early life nutrition (including early stage milks, follow on milk, and growing up milk) require the strictest food safety and quality standards. The production of powder-based nutrition involves many process stages, all which require careful integration and design to ensure the desired product quality and maximum food safety is achieved. However, despite (a) improved design of facilities and processing equipment, (b) the implementation of good manufacturing practices, (c) comprehensive sanitation programs, (d) robust biosecurity with well controlled access points for goods and people, (e) elimination of water, in particular wet cleaning procedures, in high care areas and (f) aggressive monitoring for environmental pathogens, contamination incidents in powdered infant formula (PIF) plants, and those of their suppliers, continue to occur.

Two recent outbreaks of salmonellosis in infants occurred in Europe. The first identified in 2017 originated in France (*Salmonella agona*) affected 39 infants, 37 in France, one in Spain and one in Greece. WGS was able to confirm all isolates as identical. Although the first infants fell ill in August 2017, it was 2nd December 2017 before the authorities were notified of an unusual number of *Salmonella agona* cases in France. As infants don't have a diverse diet, it quickly became apparent, on questioning the mothers, that the link between the cases was a particular company's infant formula brands. The French authorities carried out investigations at the company's factory, which manufactured these brands. All products manufactured in the implicated factory since 15th February 2017 were ordered to be withdrawn. These products have a long shelf life and the concern was that there could be contaminated material in stores and in the distribution system. Products had been distributed to 13 EU Member States and 54 countries outside the EU. Many of the Third Countries have not got the laboratory capability to identify *Salmonella agona*, or to undertake whole genome sequencing, so cases of infection are likely to have gone unreported. The implicated company was the Lactalis group which is one of the world's largest producers of dairy products, with annual sales of € 17bn ($ 21bn; £ 15bn), It has 246 production sites in 47 countries and employs 15,000 people in France alone. The implicated site, which had two driers, was in Craon France and this same site was involved in an outbreak of *Salmonella agona* with the same WGS in 2005. Corrective measures by Lactalis included dismantling of equipment and aggressive deep cleaning but despite their best efforts environmental monitoring of one tower remained positive and the decision was taken to close it completely. It is of interest that this same tower produced the contaminated product involved in the 2005 outbreak suggesting that the strain of salmonella had persisted in the production environment or in the tower itself. (Eurosurveillance, 2018).

The second outbreak affected 31 infants in France and was detected in January 2019 when Health officials in France, Belgium and Luxembourg reported *Salmonella Poona* cases in young children, all linked by WGS. Overall, 32 confirmed cases were reported in the EU: 30 in France, 1 in Belgium and 1 in Luxembourg. An assessment by EFSA and the European Centre for Disease Prevention and Control indicated that the common source of the outbreak was three rice-based infant formula products made by a factory in Spain between August and October 2018 and marketed by the French company, Lactalis (Eurosurveillance, 2019).

These outbreaks demonstrate that no infant formula company, whatever its scale, can afford to become complacent.

Disease Detectives and Forensic Microbiology

The increasing use of whole genome sequencing by the regulatory agencies of many jurisdictions, in both surveillance systems and in the investigation of outbreaks, means that companies producing contaminated product will not escape discovery (Allard et al., 2017). Pulse Field Gel Electrophoresis technology is rapidly being replaced by WGS methods. International surveillance networks sharing whole genome sequence data means that companies with globally distributed product will be identified and it is now more likely that the full extent of outbreaks will be delineated. Heretofore cases of salmonellosis in infants may not have been associated with an outbreak. With the ever-increasing rapid dissemination of adverse publicity via the digital media, brands and reputations that take years to build can be severely damaged overnight.

The epidemiologists and forensic microbiologists working for the authorities have a role to protect public health and prevent ongoing spread of outbreaks and their investigations focus on the identification of the contaminated product that is contributing to illness in infants. When they have the implicated product withdrawn, the source production facility closed and initiated sanctions against the offending company they often consider their work complete. Most of the peer reviewed scientific papers describing outbreak investigations associated with PIF stop their story at the factory gate (Weismann et al., 1977; Collins et al., 1968; Rowe et al., 1987; Brouard et al., 2007; Angula et al., 2008; Eurosurveillance, 2018 and 2019).

Often the contributing factors in the process that lead to the production of contaminated product are either never identified precisely or, if they are, they never emerge into the public domain. The umbrella term "systems failures" is often used to avoid labeling the precise root cause.

It is important to know if positive isolates are similar to previous isolates identified in the facility, which might alert a company that the pathogen has set up home in the plant. By definitively typing all isolates with WGS, it is possible to map recurring isolates to particular zones in the facility to identify where it has found a niche in the facility and where to target interventions (Russo et al., 2013; Eurosurveillance, 2017).

Contributing Factors in the Production Process

Companies are reluctant to publicise the results of in-house investigations which often demonstrate failings, and even negligence, on their part which will further damage their brands and reputation and provide ammunition to those representing ill individuals considering litigation against the companies. Contributing factors often include one, or more, of the following:

(i) Grossly contaminated raw ingredients overpowering the existing food safety management systems.

(ii) Inadequate heat treatment.

(iii) Post pasteurization contamination in the high care area, and during filling and packaging.

(iv) Contamination in the dry goods area e.g. non-dairy ingredients and packaging.

(v) Inability to seal high care zones due to bad design or old age compromising biosecurity e.g. flow of air, drains or movement of personnel and equipment.

(vi) Niches allowing powder, plus or minus moisture, to accumulate.

(vii) Inadequately trained or supervised staff.

(viii) Unsatisfactory cleaning and sanitation programs.

(ix) Lack of sufficiently comprehensive early warning environmental monitoring programs.

Company Standards and Standards of suppliers

Some PIF operations produce their own base powders whilst many are dependent on other companies to do so. Increasingly the ideal is to have total control and have everything on one site however this is not always possible and many companies are only in command of a proportion of the total process from milk to PIF. A company's brands are only as secure as the standards of its weakest supplier whether this is base powder or any other ingredient. PIF plants often boast that they are approaching pharmaceutical grade manufacturing but this claim cannot be substantiated for many powder plants. There has been a global litany of failures in powder plants demonstrating failure to maintain the highest standards. It is dangerous for companies to believe their own propaganda.

In 2016 in the US the FDA initiated a recall of 1,800 tons of milk powders produced at a Valley Milk LLC facility in the USA after the identification of *Salmonella meleagridis* in the factory environment and in product (FDA, 2016). This triggered several secondary recalls by companies who sourced the powdered milk as an ingredient for their own products creating a costly chain reaction in the market place.

In early 1997 an outbreak of *Salmonella anatum* with 22 cases in 4 countries was linked to the consumption of infant formula produced for the Dutch company Royal Numico in a production facility in France which received its base powder from two suppliers, one in France and the other in the Netherlands (Eurosurveillance, 1997).

The absence of salmonella in powders and other ingredients cannot be assured by the acceptance or rejection of batches based on microbial testing alone. A close working relationship between PIF companies and their providers is essential to yield the best results in terms of efficiency, consistent quality and above all assured food safety at every stage of the process.

Contamination of the External Environment

The emissions from spray dryers are directed through a series of cyclones to capture as much as possible of the milk powder present. Most companies undertake regular surveys of the exhaust air from the dryers to ensure that the powder capture is as effective as possible. The effectiveness of powder capture is determined to a significant extent by the nature of the material being dried, whether whole milk, skim milk or whey-based. However, mostdairy spray dryers emit some powder from their stacks; the extent of which very much depends upon the age of the facility and the capacity of the filters, with the more modern plants placing greater emphasis on waste minimisation. Less than 50 mg/m^3 is the standard for dryer emissions specified by the Best Available Techniques for the dairy processing sector but this is often exceeded particularly in plants without efficient fabric filters and plants that are in continuous production with insufficient down time for deep cleaning.

Stack losses affect the efficiency of the plant but they also result in layers of powder on the roofs and external environment. Birds are attracted by the powder and while feeding on the powder can seed it with salmonella from their faeces. Milk powder is an ideal nutrient to support the growth of salmonella so all companies must assume that their external environment is contaminated and maintain robust biosecurity.

Leakages from flanges, pumps, seals, valve glands, etc. can often seed the inside of plants with powder that can find its way into cracks and crevices providing a ideal environment for microbes further compounding the challenges when plant and equipment is old.

Renovations on Site Whilst Production is Ongoing: a Major Risk Factor

Often companies undertake major infrastructural repairs or construct new facilities on the same site when production is ongoing. This presents a major challenge to biosecurity with new people on site who are not aware of the risks, movement of equipment, generation of dust and disruption of the existing microbial flora. Moleculartyping has revealed that particular strains of Salmonella can persist for years in food processing environments and construction activity is often associated with a recurrence of product contamination incidents or recurring outbreaks. WGS now affords the possibility to identify strains in temporally spaced outbreaks as identical.

The WGS of the salmonella strain in the 2008 outbreak, linked to Malt-O Meal cereals was identical to the strain in another cereal-linked outbreak at the same company 10 years before, in 1998. CDC officials concluded that the 2008 outbreak was caused by a construction project at the manufacturing plant where workers removed a wall, which may have reintroduced salmonella into the production line (Russo et al., 2013).

The Environmental sampling results for Valley Foods LLC in 2016 indicated that the *Salmonella meleagridis* strains from 2016 were almost identical to strains found at the company in 2010, 2011 and 2013. These findings of *Salmonella meleagridis* at the company dating back several years demonstrate the existence of a persistent strain at this facility (FDA, 2016).

Sampling to Detect Positives

Companies cannot over depend on negative results from end product testing as low level intermittent contamination may not be detected even by the most stringent in house microbiological sampling. No sampling or testing plan can provide evidence of total absence of salmonella. It now behooves all major infant formula companies to use WGS to better understand the microbial ecology of their production facilities and that of their key suppliers.

Know Your Enemy

WGS now permits the identification of gene sequences that convey attributes to strains of salmonella that will influence the approach to eradication. Not all salmonellas are equally virulent or possess the same capabilities for environmental survival. Therefore it is important to "know your enemy "if you are to wage an effective war against it. WGS can help identify gene sequences that enable the strain express a range of characteristics that can make control problematic with some having the ability to persist in the factory environment for many years posing an ongoing challenge to the biosecurity measures. If you go after these salmonella you have to be sure to kill them as sub lethal stress can induce the expression of genes conveying properties that improve the strain's ability to survive in the environment. Biofilm forming ability, thermo tolerance, ability to survive desiccation, biocide resistance, reduced sensitivity to extreme pHs, increased ability to adhere to surfaces and the ability to go dormant etc. are some of the characteristics of salmonella enabling them persist (Nesss et al., 2003; Corcoran et al., 2014; Finn et al., 2013; Dhakal et al., 2019).

Culture methods may grossly under estimate the level of contamination in a facility as the possibility that strains could persist for long periods in the environment in a viable but non-culturable state exists. This poses a challenge for interpreting post pasteurisation "PCR positive but culture negative" environmental sampling results. In the past many assumed these results were due to dead bacteria, or DNA fragments, but this may not be the case. The term non-culturable is a misnomer as under the proper conditions these cells are able to resuscitate to the metabolically active and culturable state.

Many spray drying facilities and PIF plants are more than a couple of decades old and many have been designed before the full consequences of sub lethal stress have emerged. Gene transfer between pathogens and non pathogenic environmental organisms adds another dimension to the challenge to food safety a reservoir of problematic genes in the facility may present an ongoing threat preventing a facility getting the all clear after being associated with an outbreak or the production of contaminated product.

WGS technology is rapidly transforming epidemiology and food safety and PIF businesses need to use this technology to discriminate between persistent strains and sporadically introduced strains and to identify genes that may contribute to persistence.

However the most promising and far-reaching public health benefit may come from pairing a foodborne pathogen's genomic information with its geographic location and applying the principles of evolutionary biology to determine the relatedness of the pathogens.

Salmonella in Biofilms

Persistence of salmonella in PIF production environments after cleaning and sanitation maybe due to inadequate processes, acquired biocide resistance or concealment within biofilms. Many strains of salmonella readily form biofilms and cells in biofilms are less susceptible to disinfectant inactivation than cells in planktonic form as a result of the protection provided by the polysaccharide matrix. However in addition to the physical protection afforded to the bacteria by the biofilm itselfa substantial proportion of the genes of the salmonella strain may undergo up or down regulation in the transition from the planktonic to the biofilm state which may result in a inherent reduced susceptibility to biocides (Bridier et al., 2011; Condell et al., 2012; Cormican et al., 2014).

Quorum Sensing and Microbial Communities

Salmonella species are no different from other living organisms in that stressors in their environment dictate their behavior. This can be from up regulation or down regulation of their own genes. However a biofilm can contain a community of diverse bacteria who engage in cell-to-cell communication called Quorum Sensing. The possibility of pathogens and environmental organisms cooperating in a mixed microbial community to enhance survival in the environment has been suggested (Abisado et al., 2018). Furthermore when conditions are right we know that bacteria can be quiet permissive with their genetic material so it may be possible for pathogens to receive genes from environmental organisms, which has enabled them survive in hostile environments. Many of these environmental bugs are non culturable but shotgun metagenomics now present the opportunity to study their genetic make up and profile taxonomic composition and functional potential of microbial communities and to recover whole genome sequences related to environmental persistence.

Bioinformatics

Whole genome sequencing on its own is not sufficient and bioinformatic analysis is required to make sense out of the mountain of sequence data. The advent of inexpensive, ultra-high-throughput DNA sequencing instrumentation has transformed microbial whole-genome sequencing (WGS) from an expensive research exercise into a routine tool for surveillance and outbreak investigation. (Gilcrest et al., 2015). In the past a major obstacle preventing regulatory agencies' laboratories, and those used by companies, from maximizing the utility of WGS was the lack of capability to perform bioinformatic-dependent analysis. Large-scale computational capabilities, complex molecular evolutionary analyses, and dedicated bioinformatics staff were needed. The pace of WGS implementation far outpaced the number of bioinformaticians available to analyse the sequences and a bottleneck was appearing. However easy to use open access web based tools are now available making bioformatic capabilities accessible on a simple laptop to an operator with limited experience. (Oakeson et al., 2017).

Protection of Infant Health Must be Paramount

While companies might be concerned about their brands and reputations the consequences of failures could result in infant deaths so there can be no tolerance for shoddy practices. A better working relationship must be developed between companies and the regulatory agencies and consideration should be given to making reporting positive environmental results mandatory. To prevent companies stopping environmental surveillance for fear of getting positives, a minimum amount of environmental monitoring related to the throughput of the plant should be a legal requirement. Many companies use in house, or private commercial laboratories, and isolates may not be shared with the government reference laboratories. Sick infants may be the first indication that a contaminated product is on the market and if company sequences are routinely shared withthe public health authorities outbreak resolution will be rapid and further infant illness and possibly deaths will be prevented.

Both internal auditors and regulatory inspectors need to be knowledgeable of the threats to the PIF process and the tools available to perform adequate risk assessments and be competent to ask the right questions and interpret the answers.

Conclusion

Bacteria are no different from other living organisms in that their environment dictates their behavior and we now know that environmental stressors can trigger gene expression. Couple this with gene acquisition means that salmonella can have genes that contribute to persistence. WGS and bioinformatics are new tools in the armory of those charged with eradicating salmonella from production environments and should be fully utilised.

As open access public databases for sequenced pathogens continue to grow, the ability to classify nationally and internationally dispersed salmonella strains will improve. The digitization of this genetic information is making the goal of a global One Health WGS database achievable where isolates from humans, animals and the environment can be compared at the molecular level. The sharing of results in real-time means that more outbreaks will be identified, outbreaks will be recognised earlier and smaller numbers of geographically dispersed cases will be linked together.

References

Abisado R.G., Benomar S., Klaus J.R., et al., 2018. Bacterial quorum sensing and microbial community interactions [EB/OL]. mBio., 9: e02331-e02317.

Allard M. W., Bell R., Ferriera C. M., et al., 2018. Genomics of foodborne pathogens for microbial food safety [J]. Curr. Opin. Biotech., 49: 224-229.

Angulo F. J., Cahill S. M., Wachsmuth I. W., et al., 2008. Powdered Infant Formula as a source of Salmonella Infection in infants [EB/OL]. Clin. Infect. Dis., 46 (2): 268-273.

Bridier A., Briandet R., Thomas V., et al., 2011. Resistance of bacterial biofilms to disinfectants: a review [J]. Biofouling, 27 (9): 1017-1032.

Brouard C., Espié E., Weill F.X., et al., 2007. Two consecutive large outbreaks of Salmonella enterica serotype Agona infections in infants linked to the consumption of powdered infant formula [J]. Pediatr. Infect. Dis. J., 26: 148-152.

Collins R. N., Treger M. D., Goldsby J. B., et al., 1968. Interstate outbreak of Salmonella newbrunswick infection traced to powdered milk [J]. JAMA, 203: 838-44.

Condell O., Iversen C., Cooney S., et al., 2012. Efficacy of Biocides Used in the Modern Food Industry To Control Salmonella enterica, and Links between Biocide Tolerance and Resistance to Clinically Relevant Antimicrobial Compounds [J]. App. Environ. Microbiol., 78 (9): 3087-97.

Corcoran M., Morris D., De Lappe N., et al., 2014. Commonly used disinfectants fail to eradicate *Salmonella enterica* biofilms from food contact surface materials [J]. Appl. Environ. Microbiol., 80 (4): 1507-1514.

Dhakal J., Sharma C. S., Nannapaneni R., et al., 2019. Effect of Chlorine-Induced Sublethal Oxidative Stress on the Biofilm-Forming Ability of Salmonella at Different Temperatures, Nutrient Conditions, and Substrates [J]. J. Food Prot., 82: 1: 78-92.

Eurosurveillance, 1997. [EB/OL] https: //www. eurosurveillance. org/content/10. 2807/esm. 02. 03. 00190-en.

Eurosurveillance, J., et al., 2019. Outbreak of Salmonella enterica serotype Poona in infants linked to persistent Salmonella contamination in an infant formula manufacturing facility [R], France, August 2018 to February 2019. Eurosurveillance; 24; 13; 28th March.

FDA, 2016. Food regulators seize adulterated milk products for food safety violations [EB/OL] https: // www. fda. gov/NewsEvents/Newsroom/PressAnnouncements/ucm531188.htm.

Finn S., Condell O., McClure P., et al., 2013. Mechanisms of survival, responses, and sources of Salmonella in low-moisture environments [J]. Front. Microbiol., 4: 331.

Gilchrist C.A., Turner S.D., Riley M.F., et al., 2015. Whole-genome sequencing in outbreak analysis [J]. Clin. Microbiol. Rev., 28 (3): 541-562.

Jourdan-da Silva N, Fabre L, Robinson E, et al., 2018. Ongoing nationwide outbreak of Salmonella Agona associated with internationally distributed infant milk products [J]. Euro. Surveill., 23 (2): 17-00852.

Nesse L., Nordby K., Heir E., et al., 2003. Molecular analyses of Salmonella enterica isolates from fish feed factories and fish feed ingredients [J]. Appl. Environ. Microbiol., 69: 1075-1081.

Oakeson K. F., Wagnier J. M., Mendenhall M., et al., 2017. Bioinformatic Analyses of Whole-Genome Sequence Data in a Public Health Laboratory [J]. Emerg Infect Dis, 23 (9): 1441-1445.

Quince C., Walker A.W., Simpson J.T., et al., 2017. Shotgun metagenomics, from sampling to analysis [J]. Nature Biotechnology, 35: 833-844.

Rowe B., Begg N.T., Hutchinson D.N., et al., 1987. Salmonella Ealing infections associated with consumption of infant dried milk [J]. Lancet, 2: 900-903.

Russo E. T., Biggerstaff G., Hoekstra M. R., et al., 2013. A recurrent, multistate outbreak of Salmonella serotype Agona infections associated with dry, unsweetened cereal consumption, United States, 2008 [J]. J. Food Prot., 76: 227-230.

Weissman J. B., Deen R. M. A. D., Williams M., et al., 1977. An island-wide epidemic of salmonellosis in Trinidad traced to contaminated powdered milk [J]. West Indian Med. J., 26: 135-143.

Feeding Low Crude Protein Diets to Improve Efficiency of Nitrogen Use

M. D. Hanigan, S. I. Arriola Apelo, M. Aguilar

Dept. of Dairy Science, Virginia Tech

Abstract

Sustainable production of adequate quantities of food to support the human population is a world-wide goal. Under current feeding conditions in the U.S., dairy cattle convert dietary nitrogen to milk nitrogen with 25% efficiency. The remaining 75% is excreted which contributes to air and water quality problems and reduces economic performance of the industry. Efficiency could be improved to 29% if protein was fed to just meet current NRC requirements. Additional improvements are achievable given improvements in our knowledge of amino acid requirements. The current metabolizable requirement model overestimate true requirements due to lack of knowledge of the absolute amino acid requirements and due to the assumption that the partial efficiency of use of metabolizable protein for lactation is a constant 65%. Numerous data indicate the efficiency of lactational use is variable and maximally 45%. Current models also assume that responses to individual nutrients are not additive, and thus production is limited to that nutrient most deficient with respect to requirements. Several pieces of evidence also indicate this assumption is incorrect. Protein synthesis is highly regulated and the regulatory pathways independently and additively respond to the supply of energy and individual amino acids and circulating concentrations of hormones. Thus protein synthesis can respond to nutrients other than the one identified as most limiting under current theory. Addressing these deficiencies should allow formulation of diets that will achieve 35% nitrogen efficiency or greater. However, we can utilize milk urea nitrogen to establish a lower protein feeding level and monitor feeding programs today to achieve improved nitrogen efficiency. Because of genetic variation in milk urea nitrogen concentrations, one should establish a herd specific milk urea nitrogen target by systematically reducing dietary protein until a loss in production is observed.

Key words: lactation, nutrient requirements, metabolizable protein, amino acids, milk urea nitrogen

The dairy industry is coming under increased pressure from federal and local governments and from public opinion to reduce its environmental impact. Among the pollutants, nitrogen (N) excretion is a major concern. Ammonia N released from manure during storage and land application combines with sulfur dioxide and nitrogen oxides in the air to form very small particles (less than 2.5 mm). These particles cause haze and contribute to lung problems and asthma in humans. If manure is over-applied or applied to frozen or sloped land, N can migrate into surface waters and contribute to eutrophication, which can ultimately cause degradation of aquatic ecosystems and coastal hypoxia. Within the soil, manure N is converted to nitrate, nitrite, and nitrous oxide. The former 2 can migrate into aquifers causing blood oxygen exchange problems in infants (methemoglobinemia). Nitrous oxide can escape to the atmosphere contributing to decreased stratospheric ozone concentrations and greenhouse warming. In the latter case, it has greater than 300 times the greenhouse gas effect of carbon dioxide. Nitrogen oxide, nitrogen dioxide, and sulfur dioxide (derived from the sulfur containing amino acids in undigested protein) that escapes to the air creates acid when dissolved in raindrops causing acid rain, which increases the acidity of soil and surface water.

Protein, which is the source of the waste N and environmental pollution, is an expensive dietary nutrient (Table 1) representing approximately 42% of the cost of a lactating cow ration (St - Pierre, 2012). The reduction of dietary protein levels could potentially result in decreased demand for high protein ingredients, reduced price of those ingredients, and the diversion of acreage to higher yielding crops such as corn instead of growing oilseeds. This would result in increased corn supply, which would cause a reduction in the cost of low protein ingredients as well.

Table 1 Nutrient values based on central Ohio ingredient prices. From ST-Pierre and Knapp[a]

Nutrient	Sep., 2008	Aug., 2009	Oct., 2010	Sep., 2011	Sep., 2012	Nov., 2012
NE_L, $/mcal	0.145	0.103	0.121	0.166	0.194	0.115
MP, $/lb	0.292	0.466	0.277	0.280	0.498	0.636
neNDF, $/lb	-0.195	-0.257	-0.072	-0.082	-0.121	-0.041
eNDF, $/lb	-0.074	-0.045	-0.001	0.081	-0.011	0.059

[a] www. dairy. osu. edu/bdnews; MP = metabolizable protein, neNDF = non - effective NDF; eNDF = effective NDF.

In a survey carried out on 103 large scale dairies across the US [(613±46) cows; (34.5±0.3) kg of milk per cow per day], nutritionists reported feeding diets with 17.8%±0.1 % crude protein (CP) (Caraviello et al., 2006). A meta-analysis of 846 experimental diets found a similar mean CP content and conversion efficiencies for dietary and metabolizable N (based on NRC, 2001) to milk protein of 24.6% and 42.6%, respectively. Assuming the same dietary conditions (22.1 kg/d DMI and 17.8% CP) over a 10 month lactation, the national herd of 9 million dairy cattle (Livestock, Dairy, and Poultry Outlook: August 2012, LDPM-218, Dairy Economic Research Service, USDA) would excrete 1.3 million metric tons (mmt) of N per year. Efficiency could be increased to approximately 29% if animals were fed to NRC (2001). requirements which are approximately 16% CP. If a dietary protein conversion efficiency of 35% could be achieved with no change in milk protein output, excreted N would be reduced 39% to 0.51 mmt.

Dietary proteinis used to support microbial growth in the rumen. The combination of microbial protein flow from the rumen and ruminally undegraded dietary protein (RUP), which represents the majority of the metabolizable protein (MP) supply to the animal, is used for maintenance and productive functions such as milk protein synthesis. Several models, such as the NRC Nutrient Requirement models (NRC, 1989, 2001) estimate ruminal and animal N requirements and dietary supply of ruminally degradable protein (RDP) and RUP allowing users to match supply to animal requirements. These systems are commonly used in ration balancing software. Because the NRC model is widely used, it is a primary determinant of protein use in dairy diets, particularly in the US.

Provision of less RDP than required by the microbes or the animal will result in reduced milk protein yield although it willgenerally increase N efficiency. Increased N efficiency seems to be a positive effect. However, increased N efficiency under these conditions will not necessarily result in efficiency gains at the national level. If one considers that a given amount of milk is required to meet consumer demand, a loss in production per cow will require that additional cows be milked. The maintenance costs of those extra animals will quickly negate the apparent gains in efficiency per animal associated with feeding N deficient diets.

Feeding protein in excess of requirements results in the use of the surplus protein for energy needs, increased N excretion, and decreased animal efficiency (Kalscheur et al., 2006). Thus, it is important that animals be fed precisely at their requirements if maximal efficiency is to be achieved. The benefit of improved N efficiency to the dairy producer are reduced production costs associated with purchasing less dietary protein. The benefit to society is the reduced environmental impact of generating food, e.g. milk.

Metabolizeable Protein Requirements

Although we commonly state animal N requirements in terms of MP, the true requirements are for the specific amino acids (AA) resident in that protein. Because there is diversity of AA composition in the absorbed protein, stating the requirements in MP units inherently forces a certain level of over-prediction of requirements to compensate for variation in AA composition of that protein. This is perhaps most apparent when feeding diets constructed largely from corn products which are inherently low in lysine. Such a diet could be created to meet MP requirements, but animals may still respond to the addition of a protected lysine source or more protein that also provides lysine to the ration. When these types of data are mixed with all other experiments in the literature and subjected to statistical analyses to derive MP requirements, the loss in production associated with a specific amino acid deficiency forces the statistical algorithm to solve to a higher MP requirement than would be necessary if the diet contained a perfect mix of amino acids. For example, pigs can achieve efficiencies of absorbed protein deposition in muscle protein of 85% when fed a diet perfectly matched to their AA requirements (Baker, 1996) as compared to 43% efficiency of conversion of MP to milk protein in lactating cows. So, it is a given that MP requirements are greater than needed to compensate for variable AA supply. Thus, animals could successfully be fed a lower MP diet if the AA composition of that diet was better matched to AA requirements as demonstrated by Haque et al. (2012) using diets with less than 13% CP. As the cost of RUP is 3-fold greater than the cost of RDP (Knapp, 2009), being able to reduce dietary RUP is of great economic interest.

Aside from the question of balancing for AA to achieve greater efficiency, there are additional problems with the NRC (2001) MP requirement equations. Obviously, one would expect the model to predict requirements at all levels of production with the same precision. For example, if the precision of the system at 60 lbs of milk/d is plus or minus 15%, then one should expect similar precision at 80 and 90 lbs of milk. Unfortunately this is not the case. As demonstrated in Figure 1, the model over-predicts the amount of MP allowable milk at high levels of production and under-predicts at lower levels. Thus, when using the model, one may need to balance for slightly greater amounts of MP in the diet if working with high producing cows and the reverse when working with lower producing cows assuming requirements for maintenance and gestation are correct. As the MP supply predictions have been well validated, the problem appears to reside in the requirement equations.

The problem with predicting MP requirements is at least partially driven by the model assumption that the conversion of MP to milk protein, after subtraction of maintenance use, is a constant 65%. In a summary of literature data, Lapierre et al. (2007) found that the highest efficiency was 43% and it declined from there as milk protein output (and MP supply) increased. Hanigan et al. (1998) summarized publications reporting responses to post-ruminally infused casein and found a similar maximal efficiency of conversion of about 45% with an average conversion efficiency of 22%. Whitelaw et al. (1986) abomasaly infused casein at 3 different levels and observed responses at each level with efficiencies of conversion ranging from 40 for the first increment to 15% for the last increment. The reduction in efficiency at higher levels of production and thus intake protein would seem to explain the over-predictions of allowable MP at those increased levels of milk yield. This problem is well recognized by our group, as well as others, and we are working to fix this problem. Such a fix will undoubtedly be in the next release of the NRC which is likely to occur in the next 4 to 5

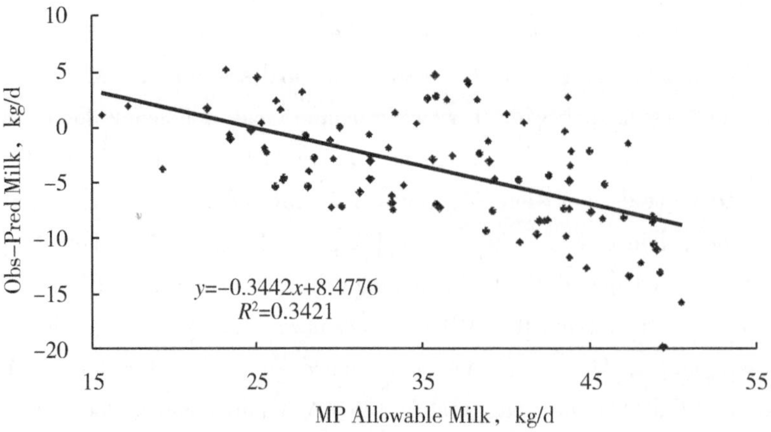

Figure 1 Residual errors (Obs-Pred) associated with predictions of metabolizable protein (MP) allowable milk yields by the NRC (2001) model. adapted from NRC (2001)

years. However, in the mean time, one should be aware of the problem.

When using the NRC model to balance a given ration at differing levels of dietary protein, one can assess the potential for saving money by removing all constraints from CP and relying on the model to solve for rations that meet RDP and MP requirements. Such an exercise is demonstrated inFigure 2 with US ingredients at prevailing prices in March of 2013. From that work, it is apparent that diets can be formulated to meet NRC requirements down to at least 15.5% CP while reducing ration costs. As protein costs increase relative to dietary energy costs, such low protein rations could become even more cost beneficial. Thus, it is important to allow the models to work without placing arbitrary limits on dietary protein content. Least cost rations, containing 15.5% CP that still meet NRC requirements, can only be achieved if arbitrary minimums are not placed on dietary CP content and a wide variety of ingredients are available.

If one assumes MP requirements are over-predicted by 2%-3% units due to lack of knowledge of true AA requirements, then the above analyses (Figure 2) would likely shift downward by a 2%-3% units. Instead of a minimum effective dietary CP content of 15.5%, one should be able to work down to a 13.5% dietary protein or less for high producing cows. Such a level of intake would achieve an efficiency of 32%-33% which comes closer to our efficiency goal. Achieving 35% efficiency (11.5% CP diets) will require very precise knowledge of AA supply and requirements so that diets can be formulated for AA as well as MP.

Amino Acid Requirements

The challenge of predicting AA supply and requirements for ruminants is much greater than formonogastric species. Flow of AA from the rumen is a function of the AA content of undigested

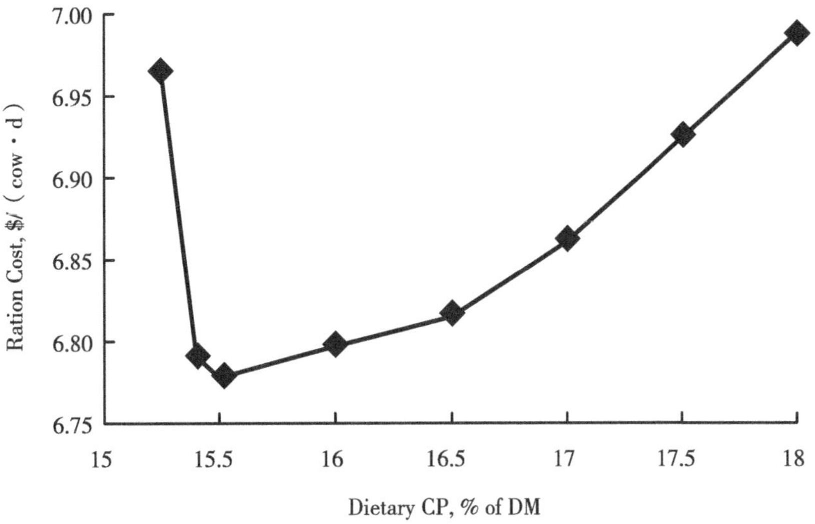

Figure 2 The cost of diets formulated to varying protein levels using the NRC (2001) model and a least cost algorithm. adapted from stewart et al. (2012) and updated using US ingredient prices from march of 2013

feed protein, microbial protein, and sloughed digestive tract cells and secretions (NRC, 2001). The difficulty of predicting each of these entities has greatly hampered our ability to derive AA requirements based on performance data as is done with swine and poultry. Most of the progress that has been made in ruminants has occurred through the use of catheterized and cannulated animals, allowing the provision of AA post-ruminally (for example Haque et al., 2012). However, this is very intensive and expensive work. To date we have amassed the most information on methionine and lysine with histidine results appearing more recently. We are far from the level of understanding that the swine and poultry people have of the remaining essential AA requirements and likely not to achieve that level of understanding any time in the near future.

As AA requirements are expressed as a percentage of MP supply, the problem with variable efficiency of MP use and over-predicting the marginal responses of milk protein to changing MP supply is partially propagated in existing AA requirement equations and likely contributing to the lack of accuracy and precision in those equations (NRC, 2001). Work at the tissue level, using multi-catheterized animals, has clearly shown that the liver and gut tissues remove a constant fraction of AA from blood presented in each pass by that tissue. Because mammary tissue does not generally remove more than half of the AA presented to it, there is significant recycling to the gut and liver resulting in additional removal. This is magnified as AA supply increases relative to energy supply, as the mammary tissue has the ability to change its removal of AA to meet its needs (Bequette et al., 2000). So if mammary tissue is presented with a good energy supply, it will be able to produce milk near its maximum potential and will increase its AA extraction efficiency to achieve this. The same will happen if energy is held constant and AA supply is reduced. Conversely if the mammary tissue is presented with inadequate energy, it will reduce its use of AA and reduce extraction from blood. In the former case, AA extraction efficiency is increased, less AA are recycled to the liver and gut, and less are catabolized. In the latter case, mammary AA extraction efficiency decreases, more AA are recycled, and catabolism increases. So, assuming a constant efficiency of post-absorptive AA use for milk protein synthesis is clearly wrong.

The above interactions between energy and AA supplies to the mammary tissue are mediated by intracellular signaling that integrates information regarding the intracellular supply of several key AA, the supply of energy in the cell (Appuhamy et al., 2009), and hormonal signals indicating overall animal status, i.e. insulin (Appuhamy et al., 2011) and probably IGF-1 (Figure 3). These signaling pathways ultimately regulate protein synthesis thus tying rates of protein synthesis to substrate supply and energy state in the animal.

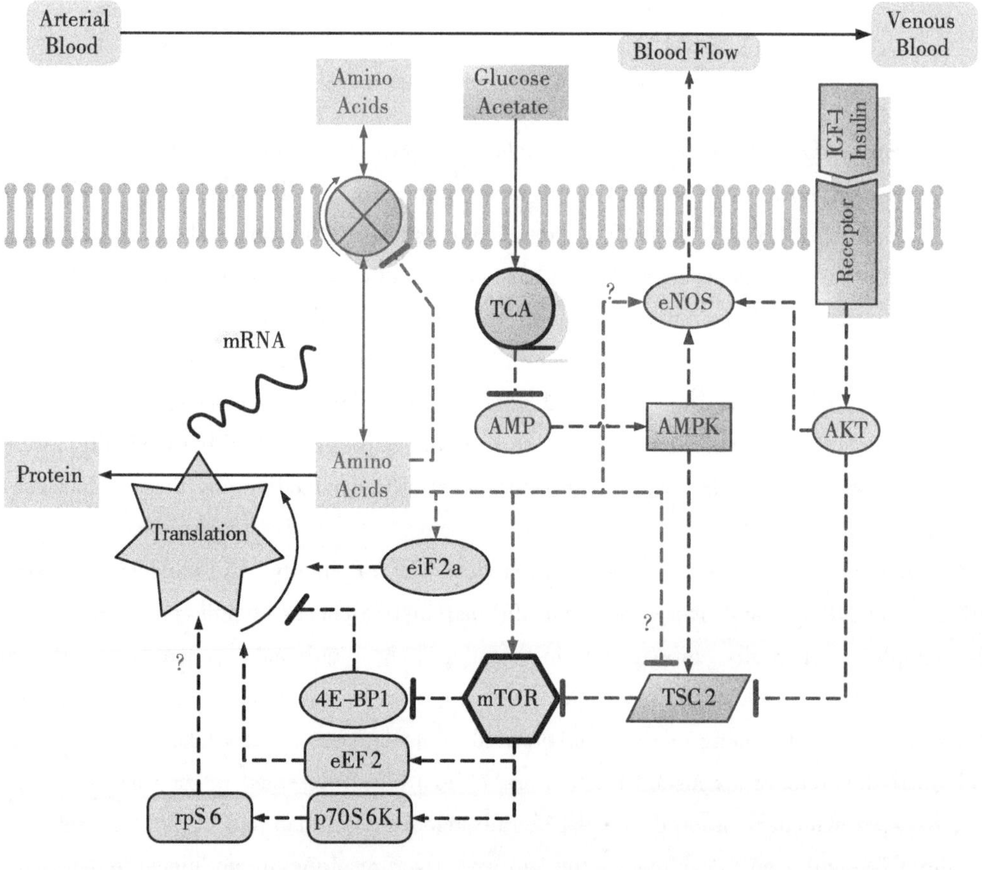

Figure 3 A partial schematic of the regulation of protein synthesis by mammalian target of rapamycin (mTOR) and associated pathways. TCA=tricarboxylic acid cycle

The mTOR pathway is the most studied of the regulatory pathways. mTOR is a serine-threonine kinase that exists in two different protein complexes: mTORC1 and mTORC2. mTORC1 phosphorylates eIF4E-binding protein 1 (4E-BP1) and ribosomal protein S6 (rpS6) kinase 1 (S6K1). When unphosphorylated, 4E-BP1 antagonizes eIF4G binding to eIF4E and inhibits translation (Mahoney et al., 2009). S6K1 phosphorylates rpS6 and eIF4B, which promote translation initiation. Another target of S6K1 is eukaryotic elongation factor (eEF) 2 kinase. This factor is inhibited by phosphorylation, resulting in activation of eEF2 and promotion of translation elongation (Dunlop and Tee, 2009). In mammary tissue of lactating cows, phosphorylation of mTOR, 4EPB1, and eEF2 was highly correlated with fractional synthesis rates of casein (Appuhamy et al., 2011) underscoring the need to better understand this mechanism with respect to nutritional effects on protein synthesis.

Appuhamy et al. (2012b) observed effects of Ile as well as Leu on mTOR phosphorylation in mammary tissue, and Moshel et al. (2006) reported that mTOR signaling in mammary epithelial cells was more sensitive to all AA than to Leu alone. Recently, it was reported that supplementation of a Val deficient diet resulted in increased milk protein synthesis in dairy cows (Haque et al., 2013). However, in the latter case, it is unknown whether this is a substrate or a cell signaling effect.

Insulin and growth factors, including IGF-1, activate the Akt pathway which subsequently activates mTORC1. In MAC-T cells, insulin stimulated phosphorylation of insulin receptor substrate 1 (IRS1), AKT, and downstream proteins (Appuhamy et al., 2011). Appuhamy et al. (2009; 2010) also observed responses of mammary cell signaling to provision of acetate and glucose via the AMP-activated protein kinase (AMPK). Activated AMPK phosphorylates Raptor, promoting the action of mTORC1 (Gwinn et al., 2008).

Because signals arising from AA supply, cellular energy status, and overall body hormonal status are integrated by mTOR which regulates the rate of milk protein synthesis, and because the tissue can adapt its AA extraction capacity to meet intracellular AA demand, the overall rate of protein synthesis and postabsorptive AA efficiency of use are functions of a combination of these regulators rather than dictated by the most limiting one.

First Limiting Nutrient Paradigm

The first limiting nutrient and AA concept is based ona hypothesis which has become known as the Law of the Minimum. Sprengel (1828) formulated this concept based on plant growth responses to soil minerals. However, the original thesis stated that a nutrient can limit plant growth, and when limiting, growth will be proportional to supply, which is clearly supported by volumes of data over the past 175 years. Von Liebig (see Paris, 1992 for a translation) subsequently restated the hypothesis in stronger terms indicating that if a nutrient was limiting for growth, responses to other nutrients could not occur (von Liebig, 1862). Mitchell and Block (1946) used von Liebig's extension of Sprengel's thesis to develop the concept of the order of limiting AA which is commonly described using the analogy of a water barrel with broken staves. Based on this formulation, if any nutrient is limiting milk production, then only the addition of that nutrient to the diet will result in a positive milk yield response, e.g. the single limiting nutrient paradigm.

In order to determine which nutrient is most limiting, one must be able to calculate the allowable milk yield from that nutrient. That calculation is quite simple if one assumesa constant transfer efficiency, as is the case in the NRC model. However, as discussed above, transfer efficiency of AA is not fixed. Because AA removal from blood is regulated in concert with needs for milk protein synthesis (Bequette et al., 2000), the efficiency of AA transfer from the gut to milk protein is variable thus complicating application of the calculations and perhaps bending the underlying somewhat. However, additive integration of signals arising from several AA, energy supply in the mammary cells, and hormonal concentrations at the cell surface to set rates of milk protein synthesis clearly violates the assumption that only one nutrient can limit production. If provision of more of one nutrient or hormone can offset the loss or deficiency of another, there are almost an infinite number of combinations of AA, energy substrates, and hormonal concentrations that will result in the very same amount of milk. This concept is demonstrated *in vivo* by the work of Rius et al. (2010a) in Figure 3. More of any one AA, while all others are held constant, will push milk protein synthesis higher regardless of which is perceived to be "first limiting". Therefore, current protein and AA requirement models for lactation inappropriately represent the underlying biology, which leads to inflated prediction errors.

The take home message from this discussion is that rations can be balanced at levels well below 15% CP, probably even below 13%, if we are able to reliably match AA supply with true animal needs. But current models of AA requirements used in field application programs appear to be incompatible with making such predictions. We are in the process of devising a new prediction scheme that will be a better representation of the biology, and thus should provide much greater accuracy allowing us to achieve N efficiencies of 35% in lactating cattle.

Milk Urea Nitrogen as a Tool to Monitor Feeding Programs

Given the errors in our prediction systems, is there any progress that can be made today in reducing protein feeding and improving efficiency? Fortunately, the answer is yes. We can use milk urea N as a tool to set minimal dietary protein levels for a herd and monitor feeding programs to ensure the desired efficiency is being achieved.

Synthesized urea is released into blood and equilibrates with body fluids including milk (Broderick and Clayton, 1997) resulting in high correlations among blood urea N, milk urea N (MUN), dietary N, and N balance in the cow (Preston et al., 1965). If protein in the diet is deficient relative to cow requirements, AA catabolism will be minimized resulting in low

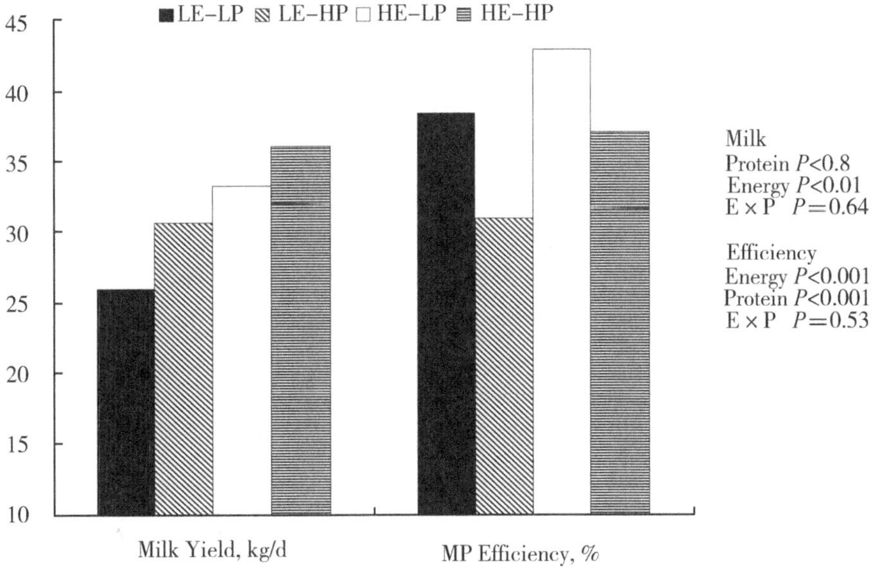

Figure 4 Milk yield and metabolizeable protein efficiency of conversion to milk protein in response to varying energy and ruminally undegraded protein supply (from rius et al., 2010). HEHP = 1. 54 MCal/kg; 11. 8% MP; HELP = 1. 54 MCal/kg, 9.5% MP; LEHP=1.45 MCal/kg, 11.8% MP; LELP=1.45 MCal/kg, 9.5% MP

urea synthesis and concentrations of urea in blood and milk. Conversely if dietary protein is in excess, AA catabolism will increase resulting in greater urea synthesis and concentrations of urea in blood and milk. Thus, protein feeding can be adjusted based on MUN concentrations to achieve maximal efficiency without compromising milk production. Because kidney urea clearance is concentration dependent, there is also a high correlation between MUN and urinary N excretion. Milk urea N is also a good indicator of ammonia emissions from dairy manure (Burgos et al., 2007). These relationships and routine measurement of MUN by milk processors and DHIA testing laboratories provides a useful tool for monitoring feeding programs and feed management practices to achieve maximum N efficiency and minimum environmental loading (NRCS, 2011).

Although MUN concentration is clearly related to protein sufficiency, there are several factors that can cause deviations from expected values. These include time of sampling, season of the year, body weight, breed, and nutritional factors. There are also significant cow effects (Wattiaux et al., 2005) that are at least partially explained by genetic variance. Given the genetic effects on MUN, it is possible that sire selection decisions within a herd may result in herd concentrations of MUN differing from the expected values based on feed management.

When the model of Kauffman and St-Pierre (2001). was used to predict MUN concentrations for individual cows in trials performed by Cyriac et al. (2008) and Rius et al.

(2010), the variance in residual MUN associated with cow was (4.1±1.1) mg/dl ($P<0.001$), indicating that individual cows can deviate considerably from the expected MUN value given a defined diet (Aguilar et al., 2012). Aguilar et al. (2012) also observed highly significant cow effects and a strong trend for a herd effect ($P<0.08$) when analyzing data from 6 herds in the state of Virginia after corrections for differences in dietary nutrients and level of production. Least squares means for MUN by herd ranged from a low of 13.6 mg/dl to a high of 17.3 mg/dl. Given that a percentage unit change in dietary CP, e.g.17% to 16%, results in a 1.1 mg/dl change in MUN, the herd with the highest MUN would have had to reduce dietary CP to 12.8% to achieve the commonly accepted MUN target of 12 mg/dl, if all other factors are held constant. Thus, it is important to recognize that not all herds can be expected to achieve the same target MUN, and herd specific calibrated targets are required if maximal efficiency is to be achieved and maintained.

Herd calibration can be achieved through an assessment of the herds feeding program, taking into account all possible factors that may affect observed variation in MUN. If the herd is well managed, fed a balanced diet that does not exceed NRC (2001) requirements for protein, and the diet has adequate energy, the prevailing MUN could serve as a calibrated target value for that herd. If the herd is overfeeding protein relative to energy supply and milk production, the ration would have to be rebalanced and fed for a period of 2 or 3 weeks before reassessing MUN. The MUN value achieved after this period of feeding to requirements reflects the calibrated NRC reference target for the herd. At this MUN level, a diet balanced properly for RDP and RUP should not precipitate a protein deficiency.

Given a reference MUN target for the herd, one can then determine if the herd will tolerate lower RDP and RUP feeding levels. Because RUP costs considerably more than RDP, it makes sense to start with RUP calibration first, although the reverse will also work. After feeding the diet balanced to NRC requirements, reduce the RUP content of the diet by 0.25% or 0.5% units while holding energy and RDP content constant. Feed the diet for a period of 2 weeks and determine if any loss in milk production or DM intake has occurred while also recording MUN content. If there is no loss in production or intake, try removing another 0.25% units from RUP and again assess production, intake, and MUN after another 2 weeks. Any loss in production will be much less than that predicted by the NRC model because of the slope bias discussed above. Once a loss is experienced, add back the last reduction in RUP and store this value as your target RUP content and MUN target level for that herd.

Having established a herd specific RUP feeding level, you can try reducing RDP content if economically favorable. Remove 0.25% or 0.5% units of RDP from the diet while holding energy content constant and RUP at the newly established level. Again feed the diet for 2 weeks and monitor DM intake, production, and MUN. Note that DM intake will generally change before milk production. If no loss in production or intake, remove another 0.25% units of RDP and repeat the process. Once a loss of DM intake or production occurs, add back the last increment of RDP and store this value as your target RDP content for that herd. Also store the new MUN value as your target MUN for the herd.

This herd specific MUN target can be used to monitor your feeding program. If MUN drops below the target value, it is highly likely that a loss in production will follow soon and corrective action should be taken immediately. Note that a drop in MUN does not provide any information regarding whether the problem is with RDP or RUP. It simply tells you the animal is short on N relative to its current level of production. You will have to determine whether it is a problem with RDP, RUP, other dietary factors such as energy and fiber, or animal health. It also important to recognize that the amount of salt in the ration will affect MUN (Spek et al., 2012), so make sure salt inclusion remains constant and similar to the level used when determining the target MUN values.

References

Aguilar, M., Hanigan, M. D., Tucker, H. A., et al., 2012. Cow and herd variation in milk urea nitrogen concentrations in lactating dairy cattle [J]. J. Dairy Sci., 95: 7261-7268.

Appuhamy, J. A., Knoebel, N. A., Nayananjalie, W. A., et al., 2012a. Isoleucine and leucine independently regulate mtor signaling and protein synthesis in mac-t cells and bovine mammary tissue slices [J]. J. Nutr., 142: 484-491.

Appuhamy, J. A. D. R. N., Bell, A. L., Nayananjalie, W. A. D., et al., 2011. Essential amino acids regulate both initiation and elongation of mrna translation independent of insulin in mac-t cells and bovine mammary tissue slices [J]. J. Nutr., 141: 1209-1215.

Appuhamy, J. A. D. R. N., Bray, C. T., Escobar, J., et al., 2009. Effects of acetate and essential amino acids on protein synthesis signaling in bovine mammary epithelial cells *in vitro* [J]. J. Dairy Sci., 92: 44.

Appuhamy, J. A. D. R. N., Knoebel, N. A., Nayananjalie, W. A. D., et al., 2012b. Isoleucine and leucine independently regulate mtor signaling and protein synthesis in mac-t cells and bovine mammary tissue slices [J]. J. Nutr., 142: 484-491.

Baker, D. H., 1996. Advances in amino acid nutrition and metabolism of swine and poultry [M]. In: E. T. Kornegay (ed.). Nutrient management of food animals to enhance and protect the environment. p 11-22. CRC Lewis Publishers, Boca Raton, FL.

Bellacosa, A., Chan, T. O., Ahmed, N. N., et al., 1998. Akt activation by growth factors is a multiple-step process: The role of the ph domain [J]. Oncogene, 17: 313-325.

Bequette, B. J., Hanigan M. D., Calder A. G., et al., 2000. Amino acid exchange by the mammary gland of lactating goats when histidine limits milk production [J]. J. Dairy Sci., 83: 765-775.

Broderick, G. A., Clayton, M. K., 1997. A statistical evaluation of animal and nutritional factors influencing concentrations of milk urea nitrogen [J]. J. Dairy Sci., 80: 2964-2971.

Burgos, S. A., Fadel, J. G., DePeters, E. J., 2007. Prediction of ammonia emission from dairy cattle manure based on milk urea nitrogen: Relation of milk urea nitrogen to urine urea nitrogen excretion [J]. J. Dairy Sci., 90: 5499-5508.

Caraviello, D. Z., Weigel, K. A., Fricke, P. M., et al., 2006. Survey of management practices on reproductive performance of dairy cattle on large us commercial farms [J]. J. Dairy Sci., 89: 4723-4735.

Clark, R. M., Chandler, P. T., Park, C. S., 1978. Limiting amino acids for milk protein synthesis by bovine mammary cells in culture [J]. J. Dairy Sci., 61: 408-413.

Cyriac, J., Rius, A. G., McGilliard, M. L., et al., 2008. Lactation performance of mid-lactation dairy cows fed ruminally degradable protein at concentrations lower than national research council recommendations [J]. J. Dairy Sci., 91: 4704-4713.

DePeters, E. J., Cant, J. P., 1992. Nutritional factors influencing the nitrogen composition of bovine milk: A review [J]. J. Dairy Sci., 75: 2043-2070.

Dunlop, E. A., Tee, A. R., 2009. Mammalian target of rapamycin complex 1: Signalling inputs, substrates and feedback mechanisms [J]. Cell. Signal., 21: 827-835.

Gwinn, D. M., Shackelford, D. B., Egan, D. F., et al., 2008. Ampk phosphorylation of raptor mediates a metabolic checkpoint [J]. Mol. Cell, 30: 214-226.

Hanigan, M. D., Cant, J. P., Weakley, D. C., et al., 1998. An evaluation of postabsorptive protein and amino acid metabolism in the lactating dairy cow [J]. J. Dairy Sci., 81: 3385-3401.

Hanigan, M. D. et al., 2000. Evaluation of a representation of the limiting amino acid theory for milk protein synthesis Modelling nutrient utilization in farm animals [M]. CABI Publishing, Wallingford, UK.

Haque, M. N., Rulquin, H., Andrade, A., et al., 2012. Milk protein synthesis in response to the provision of an "ideal" amino acid profile at 2 levels of metabolizable protein supply in dairy cows [J]. J. Dairy Sci., 95: 5876-5887.

Haque, M. N., Rulquin, H., Lemosquet, S., 2013. Milk protein responses in dairy cows to changes in postruminal supplies of arginine, isoleucine, and valine [J]. J. Dairy Sci., 96: 420-430.

Hristov, A. N., Price, W. J., Shafii, B., 2004. A meta-analysis examining the relationship among dietary factors, dry matter intake, and milk and milk protein yield in dairy cows [J]. J. Dairy Sci., 87: 2184-2196.

Jonker, J. S., Kohn, R. A., Erdman, R. A., 1998. Using milk urea nitrogen to predict nitrogen excretion and utilization efficiency in lactating dairy cows [J]. J. Dairy Sci., 81: 2681-2692.

Kalscheur, K. F., Baldwin, R. L., Glenn, VI, B. P., et al., 2006. Milk production of dairy cows fed differing concentrations of rumen-degraded protein [J]. J. Dairy Sci., 89: 249-259.

Kauffman, A. J., St-Pierre, N. R., 2001. The relationship of milk urea nitrogen to urine nitrogen excretion in holstein and jersey cows [J]. J. Dairy Sci., 84: 2284-2294.

Knapp, J. R. Accessed The good news: Feed costs have declined over the past year [EB/OL]. http://dairy.osu.edu/bdnews/bdnews.html.

Korhonen, M., Vanhatalo, A., Varvikko, T., et al., 2000. Responses to graded postruminal doses of histidine in dairy cows fed grass silage diets [J]. J. Dairy Sci., 83: 2596-2608.

Lapierre, H., Lobley, G. E., Ouellet, D. R., et al., 2007. Amino acid requirements for lactating dairy cows: Reconciling predictive models and biology [M]. In: Cornell Nutrition Conference for Feed Manufacturers, New York, 39-59.

Mahoney, S. J., Dempsey, J. M., Blenis, J., 2009. Chapter 2 cell signaling in protein synthesis: Ribosome biogenesis and translation initiation and elongation [M]. In: W. B. H. John (ed.) Progress in molecular biology and translational science No. Volume, 90: 53-107. Academic Press.

Miglior, F., Sewalem, A., Jamrozik, J., et al., 2007. Genetic analysis of milk urea nitrogen and lactose and their relationships with other production traits in canadian holstein cattle [J]. J. Dairy Sci., 90:

2468-2479.

Mitchell, H. H., Block, R. J., 1946. Some relationships between the amino acid contents of proteins and their nutritive values for the rat [J]. J. Biol. Chem., 163: 599-620.

Mitchell, R. G., Rogers, G. W., Dechow, C. D., et al., 2005. Milk urea nitrogen concentration: Heritability and genetic correlations with reproductive performance and disease [J]. J. Dairy Sci., 88: 4434-4440.

Moshel, Y., Rhoads, R. E., Barash, I., 2006. Role of amino acids in translational mechanisms governing milk protein synthesis in murine and ruminant mammary epithelial cells [J]. J. Cell. Biochem., 98: 685-700.

Noftsger, S., St-Pierre, N. R., 2003. Supplementation of methionine and selection of highly digestible rumen undegradable protein to improve nitrogen efficiency for milk production [J]. J. Dairy Sci., 86: 958-969.

NRC., 1989. Nutrient requirements of dairy cattle [M]. 6th ed. National Academy Press, Washington, DC.

NRC., 2001. Nutrient requirements of dairy cattle [M] 7th rev. Ed. National Academy Press, Washington, D. C.

NRCS., 2011. Conservation practice standard: Feed management (code 592) [S]. Natural Resources Conservation Service, Washington, D. C.

Paris, Q., 1992. The von liebig hypothesis [J]. Am. J. Agr. Econ., 74: 1019-1028.

Preston, R. L., Schnakenberg, D. D., Pfander, W. H., 1965. Protein utilization in ruminants: I. Blood urea nitrogen as affected by protein intake [J]. J. Nutr., 86: 281-288.

Rius, A. G., McGilliard, M. L., Umberger, C. A., et al., 2010. Interactions of energy and predicted metabolizable protein in determining nitrogen efficiency in the lactating dairy cow [J]. J. Dairy Sci., 93: 2034-2043.

Rulquin, H., Pisulewski, P. M., Verite, R., et al., 1993. Milk production and composition as a function of postruminal lysine and methionine supply: A nutrient - response approach [J]. Live. Prod. Sci., 37: 69-90.

Spek, J. W., Bannink, A., Gort, G., et al., 2012. Effect of sodium chloride intake on urine volume, urinary urea excretion, and milk urea concentration in lactating dairy cattle [J]. J. Dairy Sci., 95: 7288-7298.

Sprengel, C., 1828. Von den substanzen der ackerkrume und des untergrundes (about thesubstances in the plow layer and the subsoil) [J]. Journal für Technische und Ökonomische Chemie, 3: 42-99.

St - Pierre, N. Accessed The costs of nutrients, comparison of feedstuffs prices and the current dairy situation [EB/OL]. http: //dairy. osu. edu/bdnews/bdnews. html.

Stewart, B. A., James, R. E., Hanigan, M. D., et al., 2012. Technical note: Cost of reducing protein and phosphorus content of dairy rations [J]. The Professional Animal Scientist, 28: 115-119.

Stoop, W. M., Bovenhuis, H., van Arendonk, J. A. M., 2007. Genetic parameters for milk urea nitrogen in relation to milk production traits [J]. J. Dairy Sci., 90: 1981-1986.

von Liebig, J., 1862. Die chemie in ihrer anwendung auf agricultur und physiologie [M]. 7. Aufl ed. Friedrich Vieweg, Braunschweig.

Wattiaux, M. A., Nordheim, E. V., Crump, P., 2005. Statistical evaluation of factors and interactions affecting dairy herd improvement milk urea nitrogen in commercial midwest dairy herds [J]. J. Dairy Sci., 88: 3020-3035.

Whitelaw, F. G., Milne, J. S., Orskov, E. R., et al., 1986. The nitrogen and energy metabolism of lactating cows given abomasal infusions of casein [J]. British J. Nutr., 55: 537-556.

Wolfe, A. H., Patz, J. A., 2002. Reactive nitrogen and human health: Acute and long-term implications [J]. Ambio., 31: 120-125.

Wood, G. M., Boettcher, P. J., Jamrozik, J., et al., 2003. Estimation of genetic parameters for concentrations of milk urea nitrogen [J]. J. Dairy Sci., 86: 2462-2469.